Modern Regulations and Practices for Social and Environmental Accounting

Teresa Eugénio
Centre of Applied Research in Management and Economics, School of Technology and Management, Polytechnic Institute of Leiria, Portugal

Graça Azevedo
University of Aveiro, Portugal

Ana Fialho
CEFAGE, University of Évora, Portugal

A volume in the Advances in Finance, Accounting, and Economics (AFAE) Book Series

IGI Global
PUBLISHER of TIMELY KNOWLEDGE

Published in the United States of America by
IGI Global
Business Science Reference (an imprint of IGI Global)
701 E. Chocolate Avenue
Hershey PA, USA 17033
Tel: 717-533-8845
Fax: 717-533-8661
E-mail: cust@igi-global.com
Web site: http://www.igi-global.com

Library of Congress Cataloging-in-Publication Data

Names: Eugénio, Teresa, 1974- editor. | Azevedo, Graça Maria do Carmo, editor. | Fialho, Ana, 1966- editor.
Title: Modern regulations and practices for social and environmental accounting / Teresa Eugénio, Graça Azevedo, and Ana Fialho, editor.
Description: Hershey, PA : Business Science Reference, [2022] | Includes bibliographical references and index. | Summary: "This book researches the needs in terms of accounting regulations, norms, organizational practices, and the challenge of education that can lead to the implementation of accounting and reporting practices that contribute to the achievement of the sustainable development goals"-- Provided by publisher.
Identifiers: LCCN 2021054736 (print) | LCCN 2021054737 (ebook) | ISBN 9781799894100 (hardcover) | ISBN 9781799894117 (paperback) | ISBN 9781799894124 (ebook)
Subjects: LCSH: Social accounting. | Environmental accounting. | Sustainable development reporting.
Classification: LCC HD60.3 .M64 2022 (print) | LCC HD60.3 (ebook) | DDC 658.4/083--dc23/eng/20211117
LC record available at https://lccn.loc.gov/2021054736
LC ebook record available at https://lccn.loc.gov/2021054737

This book is published in the IGI Global book series Advances in Finance, Accounting, and Economics (AFAE) (ISSN: 2327-5677; eISSN: 2327-5685)

British Cataloguing in Publication Data
A Cataloguing in Publication record for this book is available from the British Library.

All work contributed to this book is new, previously-unpublished material. The views expressed in this book are those of the authors, but not necessarily of the publisher.

For electronic access to this publication, please contact: eresources@igi-global.com.

Advances in Finance, Accounting, and Economics (AFAE) Book Series

Ahmed Driouchi
Al Akhawayn University, Morocco

ISSN:2327-5677
EISSN:2327-5685

Mission

In our changing economic and business environment, it is important to consider the financial changes occurring internationally as well as within individual organizations and business environments. Understanding these changes as well as the factors that influence them is crucial in preparing for our financial future and ensuring economic sustainability and growth.

The **Advances in Finance, Accounting, and Economics (AFAE)** book series aims to publish comprehensive and informative titles in all areas of economics and economic theory, finance, and accounting to assist in advancing the available knowledge and providing for further research development in these dynamic fields.

Coverage

- Economic Downturn
- Economics of Risks, Uncertainty, Ambiguity, and Insurance
- Microeconomics
- Economics of Migration and Spatial Mobility
- Applied Accounting
- Environmental and Social Accounting
- Public Finance
- Economics of Natural and Environmental Resources
- Managerial Accounting
- Accounting information systems

IGI Global is currently accepting manuscripts for publication within this series. To submit a proposal for a volume in this series, please contact our Acquisition Editors at Acquisitions@igi-global.com or visit: http://www.igi-global.com/publish/.

The Advances in Finance, Accounting, and Economics (AFAE) Book Series (ISSN 2327-5677) is published by IGI Global, 701 E. Chocolate Avenue, Hershey, PA 17033-1240, USA, www.igi-global.com. This series is composed of titles available for purchase individually; each title is edited to be contextually exclusive from any other title within the series. For pricing and ordering information please visit http://www.igi-global.com/book-series/advances-finance-accounting-economics/73685. Postmaster: Send all address changes to above address.

Titles in this Series

For a list of additional titles in this series, please visit: www.igi-global.com/book-series

Impact of Open Innovation on the World Economy
Orlando Lima Rua (Polytechnic of Porto, Porugal)
Business Science Reference • © 2022 • 306pp • H/C (ISBN: 9781799886655) • US $225.00

Handbook of Research on Challenges in Public Economics in the Era of Globalization
Şahin Akkaya (Istanbul University, Turkey) and Başak Ergüder (Istanbul University, Trkey)
Business Science Reference • © 2022 • 480pp • H/C (ISBN: 9781799890836) • US $295.00

Handbook of Research on Green, Circular, and Digital Economies as Tools for Recovery and Sustainability
Patricia Ordóñez de Pablos (The University of Oviedo, Spain) Xi Zhang (Tianjin University, China) and Mohammad Nabil Almunawar (Universiti Brunei Darussalam, Brunei)
Business Science Reference • © 2022 • 418pp • H/C (ISBN: 9781799896647) • US $295.00

Handbook of Research on the Significance of Forensic Accounting Techniques in Corporate Governance
Suleman Sherali Kamwani (A'Sharqiyah University, Oman) Elisabete S. Vieira (University of Aveiro, Portugal) Mara Madaleno (University of Aveiro, Portugal) and Graça Azevedo (University of Aveiro, Portugal)
Business Science Reference • © 2022 • 417pp • H/C (ISBN: 9781799887546) • US $265.00

COVID-19's Impact on the Cryptocurrency Market and the Digital Economy
Nadia Mansour (University of Sousse, Tunisia & University of Salamanca, Spain) and Salha Ben Salem (University of Monastir, Tunisia)
Business Science Reference • © 2022 • 320pp • H/C (ISBN: 9781799891178) • US $195.00

Handbook of Research on Energy and Environmental Finance 4.0
Abdul Rafay (University of Management and Technology, Pakistan)
Business Science Reference • © 2022 • 545pp • H/C (ISBN: 9781799882107) • US $295.00

Building the High-Performance Finance Function
André de Waal (Finance Function Research and Development Center BV, The Netherlands & HPO Center BV, The Netherlands) Eelco Bilstra (Finance Function Research and Development Center BV, The Netherlands) and Jacques Bootsman (Finance Function Research and Development Center BV, The Netherlands)
Business Science Reference • © 2022 • 298pp • H/C (ISBN: 9781799869290) • US $215.00

IGI Global
PUBLISHER of TIMELY KNOWLEDGE

701 East Chocolate Avenue, Hershey, PA 17033, USA
Tel: 717-533-8845 x100 • Fax: 717-533-8661
E-Mail: cust@igi-global.com • www.igi-global.com

Editorial Advisory Board

List of Reviewers

Table of Contents

Detailed Table of Contents

Conventional accounting practices are not sufficient to provide the necessary information demanded by the stakeholders. By environmental accounting and reporting, valuable information regarding the effects of activities of companies on the environment are identified and disclosed. Environmental accounting and reporting have numerous advantages, such as providing a basis for more efficient decision-making processes, optimization of company value, improved corporate image, and increased customer loyalty and investor trust. On the other hand, this chapter mainly focuses on the challenges and threats to environmental accounting and reporting and discusses the possible solutions to the problems identified and recommendations provided by prior literature.

This chapter contributes to the knowledge and understanding of recent trends in social and environmental accounting with the Sustainable Development Goal of the UN Agenda 2030. This study has two specific objectives: 1) to analyse the scientific production that crosses the area of SEA with SDG achievement, quantifying the articles published by year and by title of origin, by citation and research network by author, and by journal impact and 2) to identify publications by author, affiliation, and country and by type, subject, and funder. It is important to understand how sustainability accounting research might contribute to responsible business management towards SDG. A bibliometric analysis of articles published in journals in the Scopus database was carried out in the period between 2016 and 2021. The results of this study describe the current state of research in SEA and SDG achievement and provide a framework that can guide researchers.

of a simplified NFI standard and/or reporting format for SMEs; if that is an effective means of limiting the burden on SMEs; and, finally, in case of implementing it, if it should be mandatory or voluntary. Data were obtained from the stakeholders' answers provided during the public consultation conducted by the European Union about possible revisions of the Directive 2014/95/EU, from February to June 2020. The results show that the most controversial topic is the mandatory or voluntary nature of the simplified standard for SMEs. Furthermore, there are different opinions about the usefulness and effectiveness of a simplified standard for SMEs, particularly between capital providers versus other entities.

The objective of the chapter is to understand the multidimensional concept of integrated thinking and propose a set of indicators that enable organizations to identify integrated thinking. The chapter is characterized as qualitative, descriptive, and exploratory. Seventeen semi-structured interviews were conducted with researchers who have developed dissertations and/or theses on integrated reporting. The results were analyzed from content analysis and allowed comparing the dimensions proposed for integrated thinking. The results show that the concept of integrated thinking is multidimensional and has four dimensions: 1) connectivity and systemic view, 2) integrated planning, 3) communication and cooperation, 4) understanding stakeholders. Based on that, it was possible to develop a set of 27 indicators. The findings contribute to the academic literature on integrated reporting, filling research gaps, and the development of integrated thinking indicators contributes to organizational practice, as they enable the identification of integrated thinking.

This study aimed to analyze the perception of the International Integrated Reporting Council (IIRC) and social actors about the external assurance process involving integrated reporting, based on Zilber's institutionalization model. Two documents made available for public consultation by the IIRC, as well as the 62 documents with responses from social actors, were analyzed by means of the content analysis technique. The results indicate that the IIRC seeks to institutionalize its guidelines, as it has a favorable stance for believing that the theme is substantial, even though the actors with an interest in the subject demand guidance due to the subjectivity involving external assurance and some contents of Framework 1.0. Thus, Framework 1.0 for IR and its external assurance must evolve in view of the needs exposed by the social actors. The opinions of the social actors indicate that they seek to meet their own specific interests, rather than critically contributing to adjustments in the meanings of external IR assurance.

A regulatory policy based on the idea of smart regulation takes the view that legislation and regulations serve the economy, and therefore, laws must be intended to create the conditions to improve national economies and economic growth, as laid down by the OECD and the European Union itself. Such a regulatory policy is informed by a set of principles which apply to both regulatory institutions and regulations, such as transparency, cooperation, accountability, efficiency, effectiveness, participation, and openness. This regulatory approach takes the view that regulations are of better quality when they have been influenced by citizens. As a consequence, smart regulation results in participatory democracy. Citizens collaborate with the government in the lawmaking process at an early stage and throughout the lifecycle of a legislative act by means of consultations. This chapter focuses on consultation as the principal mechanism for participatory democracy in the context of smart regulation law-making in the European Union.

This chapter analyzes disclosures on anti-money laundering made by the largest banks in Germany, France, and the United Kingdom. Directive 2014/95/EU has transformed the way banks with more than 500 employees disclose non-financial information. The findings show that between 2013 and 2017 compliance has increased, while understandability has remained the same. Despite the fact that a common regulation is applied, this research highlights differences and similarities in disclosures of non-financial information among banks. It adds an overview of the non-financial reporting in the banking sector to the literature, which is not limited to a specific country.

The study aims to identify changes in personal financial behavior in the midst of the COVID-19 pandemic. Data from 796 participants of the Extension Project ‘Financial Education in Times of COVID-19’, between April/2020 and February/2021, were analyzed. It is characterized as qualitative, quantitative, and descriptive research with a survey. The results indicate that 66% of respondents are women, 47% are married, 54% are between 25 and 44 years old, and 32% are financiers. Regarding financial behavior during the pandemic, 52% of the participants make a shopping list to go to the supermarket and search for cheaper brands, and 56% noticed an increase in expenses. Respondents changed their behavior and expanded their control over finances, starting to reflect more when buying. The research contributes, highlighting the importance of education and the use of money in facing financial crises. It enriches the literature regarding education and financial behavior in times of pandemic.

Preface

ABOUT THE SUBJECT

Modern Regulations and Practices for Social and Environmental Accounting discusses social and environmental accounting issues such as regulations, norms, organizational practices, and the challenges of education. There has been an increased interest in social and environmental issues in recent years as more consideration is given to the idea of sustainability and social accounting. Gray et al. (1996) argue that social and environmental accounting and reporting play a relevant role in organizations' context as tools for analyzing the sustainability performance of the organizations and note that these have been relevant subjects in the academic literature. The issue of social and environmental accounting has gained a huge international relevance.

Social accounting can be considered a straightforward manifestation of corporate enforcement to legitimize, explain, and justify the organization's activities or an ethically desirable component of any well-functioning democracy. According to Mishra (s/d), social accounting is concerned with the study and analysis of accounting practice of those activities of an organization. The concept of socialistic pattern of society, civil rights movements, environmental protection, and ecological conservation groups, increasing awareness of society towards corporate social contribution, etc. Have contributed towards the growing importance of social accounting.

Social accounting can be three things: a fairly straightforward manifestation of corporate enforces to legitimate, explain, and justify the organization activities; an ethically desirable component of any well-functioning democratic and, just possibly, one of the few available mechanisms to address sustainability that does not involve the extinction of the species (Gray et al., 2014).

Environment Accounting, as Gray (1993) said, introduces and explains environmental issues as they relate to accountants today, covering a broad range of topics including Why environmental issues need to be at the heart of accounting practice; The development and implementation of an environmental policy; Environmental audit and management systems; Controlling costs of energy and waste; Environmentally aware investment, budgeting, and appraisal.

Some examples of environmental disclosures considered are any statement and information related to environmental policies, environmental management system (including ISO 14001 and Eco-Management and Audit Scheme – EMAS), pollution from business operations, pollution arising from the use of the product, discussion of specific environmental laws and regulations affecting company operations and products, prevention or repair of damage to the environment resulting from the processing of natural resources; conservation of natural resources and recycling activities, sustainability environmental aesthetics, conservation of energy in the conduct of business operations; the energy efficiency of products; and other environmental matters.

The increase of interest in social and environmental issues has been followed by an increase in academic writing and publications. In fact, additional study is required to better understand the relevancy of social and environmental accounting in today's modern business world.

Therefore, it would be fruitful to investigate and disseminate the results of theoretical and empirical studies that reveal different aspects of Social and environmental accounting and corporate social responsibility phenomenon in order to obtain insightful knowledge on how to promote effectively the social and environmental accounting and reporting play a relevant role in this context.

This book brings together several coherent chapters that allow, on a scientific basis, to discuss internationally the new directions in social and environmental accounting and reporting.

EXPECTATIONS

Covering a range of topics such as non-financial reporting and corporate social responsibility, Editors expect that this reference work is ideal for industry professionals, researchers, academicians, managers, practitioners, instructors, students, and all others interested in the subject of social and environmental accounting. This is a current issue where research is urgently needed in terms of regulation, norms, organizational practices, and the challenge of education. It is expected that new studies will trigger future developments in this area, fostering new perspectives and approaches, with the possibility of outputs/consequences/practical results that converge towards a fairer society and a more awake society for environmental and social protection.

On the other hand, research in this area may conduce to the implementation of accounting and reporting practices that contribute to the achievement of the Sustainable Development Goals.

This book aims to achieve the following main objectives:

Discuss the challenges and threats of social and environmental accounting
Identify the opportunities of corporate social responsibility
Relate Sustainable Development Goals with social and environmental accounting and reporting
Driving developments in education for sustainability
Analyze the models of non-financial information
Discuss and disseminate non-financial assurance
Analyzing trends in disclosure initiatives

ORGANIZATION OF THE BOOK

This book is organized into 14 chapters. A brief description of each of the chapters follows:

Chapter 1, "Challenges and Threats to Environmental Accounting and Reporting," based on a review of current literature, discusses the perspectives regarding challenges and threats to environmental accounting and reporting as well as the solutions and recommendations provided.

Chapter 2, "Bibliometric Analysis of Social and Environmental Accounting Research and United Nations SDG Achievement," contributes to the knowledge and understanding of more recent trends on research that crosses the area of social and environmental accounting (SEA) with the Sustainable Development Goals (SDGs) of the United Nations Agenda 2030 achievement.

Chapter 3, "Stranded Assets in the Context of the Low Carbon Economy: A Bibliometric Study," aims to understand technical characteristics of scientific research whose theme is stranded assets. Specifically, it seeks to identify prominent authors in the study of stranded assets with articles published in international journals; to know the main occurrences of research concentration related to the stranded assets theme, and identify trends that drive research on the impacts of stranded assets in the context of the low carbon economy.

In Chapter 4, "Supply Chain Pressure on Environmental Management Systems Adoption: A Systematic Literature Review," through a systematic literature review, the past and current research regarding the effects of supply chain pressure on the adoption of EMS is analyzed and synthesized. As climate change is becoming more visible, individuals, governments, and firms around the globe increasingly try to mitigate its ecological impact. For companies, the implementation of (certified) Environmental Management Systems, can improve their environmental performance.

Chapter 5, "The Reporting of Non-Financial Information by SMEs: Assessing the Answers to the Project of the Revised Directive 2014/95/EU," analyses the stakeholders' perspective about three core aspects related to small and medium enterprises (SMEs) and non-financial information (NFI) disclosures: the usefulness of a simplified NFI standard and/or reporting format for SMEs, if that is an effective means of limiting the burden on SMEs, and, finally, in case of implementing it, if it should be mandatory or voluntary.

Chapter 6, "Multidimensional Concept of Integrated Thinking and Proposal of Indicators for Its Identification," consists of understanding the multidimensional concept of Integrated Thinking and proposes a set of indicators that enable organizations to identify Integrated Thinking. This chapter contributes to organizational practice by proposing a set of indicators to identify and map Integrated Thinking, which can be applied by organizations.

Chapter 7, "Institutionalization of the Process for External Assurance on Integrated Reporting: Perceptions of Both the IIRC and Social Actors," analyzes the perception of both the IIRC and social actors about the process for external IR assurance, based on Zilber's institutionalization model (2002, 2006). In this study, institutionalization is approached as a process in which social actors, by assigning meanings, create and recreate institutions, that is, social or organizational practices.

Chapter 8, "Future Perspectives and Evolution of Citizen Participation in Government," analyzes and evaluated the practice of the aforementioned mechanisms, highlighting the problems and shortcomings that prevent and hinder their complete effectiveness or distract them from the main objective for which they were conceived. Ultimately, the legal justification for the application of participatory democracy mechanisms in standard-setting procedures has several reasons and is the product of a synergy of factors that are both cause and effect of the normative outcome.

Chapter 9, "Non-Financial Reporting: A Comparative Analysis with a Focus on the Combat Against Money Laundering – Evidence From Major European Banks," looks at the disclosure of non-financial information by eight European banks from France, Germany, and the United Kingdom. It analyses the framework for disclosure, the compliance level interns of the EU regulation, and some characteristics of the information disclosed by banks, namely understandability, and comparability.

Chapter 10, "Changes in Personal Financial Behavior Amid the COVID-19 Pandemic in Brazil," discusses the changes in personal financial behavior with the COVID-19 pandemic, as this disease has not only affected the countries' health systems but also their economies, increasing unemployment and causing many families to change their financial habits. This chapter also highlights the importance of the concepts of financial education.

Chapter 11, "Gender Visibility of Female Directors in the Local Licensed Commercial Banks in Sri Lanka," exposes and portrays the investigation of female occupational compositions, trends in designation portfolio, and female visibility in the Board of Directors and Corporate Management in Local licensed commercials banks in Sri Lanka. This inquiry directs towards understanding common career advancement influences for female managers towards increasing their visibility in reaching executive positions, despite the debatable composition in equal representation by gender.

Chapter 12, "Integrated Reporting Compliance in Portuguese Companies," evaluates integrated reporting by companies in Portugal and assesses whether it is in line with the essentials prescribed by IIRC and the European Commission - COM 2017/C 215/01. This topic has special importance nowadays, given the current debate on future guidelines for the reporting of non-financial information and the possibility of the publication of International Non-financial Reporting Standards by the IASB.

Chapter 13, "Sustainable Development Goals-Related Disclosures: University of Minho's Webpage and Sustainability Reports," aims to analyze the SDGs-related disclosures performed by the University of Minho in its webpage and its sustainability reports published since 2010. The specific objectives are to analyze: how UMinho discloses information related to SDGs in its webpage; the linkage to SDGs in its sustainability reporting, after the 2030 Agenda approval (in 2015); and whether sustainability reports' content was already connected to SDGs themes during the previous period (2010-2014) - before the publication of 2030 Agenda.

Chapter 14, Organizational Virtuousness and Workplace Well-Being in Organizational Citizenship Behaviors: A Case Study in Intermunicipal Community of Terras de Trás-os-Montes," intends to evaluate the organizational virtuousness, affective well-being at work and organizational citizenship behaviors, following the studies developed by Cameron et al. and by Rego et al. The study was applied in the municipalities of the Intermunicipal Community (CIM) – Trás-os-Montes and seven hundred and sixty employees were surveyed.

Teresa Eugénio
Centre of Applied Research in Management and Economics, School of Technology and Management, Polytechnic of Leiria, Portugal

Graça Azevedo
Center for Research in Accounting and Taxation, University of Aveiro, Portugal

Ana Fialho
CEFAGE, University of Évora, Portugal

REFERENCES

Gray, R. (1993). Current practice in environmental reporting. *Social and Environmental Accounting, 13*(1), 6–8.

Gray, R., Brennan, A., & Malpas, J. (2014). New accounts: Towards a reframing of social accounting. *Accounting Forum, 38*(4), 258–273.

Gray, R., Owen, D., & Adams, C. (1996). *Accounting & accountability: changes and challenges in corporate social and environmental reporting*. Prentice Hall.

Chapter 1
Challenges and Threats to Environmental Accounting and Reporting

Derya Üçoğlu
https://orcid.org/0000-0001-5510-3574
Istanbul Bilgi University, Turkey

ABSTRACT

Conventional accounting practices are not sufficient to provide the necessary information demanded by the stakeholders. By environmental accounting and reporting, valuable information regarding the effects of activities of companies on the environment are identified and disclosed. Environmental accounting and reporting have numerous advantages, such as providing a basis for more efficient decision-making processes, optimization of company value, improved corporate image, and increased customer loyalty and investor trust. On the other hand, this chapter mainly focuses on the challenges and threats to environmental accounting and reporting and discusses the possible solutions to the problems identified and recommendations provided by prior literature.

INTRODUCTION

Since the mid-1980s, environmental accounting has become an important issue for companies, accounting regulatory bodies, academics, professional associations, and other related parties (Mata et al., 2018). Interest in environmental accounting originated from the growing societal awareness that led to an expectation of a corporate response on ecological issues (Zrnic et al., 2020).

The need for environmental accounting and reporting is also related to the fact that the environment is under threat. These threats arise from air and water pollution, anthropogenic actions that have contributed to climate change and an increase in greenhouse gases, loss of rainforests, species vanishing, and degraded agricultural soils. As well as individuals, organizations have a duty to protect the environment, conserve natural resources and maintain biodiversity (Jones, 2010).

DOI: 10.4018/978-1-7998-9410-0.ch001

As business activities have critical roles in the above-mentioned environmental threats, environmental accounting encourages companies to make better decisions by considering their effects on the environment and the interests of all stakeholders (Carandang & Ferrer, 2020). Likewise, Lindenmayer (2017) states that environmental accounting can be used as a tool for making better decisions, minimizing resource mismanagement risks and potential collapse of impacted ecosystems (Liu et al., 2018).

The extent of activities of companies and their concern about the impacts on the environment attract public interest, and the stakeholders expect companies to conserve the environment and minimize the possible damages that could be caused by their actions (Wellalage & Kumar, 2021). Although some industries, such as manufacturing businesses, are considered the leading causes of environmental changes, many other companies in the trade and service sectors also contribute to environmental damages. For instance, if financial institutions had not employed green banking activities such as green credit cards, more paper consumption, and higher environmental pollution levels might have occurred (Dhar & Chowdhury, 2021).

Several researchers consider environmental disclosures one of the most important factors of sustainable development. When companies declare their environmental activities, due to the increase in environmental transparency, loyalty and trust of customers increase, and a favorable impact on the company's sustainability is observed (Khosroshahi et al., 2021).

Based on a careful review of current literature, this chapter will discuss the perspectives regarding challenges and threats to environmental accounting and reporting as well as the solutions and recommendations provided.

ENVIRONMENTAL ACCOUNTING AND REPORTING

In the United Nations' Conference on the Environment (Earth Summit in Rio de Janeiro) in 1992, an action plan on sustainable development was approved, and the concept of environmental accounting was introduced regarding the implementation of related policies (Council of Europe, 2004). According to the European Environment Agency, "the green growth, in the long term, is the only way to sustain economic growth". Therefore, the enterprises need to maintain a balance between economic growth and protecting the environment (Ribeiro et al., 2016).

For this purpose, the Kyoto Protocol was signed in 2005 by 175 countries to preserve the climate and reduce carbon emissions. The Paris Agreement, which replaced the Kyoto Protocol (Elmadhoun & Reddy, 2019), is a legally binding international treaty signed in 2015, and 196 parties agreed on the long-term temperature goal of limiting global warming to 1.5°C by reducing greenhouse gas emissions (United Nations, n.d.)

Accordingly, the World Economic Forum published the Global Risks Report 2022, which identified climate change as one of the most significant global risks and mentioned an average temperature increase of 1.8°C even in the optimistic scenario. If the net-zero goal is not achieved by 2050, irreversible consequences might be faced.

Therefore, considering all the risks, the companies have an ethical responsibility to society and their shareholders for conserving natural resources and restoring the damages to ecosystems. If the companies want to survive for long years, they need to ensure the sustainable consumption of natural resources (Weidman et al., 1994).

Traditional cost accounting does not include the effect of human beings on the environment (Jones, 2010). It is insufficient to provide necessary information as environmental costs and activities that affect the environment are ignored (Arowoshegbe & Emmanuel, 2016). For instance, the consumption of natural resources or environmental changes such as air pollution resulting from production processes is not fully reported in conventional accounting practices (Rounaghi, 2019). This challenge is related to traditional accounting having a capitalist orientation, a narrow business-oriented view, and being mainly based on monetary values and numerical quantification (Jones, 2010).

Similarly, the financial reporting framework focuses on companies' past financial position and performance. However, stakeholders such as shareholders and creditors also give importance to the future prospects of companies in addition to the financial information disclosed (Needles et al., 2016). So, environmental accounting has emerged as a new branch of accounting due to the limitations of traditional accounting approaches and the issue of corporate responsibility for sustainable development (Sarkar & Ahmed, 2020).

Environmental accounting mainly focuses on the environmental effect of the activities of companies, and environmental reporting provides financial, non-financial quantitative information, and qualitative information such as descriptive narratives (Harrison & Lamberton, 2007). For instance, non-financial disclosure includes information about formal environmental policies and environmental management systems. On the other hand, financial information disclosure is about environmental expenditures, environmental assets and liabilities, and environmental contingencies (Moneva & Cuellar, 2009).

Environmental accounting is a subcategory of social responsibility accounting, which is related to disclosing organizations' social and environmental impacts. Environmental accounting mainly concentrates on the environmental performance of companies (Arowoshegbe & Emmanuel, 2016) by introducing environmental costs in the decision-making processes and reporting other environmental information to the public (Sarkar & Ahmed, 2020), and includes waste and energy reporting, social and environmental reporting, accounting for environmental assets and liabilities, and environmental impact assessment (Gulluscio et al., 2020).

Socially responsible companies emphasize public interest by eliminating any practice that causes damage to the environment. Rather than focusing only on incurring high profit, such enterprises encourage community growth and development and have a triple bottom line involving profit, people, and the planet (Baker et al., 2012).

Social costs can be defined as the environmental effect of activities of companies on the public. For instance, any damage to people's health or agriculture and forests may have occurred due to a company's activities (Baldarelli & Nesheva-Kiosseva, 2017). Similarly, environmental disasters caused by the actions of companies, such as oil spills due to an explosion at a drilling rig, cause significant damage to the ecosystem and unfavorable financial implications for the companies (Alazzani & Wan-Hussin, 2013). Deepwater Horizon oil spill in the Gulf of Mexico, Exxon Valdez oil spill, Brumadinho, and Mariana dam disasters are only a few examples of environmental disasters. In such cases, unless there is evidence of a causal link between a company's activities and the environmental damage, it will not create a burden on the company. As a result, the cost is on the society, whereas the company takes the entire profit. In this sense, environmental accounting helps identify the party causing the damage but not paying for it (Baldarelli & Nesheva-Kiosseva, 2017) and contributes to the financial valuation of disasters so that fines can be paid. According to Enahoro (2009), when environmental costing is avoided, financial information reported to users cannot be considered complete or correct concerning the concept of true and fair view[1] (Carandang & Ferrer, 2020).

"Businesses are increasingly recognizing that maximizing shareholder value involves more than profits," said Chloe Xie from MIT Sloan School of Management (Vereckey, 2021). Consumers, investors, and other stakeholders demand more information regarding companies' environmental and social performance (Hossain, 2019). As for companies, optimizing their values by disclosing environmental issues to the public to ensure long-term sustainability and improving corporate image is essential (Ekundayo & Josiah, 2020). Furthermore, by applying environmental accounting, environmental costs that are incurred by companies when providing goods and services can be eliminated or at least reduced by business decisions ranging from investment in greener technologies, redesigning processes, products, or services, or making operational changes (U.S. Environmental Protection Agency, 1995).

The environmental accounting adoption by European companies is encouraged by European Union's Directive 2014/95/EU issued on 22 October 2014. The Directive points out the importance of businesses disclosing information on sustainability such as social and environmental factors to increase both investor and consumer trust. The non-financial information disclosure is also expected to combine profitability with environmental protection (Official Journal of the European Union, 2014).

Environmental accounting and reporting help companies achieve a competitive advantage, as customers increasingly prefer products and services provided through green manufacturing processes (Baldarelli & Nesheva-Kiosseva, 2017). In addition to its contribution to the reputation of companies, the companies engage in environmental reporting because of different factors such as organizational performance, company size, the pressure of stakeholders and legislation (Mata et al., 2018), the attitude of management, and public's expectation towards the environmental performance of entities (Ribeiro et al., 2016). Management of some companies is concerned with environmental reporting due to the motivation of being evaluated positively by the stock market participants and the focus on social acceptance (Nishitani et al., 2020).

Environmental accounting also helps companies cope with threats related to organizational legitimacy by showing their positive impact on society which creates value for the businesses (Brown & Fraser, 2006). So, environmental reporting seeks to highlight the relationship between financial and social performance and encourages corporate social responsibility in business practices (Baxi & Ray, 2009).

There are many studies in the literature (Nguyen, 2019; Lin & Dong, 2018; Cai et al., 2016; Fuadah et al., 2019; Al-Hadi et al., 2019; Poddi & Vergalli, 2016; Ghoul et al., 2011; Yeh et al., 2019; Oikonomou et al., 2014; Salvi et al., 2018; Jha & Cox, 2015) related to the level of environmental accounting and reporting practices. These studies identified that companies with better environmental performance experienced favorable effects on firm value, lower probability of filing for bankruptcy when the company is in financial distress, better corporate financial performance, contribution to reducing the cost of equity, cost of debt, and credit spreads (Anh et al., 2020).

Environmental Costs

Environmental remediation costs are any expenses or costs incurred to remove, clean up, remediate, prevent, or minimize damages and injuries resulting from the actions of any person or entity, or to achieve compliance with the provisions of any Environmental Laws (Law Insider, 2021).

Based on current legislation and available technology, environmental remediation obligations can be estimated using the company management's judgments about remediation methods. In some instances, even if a company does not have a legal obligation, ethical responsibility to society should be considered a liability regarding environmental damage. Suppose the liability cannot be estimated reliably or is not probable, or there are difficulties concerning shared responsibility. In that case, the companies should at least disclose the environmental loss in the notes to the financial statements. The presence of other potentially responsible parties, uncertainty caused by changing legislation and remediation technology, and complexity of such technology create difficulties concerning accounting for remediation costs (Weidman et al., 1994), although it does not exempt companies from seeking new strategies for recognizing environmental contingencies.

On the other hand, environmental costs, in general, refer to the expenses related to environmental management and measures for protecting the environment (Taygashinova & Akhmetova, 2019). Environmental costs can be classified under two headings:

- Environmental protection costs: These costs are related to environmental management systems and environmental protection (Taygashinova & Akhmetova, 2019). For instance, expenditures related to waste and wastewater management, biodiversity protection, landscapes, air, climate, ozone layer, soil remediation, surface water, and groundwater can be classified as costs for environmental protection activities (Eurostat, 2021).
- Costs of environmental load: If the environment is already damaged, any cost incurred for compensation can be considered the cost of environmental load. Any penalties charged, contributions for waste or insurance, and taxes are examples of such costs (Taygashinova & Akhmetova, 2019).

Other than this classification, environmental costs can be classified as internal and external environmental costs, financial and non-financial costs (Gurarda, 2015), or conventional, hidden, contingent, societal, and relationship/image costs (U.S. Environmental Protection Agency, 1995).

Circular Economy

The circular economy aims to maintain the values of materials, products, and resources as long as possible and manage wastes by reducing, reusing, and recycling activities conducted during production, circulation, and consumption processes (Opferkuch et al., 2021).

The European Commission has urged identifying priorities for the circular economy, which encourages longer product lifespans, high efficiency of resources, recycling, zero waste emissions, and eco-design. Environmental reporting is perceived as a supporting tool for solving environmental problems and adapting the circular economy model to companies (Dagiliene et al., 2020).

On the other hand, the study by Dagiliene et al. (2020) provided evidence that large European manufacturing companies still do not report much information regarding the circular economy, and they mainly focus on material reduction and water and air emissions.

Carbon Accounting

As a part of environmental accounting, carbon accounting provides the material value of the unfavorable impact of carbon emissions so that companies can quantify their climate impact (Elmadhoun & Reddy, 2019). Based on the previous literature, Stechemesser & Guenther (2012) defined carbon accounting as the monitoring, evaluation, and recognition of greenhouse gas emissions and the impact of these emissions on ecosystems' carbon cycle.

The Climate Disclosure Standards Board (CDSB), an international consortium of business and environmental non-governmental organizations (NGO), has been established to advance and align environmental information disclosure in a mainstream reporting model for equating natural capital and financial capital. The scope of the CDSB Framework for reporting environmental and climate change information, first released in 2010, has been expanded to cover environmental information and natural capital in addition to climate change and greenhouse gas (GHG) emissions. The framework includes guiding principles and environmental information reporting requirements (CDSB Framework, 2019).

Although there is pressure towards decarbonization by climate-conscious investors, governments, and the public, the transition to low-carbon technologies still lacks the necessary investment channels, especially in developing countries. Unfortunately, climate disclosures are insufficient for reallocating capital flows towards low-carbon investments. Still, they can only be considered a minor policy tool used to satisfy the need for protecting the financial system from climate-related risks (Ameli et al., 2021).

Biodiversity and Extinction Accounting

Biodiversity and extinction accounting is related to capturing necessary information for biodiversity management by preparing reports on species, the impacts of the organizations on species and their habitats, the measures taken to reduce the unfavorable organizational impacts on the ecosystem, and determining the biodiversity-related performance (Skouloudis et al., 2019; Atkins & Maroun, 2020).

COVID-19 pandemic and the possibility of future zoonotic pandemics resulting from biodiversity loss drew more comprehensive attention to environmental issues. Unfortunately, biodiversity and extinction accounting, an extension of corporate social responsibility, had attracted little consideration (Roberts et al., 2021) until the last years.

But recently, there have been more studies on biodiversity and extinction accounting. For instance, Houdet et al. (2020) proposed new double-entry bookkeeping conventions for facilitating biodiversity accounting and disclosing net biodiversity impact.

Sustainability Reporting and Integrated Reporting

Sustainability focuses on the economic, environmental, and social impacts of organizations and their interactive effects. Long-term corporate sustainability is crucial for the sustainability of society and the world economy. In this respect, sustainability accounting and reporting are related to designing accounting and information management methods for reporting high-quality information to ensure that organizations are moving towards sustainable development (White, 2016).

According to the international accounting and reporting principles, the companies prepare annual reports to provide relevant and faithfully presented financial information (Deloitte - Conceptual Framework for Financial Reporting, 2018). The integrated reports combine financial and non-financial information regarding economic performance, environmental performance, social, and other sustainability issues (INTOSAI WGEA, n.d.) for improving a company's sustainable approach. Before integrated reporting, social or sustainability reports were used to disclose an organization's relations with the social and natural environment (Paternostro, 2020). But integrated reporting introduced disclosing financial, social, environmental, and governance components for value creation (Soriya & Rastogi, 2021).

The differences and similarities between annual reports, sustainability reports, and integrated reports are provided in Table 1 concerning specific issues.

Figure 1. Integrated reporting framework
Source: (INTOSAI WGEA, n.d.)

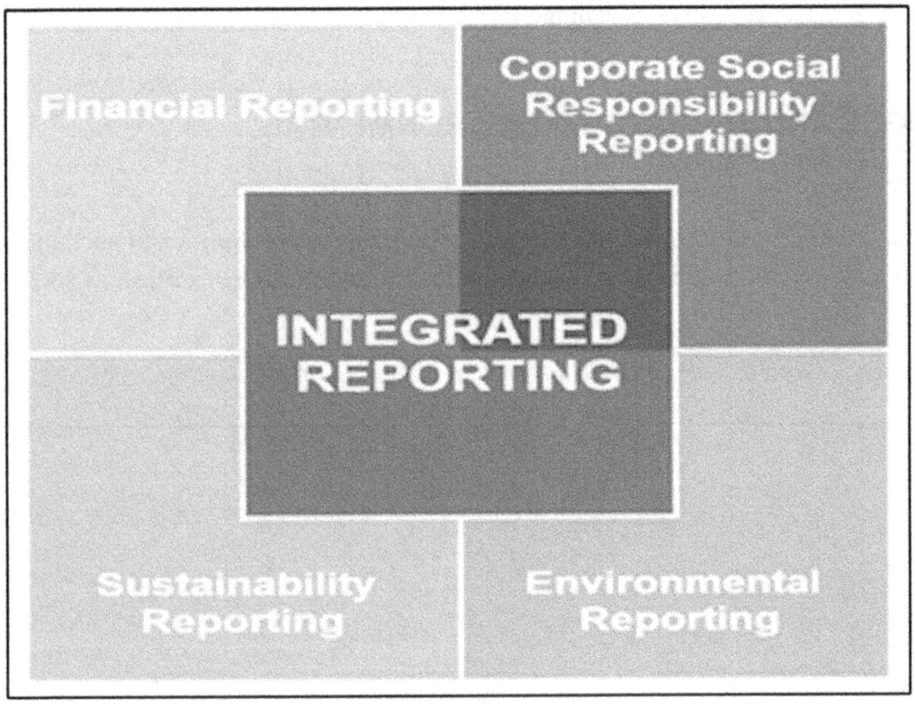

Table 1. Differences and similarities between annual, sustainability, and integrated reports

	Annual Reports	Sustainability Reports	Integrated Reports
Targeted stakeholders	Narrow (specific stakeholders such as investors and shareholders)	Wide (several different groups of stakeholders)	Narrow (mainly investors and providers of financial capital)
Mandatory/ voluntary	Mandatory	Mostly voluntary	Mostly voluntary
Orientation	Short-term and past performance-oriented	Long-term and value creation oriented	Strategic, long-term focus and future orientation
Content	Financial	Non-financial	Financial and non-financial
Regulation or guidelines	National and international laws and GAAP (or IFRS)	Global reporting initiative (GRI)	International Integrated Reporting Council (IIRC) framework
Industry-specific guidelines	Low	Medium	Low
Comparability	High	Medium (standards created by GRI)	Low
Assurance level	High	Low (non-financial information is more challenging to assure)	Low (non-financial information is more challenging to assure)
Scope	Financial reporting entity (one company or group of companies – individual or consolidated financial statements)	Broader than the reporting entity – supply chain, Life Cycle Assessment (LCA) approach	Broader than financial reporting entity – supply chain, Life Cycle Assessment (LCA) approach

Source: (Fasan, 2013; Needles, 2016; Speziale, 2019)

The perceived reasons for voluntarily adopting integrated reporting are to increase the awareness of sustainability, provide higher transparency and data accuracy, enhance the brand and reputation of the company, and create a competitive advantage (Eccles & Krzus, 2014).

CHALLENGES AND THREATS TO ENVIRONMENTAL ACCOUNTING AND REPORTING

A review of the literature reveals many challenges and threats to environmental accounting and reporting. This chapter summarizes these threats and challenges under three headings, as shown in Figure 2.

Figure 2. Challenges and threats to environmental accounting and reporting

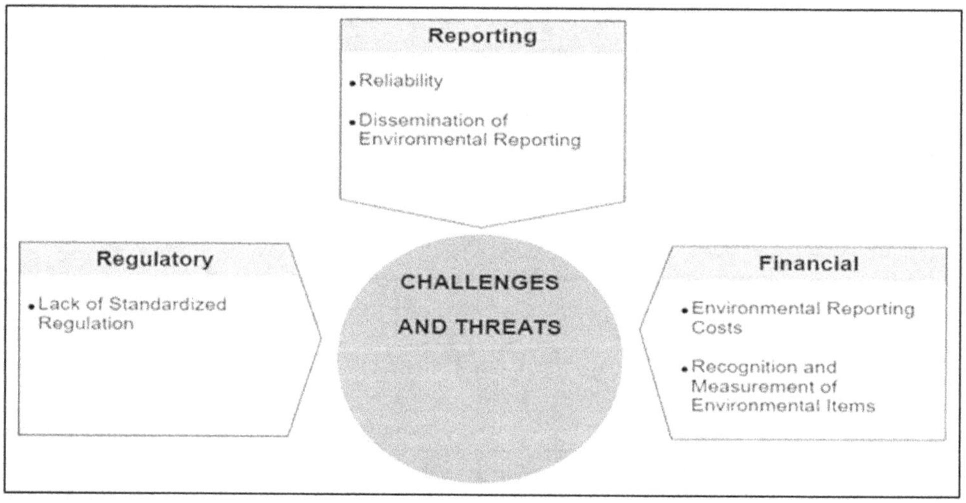

Lack of Standardized Regulation

Annual reports are a means of communication between the companies and the stakeholders. Therefore, some companies use these reports to include their views on environmental issues, whereas others publish separate reports on environmental sustainability (Miklosik & Evans, 2021).

Although companies tend to be more transparent in reporting, the reports become longer as additional information is provided, and readers may find more comprehensive reports harder to manage and interpret (MacLean & Rebernak, 2007). Furthermore, the lack of a unified conceptual framework results in companies presenting different environmental reports based on biased reporting systems. In the absence of proper environmental accounting and reporting standards, the disclosure policies may continue to be deficient (Herath, 2005). Also, the quality of environmental reporting varies across larger and smaller companies and countries with different regulatory environments (Leung, 2021). Another debate for environmental accounting and reporting is whether a regulator or self-regulation is more appropriate (Herath, 2005).

Although having similar goals like providing a single report, different institutions work on advancing environmental information disclosure for sustainability.

Table 2. Initiatives for environmental and sustainability reporting

Initiatives	Establishment Year	Mission	Action
CDP Disclosure Insight Action	2000	Focusing investors, companies, cities, and governments on building a sustainable economy by measuring and acting on their environmental impact	Runs the global environmental disclosure system to support companies, cities, states, and regions to measure and manage their risks and opportunities associated with climate change, water security, and deforestation
Climate Disclosure Standards Board (CDSB)	2007	Advancing and aligning the global mainstream corporate reporting model to equate natural capital with financial capital	Offers companies a framework for reporting environmental information with the same rigor as financial information
Global Reporting Initiative (GRI)	1997	Promoting economic, environmental, and social sustainability as a global standard-setter for impact reporting	Creates the global common language for organizations to report their impacts and take responsibility for a sustainable future
International Integrated Reporting Council (IIRC)	2010	Creating a globally accepted framework for a process that results in communications by an organization about value creation over time	
International Sustainability Standards Board (ISSB)	2021	Delivering a comprehensive global baseline of sustainability-related disclosure standards that will provide investors and other capital market participants with information about companies' sustainability-related risks and opportunities for helping them in making informed decisions	A new standard-setting board that will meet the demands for high-quality, transparent, reliable, and comparable reporting on climate and other environmental, social, and governance (ESG) matters
Sustainability Accounting Standards Board (SASB)	2011	Enabling businesses around the world to identify, manage and communicate financially-material sustainability information to their investors	
United Nations Environment Programme Finance Initiative (UNEP FI)	1992	Associating the United Nations and the financial sector globally regarding environmental, social, and governance (ESG) challenges	Contributes the perspectives of financial institutions to the various United Nations and global activities on sustainable finance
Value Reporting Foundation (VRF)	2021 - by the merger of the International Integrated Reporting Council (IIRC) and the Sustainability Accounting Standards Board (SASB)	Advancing progress towards a globally accepted corporate reporting system	A new foundation that will support key bodies such as the IFRS Foundation and work with colleagues around the world to drive progress towards a comprehensive corporate reporting system

Source: (Official websites of organizations)

Other organizations also provide reporting approaches such as the European Commission, International Organization of Standardization, Organization for Economic Co-operation & Development, United Nations, World Economic Forum, and associations of chartered accounts of different countries

(Opferkuch, 2021). The guides, standards, or reporting approaches described by these organizations may create a standards overload for companies on which framework to follow and report their impact on the environment.

For instance, even the materiality approaches in sustainability reporting provided by GRI and SASB are different. GRI uses a two-dimensional matrix for visualizing materiality degrees and focuses on the "significance of the reporting organization's economic, environmental and social impacts". On the other hand, the SASB approach of materiality emphasizes issues "that are likely to affect the financial condition or operating performance of companies" (Jorgensen et al., 2021). In addition, some of these standards are criticized for not being sufficient in addressing biodiversity loss and species extinction (Roberts, 2021).

The study by Beck et al. (2010) showed that the content of environmental reporting both changed by time and from country to country depending on the standards and regulations published. Considering the British and German companies included in the content analysis, the authors also identified that similar disclosure content patterns might have been observed because of lower creativity in environmental reporting. In addition, the disclosure quality was relatively low due to environmental reporting being used to create a positive image for the company rather than a means of decision-making.

Reliability of Disclosure

An essential challenge to environmental reporting is the reliability of information disclosed. Environmental disclosures made by companies mainly operating in environmentally sensitive industries such as utilities, chemicals, energy, mining, and resources, are likely to be perceived as less reliable. Such companies may disclose softer or misleading information about their environmental practices to affect stakeholders' perceptions. Therefore, stakeholders may not be able to make a differentiation between credible and unreliable information (Radhouane et al., 2018a). The study by Dobbs & Van Staden (2016) provided evidence on companies reporting good news rather than bad news. As little bad news is disclosed, the content of environmental reporting is not of high quality, and the preparers do not seem to report what they believe is essential. On the other hand, the stakeholders believe that corporate social responsibility reporting, including environmental information, is assured due to misinterpreting the scope of a standard financial audit. This study also identified that most of the companies do not have highly formalized internal systems for social and environmental reporting, do not use structured reporting frameworks, and only very few companies have the information assured.

In this sense, environmental reporting seems to be used as a tool to change public perceptions instead of showing the actual environmental performance of companies. Companies that have lower environmental performance may face more public pressure. Thus, they may be more motivated to increase the number of environmental disclosures so that legitimacy can be obtained (Braam et al., 2016). Similarly, the study by Michelon et al. (2016) provided evidence on companies with poor environmental management making comprehensive environmental disclosures on company websites (Sadiq & Tran, 2020).

When the companies are obliged to make environmental disclosures by law, the parameters for reporting are well defined. On the other hand, when voluntary reporting is considered, the companies decide what to disclose to the public. In such cases, the scope and content of environmental reporting and the extent of disclosures may depend on different factors such as the strategic intent of the companies, ownership structure (Baxi & Ray, 2009), the independence of the board of directors (Amosh & Khatib, 2021), the industry and the country in which the company operates. According to the study by Waldman et al. (2006), even a country's societal culture may impact the social responsibility values of top managers

(Alipour et al., 2019). Similarly, Baldini et al. (2018) mentioned the influence of the cultural system on environmental disclosure practices and the impact of the political system consisting of legal framework and corruption, the density of labor unions, and the unemployment rate.

Laine et al. (2017) concluded that they "remain skeptical regarding the provision of environmental expenditure and investment figures, which despite their seemingly accurate and precise form, appear to include mainly arbitrary and subjective allocations". Although the accounting figures are beneficial, they are at least partially the results of professional judgment similar to the ones disclosed in environmental reports. In addition, if there is an unregulated setting, it might not be easy to determine whether a true and fair view of environmental performance is disclosed (Braam et al., 2016).

Environmental reporting will not add credibility to the reporting process and will continue to be perceived as a public relations tool unless the environmental information disclosed is assured by third parties (Jones & Solomon, 2010).

Dissemination of Environmental Reporting

A further problem is the dissemination of environmental accounting and how to develop approaches for helping spread environmental reporting and accounting (Schaltegger et al., 2017). While companies strive to make progress in environmental reporting, if environmental accounting and reporting are not linked to core business strategies, the managers may be unwilling to provide these reports and overlook the benefits of strategic issues (MacLean & Rebernak, 2007).

Thus, environmental accounting suffers from an identity crisis on whether it should be a part of the traditional accounting system or should be designed as a different system (Mathews, 1995). Also, there are differences between developed and developing countries regarding environmental reporting awareness and disclosures.

Developing countries have lower gross national income (GNI) per capita, gross domestic product (GDP) per capita, and human development index (HDI). Developing countries that constitute over three-quarters of all countries go through more severe environmental problems, although they benefit less from resources and nature (Qian et al., 2021). In developed countries, environmental accounting and reporting are widely known, whereas, in developing countries like Bangladesh, companies and individuals are less informed about environmental reporting (Sarkar & Ahmed, 2020). Also, due to the inadequate research on environmental reporting in developing countries (Qian et al., 2021), it is not possible to identify the level and quality of environmental reporting in such countries.

Environmental Reporting Costs

Although having many advantages, environmental reporting has several costs for businesses. One main challenge is the perception that environmental accounting practices may create a decrease in companies' profits because of compliance costs and penalties that could be received for harming the environment (Herath, 2005). Increased transparency causes high costs regarding certifications, purchasing necessary equipment, essential laboratory infrastructure, and conducting experiments. But if companies have more increased transparency, it may have a favorable impact on customer behavior. On the other hand, if what is reported is evidence of poor environmental performance, then the customers may not be willing to buy the products or services of such companies (Khosroshahi et al., 2021). So, in addition to implementation costs, a business may experience negative consequences, such as a decrease in sales, due to reporting environmental issues.

Recognition and Measurement of Environmental Items

Although companies adopt their environmental policies, they face challenges in defining, identifying, measuring, classifying, and reporting environmental costs (Singh et al., 2016). For the concerned authorities and the management of the companies, the identification of environmental costs is perceived as a challenging task (Khan, 2020). For instance, when two companies operating in the same industry have their factories in the exact location, it might be challenging to assess the amount of air pollution created by each factory.

Sometimes, it may not be easy to determine whether a cost is environmental as some costs may be partly environmental, or they may fall into a gray zone. Suppose a company has clean technology equipment used in production that can be considered environmental. In such a case, is it possible to classify the costs related to the equipment as environmental costs? In other cases, health and safety costs may be confused with environmental costs (U.S. Environmental Protection Agency, 1995).

Another problem is related to the lack of a standard tool for the measurement of environmental costs. As the companies have the flexibility of accounting for environmental costs in a way relevant to their company (Khan, 2020), the amounts disclosed to the public by different companies may not be comparable due to using different bases for recognition and measurement.

Due to not having observable prices, assigning monetary values to natural resources is a complicated process (Herath, 2005). Besides, determining the environmental losses and gains and whether the environmental damages can be repaired or are irreversible (Herath, 2005) are other important issues.

In addition to environmental costs, the companies may have difficulty identifying and measuring their environmental assets and liabilities. If a company has environment-friendly activities and processes, then due to less contamination, waste reduction, enhanced safety, and increased productivity and efficiency, the company may recognize environmental assets. On the other hand, the activities of a company may result in future costs of cleaning a site which needs to be accounted as an environmental liability. Identifying and measuring such items are other challenges for the entities (Khan, 2020). For instance, Maurice (2012) determined that there are hardly any experts on identifying and valuing environmental liabilities in many organizations (Chen et al., 2014).

International Financial Reporting Interpretations Committee Interpretation 3: Emission Rights was withdrawn in June 2005, which led to the discussions on how to account for carbon allowances (He et al., 2021). So, when the environmental costs and liabilities are assessed according to International Financial Reporting Standards (IFRS), the companies need to consider whether they have a legal or constructive obligation concerning environmental damage prevention, reduction, or restoration. Besides, another challenge occurs in recognizing environmental items, as periodicity and matching principles result in artificial and short-term periods. While on the contrary, environmental problems and costs associated with these problems are long-term by nature. Therefore, when biodiversity is harmed because of a company's activities, it is controversial whether the matching principle may be applied or not, as this principle is blind to such costs (Bracci & Maran, 2013)

Chen et al. (2014) conducted a study on environmental liability disclosure. The findings have identified exciting results. The average environmental liability amounts of companies that have newly begun disclosing such information are lower than those already making disclosures. The authors believe that disclosing environmental liabilities might have been used as a manipulative image management tool. Rather than being transparent, the companies may be motivated to prevent stakeholders' assessment of overstated environmental liabilities. On the contrary, Maurice (2012) suggested that better disclosures for environmental liabilities might have resulted from the adoption of a formal environmental management system (Chen et al., 2014).

Therefore, according to Ghosh & Wolf (2021), environmental accounting is an unreliable resource for reporting and can even be used as a tool to disguise political interests.

SOLUTIONS AND RECOMMENDATIONS

When the top management is willing to adopt environmental accounting, a new way of looking at environmental management can be raised as environmental accounting integrates environmental information into business decisions (U.S. Environmental Protection Agency, 1995). In addition to management commitment, creating employee awareness, improving employee knowledge and expertise, assuring the coordination between all departments, integrating corporate strategy into environmental issues, and benefiting from management accounting practices can contribute to environmental accounting and reporting issues (Solovida & Latan, 2017). Thus, even environmental costs considered a burden for companies can be offset by generating additional revenue from selling by-products or recycled products (Singh et al., 2016).

The regulatory agencies may provide a common framework, including policies for environmental disclosures (Khan, 2020). Similarly, the study by Helfaya et al. (2019) has concluded that high-quality environmental reporting can be achieved by authoritative, powerful pressure on companies, and the regulatory bodies should publish the principles for environmental reporting and conduct continuous audits considering the rules and regulations. In this way, besides quantitative information and public disclosures on the environment, the companies may be encouraged to disclose more detailed qualitative information.

There have been some positive developments in this direction. Erkki Liikanen, Chair of the IFRS Foundation Trustees, has announced the establishment of the International Sustainability Standards Board (ISSB) on 3 November 2021, at the UN climate-change summit's Finance Day Presidency event titled 'A Financial System for Net Zero'. ISSB is expected to develop and publish standards to provide a basis for sustainability disclosures that will assure high-quality and globally comparable sustainability information. In his speech, Liikanen also delivered the commitment to consolidating the VRF and the CDSB as a part of the IFRS family. The VRF, the CDSB, the World Economic Forum, and the International Accounting Standards Board (IASB), supported by IOSCO, will publish the requirements related to general sustainability and climate-related disclosures. These initiatives are intended to create a global sustainability disclosure standard-setter that the financial markets lack (IFRS, 2021). In this way, the challenges related to the lack of standardization, dissemination of environmental reporting, and issues concerning recognition and measurement of environmental issues might be solved.

The European Union also has a new sustainable taxonomy focusing on sustainable activities contributing to climate change mitigation. Moreover, the UK Department for Work and Pensions asks trustees of pension funds to consider any risks and opportunities regarding material environmental issues. Current and future legislative changes can make a favorable impact on companies for environmental reporting practices. But the policymakers should give more importance to determining the factors that drive companies to make voluntary disclosures rather than compulsory ones (Brooks & Schopohl, 2021).

Due to the requests by investors and other market participants, the US Securities and Exchange Commission (SEC) has been developing mandatory climate disclosures for providing comparable and consistent information to the public (Solum et al., 2021).

Although these developments are valuable, voluntary reporting practices may be examined and systematized to solve potential problems before mandatory reporting is applied widespread. For instance, with the efforts of accountants, economists, and other experts, credible valuation procedures may be developed so that reliable values can be reported (Herath, 2005). Moreover, the study by Senn & Giordano-Spring (2020) has pointed out that it might be challenging to enforce regulations for environmental reporting when weak definitions that could be a significant limitation to relevance and reliability are provided.

Although the investors are looking for consistent and comparable metrics for environmental reporting, such issues are dynamic, interconnected, complex, and uncertain. The interviews by IIRC and Kirkshhoff (2020) have identified that mandatory reporting requirements for achieving disclosure standardization may reduce the unique nature and richness of information and provide fewer benefits (Adams & Abhayawansa, 2021).

Another critical issue related to environmental reporting is the identification of costs. It is not easy to identify which costs are environmental costs and which are not, but the correct classification of costs is not the key to successfully implementing environmental accounting. The main aim of environmental accounting is to supply relevant information to the stakeholders. In this sense, a company may disclose its definition of environmental cost and how the costs were classified considering its goals and the intended uses related to environmental accounting (U.S. Environmental Protection Agency, 1995). In addition, the companies should not focus only on reporting quantified indicators based on economic events. Still, they should find ways to account for environmental issues and develop and apply appropriate narratives presenting the interaction between a company and the ecologic system (Boyce, 2000). For this purpose, to report financial and non-financial environmental information, ISSB may develop principle-based standards to achieve comparability and flexibility similar to International Financial Reporting Standards.

The information needed for environmental accounting cannot be collected only by the efforts of an accountant. Environmental accounting and reporting practices need an organizational change in which operational teams consisting of engineers, chemists, system designers, production managers, operators, purchasing specialists, and accountants work together, and an official environmental management system is employed (Rounaghi, 2019).

On the other hand, external assurance for enhancing the credibility of reports may create additional problems due to the increased cost of the audit, the complexity of environmental reporting, the nature of the assurance process, the level of independence of the environmental consultants/assurors, and the lack of stability in methods and standards (Jones & Solomon, 2010; Senn & Giordano-Spring, 2020). Moreover, as environmental management systems and processes are not very developed, it is not easy to conduct an efficient audit (Jones & Solomon, 2010). However, Maroun (2019) has identified that external assurance contributed significantly to the credibility and reliability of integrated and sustainability reports.

Although the current assurance standards are not designed for integrated or sustainability reports (Maroun & Atkins, 2015), considering the potential benefits, the companies should have their sustainability, and integrated reports, including environmental information assured. As auditing non-financial information becomes widespread, it might even be a part of a financial audit, and the problem of lack of stability in auditing procedures and standards might be solved.

FUTURE RESEARCH DIRECTIONS

Environmental reporting is criticized for its selective nature, for reporting favorable environmental actions, and for not disclosing unfavorable events. This ethical issue addresses the problem of how the stakeholders perceive voluntary environmental disclosures (Radhouane et al., 2018b). For this purpose, it may be interesting to conduct a study that compares the perceptions of different stakeholders about environmental information reported and how they interpret them. Since the companies probably design the content of their environmental disclosures considering their potential impact on stakeholders, their views on environmental reporting may be critical for encouraging companies to disclose both favorable and unfavorable information.

In addition, as there is inadequate research in the literature about developing countries, the extent and quality of environmental information disclosed in developing countries, the factors affecting the level of voluntary disclosures, and the impact of external assurance of environmental reporting on investor decisions may be explored.

CONCLUSION

Climate change is rapidly happening. According to the records, 2020 was the hottest year, and in 2021, floods destroyed Germany, Belgium, the Netherlands, Switzerland, and China, while severe high temperatures were recorded in Finland, Turkey, and America. Although environmental issues and climate change were not perceived as major threats by most investors and companies, considering the clear message of nature, reforms need to be implemented rapidly (CDP Strategy 2021-2025).

In this respect, traditional accounting practices can be considered essential factors for the occurrence of environmental crises, as they concentrate on financial capital and economic considerations rather than environmental issues. Moreover, traditional accounting focuses on the entities with a short-term view, ignoring the long-term effects of managerial decisions on other organizations or environmental considerations (Tregidga & Laine, 2021). Therefore, environmental accounting and reporting gained importance to meet the demands of stakeholders and provided companies the opportunity to enhance legitimacy and trust (Amosh & Khatib, 2021).

When the companies' awareness of protecting and conserving the environment is weak, and companies do not have adequate resources and environmental management systems, they are more likely to disclose information that might be useful in avoiding legal obligations and taking any risks. Moreover, a voluntary disclosure may include positive environmental information and consciously conceal negative information that might harm the company image (Liu & Liu, 2021).

Environmental reporting helps companies take more responsibility towards the environment and is an effective organizational control mechanism as long as the information provided is comprehensive, complete, and reflects the actual environmental performance of companies (Van Staden et al., 2011).

Although reporting environmental information is a critical issue and has attracted the attention of different stakeholders, it seems to be used as a tool of impression management to show that companies are concerned about the environment. Therefore, self-reported environmental information should be interpreted carefully, considering its reliability (Dobbs & Van Staden, 2016).

To overcome the challenges mentioned, principle-based standards for reporting environmental information should be developed and should apply to companies worldwide. In addition, environmental reporting cannot provide reliable information for decision-making unless it is assured. Therefore, environmental auditing should be a part of financial audits, and international standards on environmental auditing should be published for auditing non-financial information.

REFERENCES

Adams, C. A. & Abhayawansa, S. (2021). Connecting the COVID-19 pandemic, environmental, social and governance (ESG) investing and calls for 'harmonisation' of sustainability reporting. *Critical Perspectives on Accounting*, *102309*, 1-13. doi:10.1016/j.cpa.2021.102309

Alazzani, A., & Wan-Hussin, W. N. (2013). Global Reporting Initiative's environmental reporting: A study of oil and gas companies. *Ecological Indicators*, *32*, 19–24. doi:10.1016/j.ecolind.2013.02.019

Alipour, M., Ghanbari, M., Jamshidinavid, B., & Taherabadi, A. (2019). Does board independence moderate the relationship between environmental disclosure quality and performance? Evidence from static and dynamic panel data. *Corporate Governance: The International Journal of Business in Society*, *19*(3), 580–610. doi:10.1108/CG-06-2018-0196

Ameli, N., Kothari, S., & Grubb, M. (2021). Misplaced expectations from climate disclosure initiatives. *Nature Climate Change*, *11*(11), 917–924. doi:10.103841558-021-01174-8

Amosh, H. A. & Khatib, S. F. A. (2021). Ownership structure and environmental, social and governance performance disclosure: the moderating role of the board independence. *Journal of Business and Socio-economic Development*, 1-18. doi:10.1108/JBSED-07-2021-0094

Anh, N. H., Soa, N. L., & Hanh, H. H. (2020). Environmental accounting practices and cost of capital of enterprises in Vietnam. *Cogent Economics & Finance*, *8*(1), 1–17. doi:10.1080/23322039.2020.1790964

Arowoshegbe, A. O., & Emmanuel, U. (2016). Accounting for social and environmental challenges: A theoretical perspective. *Saudi Journal of Business and Management Studies*, *1*(3), 143–148. doi:10.21276jbms.2016.1.3.10

Atkins, J., & Maroun, W. (2020). The Naturalist's Journals of Gilbert White: Exploring the roots of accounting for biodiversity and extinction accounting. *Accounting, Auditing & Accountability Journal*, *33*(8), 1835–1870. doi:10.1108/AAAJ-03-2016-2450

Baker, C. R., Cohanier, B., & Pederzoli, D. (2012). Corporate social and environmental reporting in the large retail distribution sector. *Procedia Economics and Finance*, *2*, 209–218. doi:10.1016/S2212-5671(12)00081-0

Baldarelli, M. G., & Nesheva-Kiosseva, N. (2017). Environmental accounting: Conceptual framework. In S. O. Idowu & R. Schmidpeter (Eds.), *Environmental Accounting and Reporting: Theory and Practice* (pp. 35–67). Springer International Publishing. doi:10.1007/978-3-319-50918-1_2

Baldini, M., Dal Maso, L., Liberatore, G., Mazzi, F., & Terzani, S. (2018). Role of country- and firm-level determinants in environmental, social, and governance disclosure. *Journal of Business Ethics*, *150*(1), 79–98. doi:10.100710551-016-3139-1

Baxi, C. V., & Ray, R. S. (2009). Corporate social & environmental disclosures & reporting. *Indian Journal of Industrial Relations*, *44*(3), 355–375.

Beck, A. C., Campbell, D., & Shrives, P. J. (2010). Content analysis in environmental reporting research: Enrichment and rehearsal of the method in a British–German context. *The British Accounting Review*, *42*(3), 207–222. doi:10.1016/j.bar.2010.05.002

Boyce, G. (2000). Public discourse and decision making: Exploring possibilities for financial, social and environmental accounting. *Accounting, Auditing & Accountability Journal*, *13*(1), 27–64. doi:10.1108/09513570010316135

Braam, G. J. M., Uit de Weerd, L., Hauck, M., & Huijbregts, M. A. J. (2016). Determinants of corporate environmental reporting: The importance of environmental performance and assurance. *Journal of Cleaner Production*, *129*, 724–734. doi:10.1016/j.jclepro.2016.03.039

Bracci, E., & Maran, L. (2013). Environmental management and regulation: Pitfalls of environmental accounting? *Management of Environmental Quality*, *24*(4), 538–554. doi:10.1108/MEQ-04-2012-0027

Brooks, C., & Schopohl, L. (2021). Green accounting and finance: Advancing research on environmental disclosure, value impacts and management control systems [Editorial]. *The British Accounting Review*, *53*(1), 1–5. doi:10.1016/j.bar.2020.100973

Brown, J., & Fraser, M. (2006). Approaches and perspectives in social and environmental Accounting: An overview of the conceptual landscape. *Business Strategy and the Environment*, *15*(2), 103–117. doi:10.1002/bse.452

Carandang, J. C., & Ferrer, R. C. (2020). Effect of environmental accounting on financial performance and firm value of listed mining and oil companies in the Philippines. *Asia-Pacific Social Science Review*, *20*(1), 117–134.

CDP. (n.d.). *About Us*. Retrieved from https://www.cdp.net/en/info/about-us

CDP. (n.d.). *Accelerating the Rate of Change: CDP Strategy 2021-2025*. Retrieved from https://cdn.cdp.net/cdp-production/comfy/cms/files/files/000/005/094/original/CDP_STRATEGY_2021-2025.pdf

Chen, J. C., Cho, C. H., & Patten, D. M. (2014). Initiating disclosure of environmental liability information: An empirical analysis of firm choice. *Journal of Business Ethics, 125*(4), 681-692. doi:10551-013-1939-0 doi:10.1007/s

Climate Disclosure Standards Board. (2019). *CDSB Framework for reporting environmental & climate change information*. Retrieved from https://live-cdsb.pantheonsite.io/sites/default/files/cdsb_framework_2019_v2.2.pdf

Climate Disclosure Standards Board. (n.d.). *About the Climate Disclosure Standards Board*. Retrieved from https://www.cdsb.net/our-story

Council of Europe Parliamentary Assembly. (2004). *Environmental accounting as a sustainable development tool*. Retrieved from https://assembly.coe.int/nw/xml/XRef/Xref-XML2HTML-en.asp?fileid=17198&lang=en

Dagiliene, L., Frendzel, M., Sutiene, K., & Wnuk-Pel, T. (2020). Wise managers think about circular economy, wiser report and analyze it. Research of environmental reporting practices in EU manufacturing companies. *Journal of Cleaner Production, 274*, 1–14. doi:10.1016/j.jclepro.2020.121968

Deloitte. (n.d.). *Conceptual Framework for Financial Reporting 2018*. Retrieved from https://www.iasplus.com/en/standards/other/framework

Dhar, S., & Chowdhury, M. A. F. (2021). Impact of environmental accounting reporting practices on financial performance: Evidence from banking sector of Bangladesh. *International Journal of Asian Business and Information Management, 12*(1), 24–42. doi:10.4018/IJABIM.20210101.oa2

Dobbs, S., & Van Staden, C. (2016). Motivations for corporate social and environmental reporting: New Zealand evidence. *Sustainability Accounting, Management and Policy Journal, 7*(3), 449–472. doi:10.1108/SAMPJ-08-2015-0070

Eccles, R. G., & Krzus, M. P. (2014). *The integrated reporting movement: Meaning, momentum, motives, and materiality*. John Wiley & Sons.

Ekundayo, G., & Josiah, M. (2020). Environmental accounting disclosure: A critical examination of literature. *British Journal of Economics, Finance and Management Sciences, 17*(2), 34–45.

Elmadhoun, I. R. I., & Reddy, G. N. (2019). The role of carbon accounting in managing the climate change risks: Literature review. *International Journal of Scientific Research, 9*(11), 309–315. doi:10.21275/SR20921154637

Eurostat. (2021). *Environmental protection expenditure accounts (env_ac_epea)*. Retrieved from https://ec.europa.eu/eurostat/cache/metadata/en/env_ac_epea_esms.htm

Fasan, M. (2013). Annual Reports, Sustainability Reports and Integrated Reports: Trends in Corporate Disclosure. In C. Busco, M. L. Frigo, A. Riccaboni, & P. Quattrone (Eds.), *Integrated Reporting: Concepts and Cases that Redefine Corporate Accountability* (pp. 41–57). Springer International Publishing Switzerland. doi:10.1007/978-3-319-02168-3_3

Ghosh, R., & Wolf, S. (2021). Hybrid governance and performances of environmental accounting. *Journal of Environmental Management, 284*(111995), 1–11. doi:10.1016/j.jenvman.2021.111995 PMID:33515840

Global Reporting Initiative. (n.d.). *About GRI*. Retrieved from https://www.globalreporting.org/about-gri/

Gulluscio, C., Puntillo, P., Luciani, V., & Huisingh, D. (2020). Climate change accounting and reporting: A systematic literature review. *Sustainability, 12*(5455), 1–31. doi:10.3390u12135455

Gurarda, S. (2015). Environmental Management Accounting. In U. Akkucuk (Ed.), *Handbook of Research on Developing Sustainable Value in Economics, Finance, and Marketing* (pp. 278–296). IGI Global. doi:10.4018/978-1-4666-6635-1.ch016

Harrison, J. L., & Lamberton, G. (2007). Who's doing environmental accounting? A survey of NSW dairy farms. *Rural Society, 17*(2), 182–195. doi:10.5172/rsj.351.17.2.182

He, R., Luo, L., Shamsuddin, A., & Tang, Q. (2021). Corporate carbon accounting: A literature review of carbon accounting research from the Kyoto Protocol to the Paris Agreement. *Accounting & Finance*, 1-38. doi:10.1111/acfi.12789

Helfaya, A., Whittington, M., & Alawattage, C. (2019). Exploring the quality of corporate environmental reporting: Surveying preparers' and users' perceptions. *Accounting, Auditing & Accountability Journal*, *32*(1), 163–193. doi:10.1108/AAAJ-04-2015-2023

Herath, G. (2005). Sustainable development and environmental accounting: The challenge to the economics and accounting profession. *International Journal of Social Economics*, *32*(12), 1035–1050. doi:10.1108/03068290510630999

Hossain, M. M. (2010). Environmental accounting challenges of selected manufacturing enterprises in Bangladesh. *Open Journal of Business and Management*, *7*(02), 709–727. doi:10.4236/ojbm.2019.72048

Houdet, J., Ding, H., Quetier, F., Addison, P., & Deshmukh, P. (2020). Adapting double-entry bookkeeping to renewable natural capital: An application to corporate net biodiversity impact accounting and disclosure. *Ecosystem Services*, *45*(101104), 1–14. doi:10.1016/j.ecoser.2020.101104

IFRS. (2021). *Global sustainability disclosure standards for the financial markets*. Retrieved from https://www.ifrs.org/news-and-events/news/2021/11/global-sustainability-disclosure-standards-for-the-financial-markets/

International Integrated Reporting Council. (n.d.). *The International <IR> Framework*. Retrieved from https://www.integratedreporting.org/wp-content/uploads/2013/12/13-12-08-THE-INTERNATIONAL-IR-FRAMEWORK-2-1.pdf

International Sustainability Standards Board. (n.d.). *International Sustainability Standards Board*. Retrieved from https://www.ifrs.org/groups/international-sustainability-standards-board/

INTOSAI WGEA. (n.d.). *INTOSAI WGEA research project - Environment and sustainability reporting: Extended outline*. Retrieved from https://www.wgea.org/media/1275/a_2_sc11-sustainability-reporting-eo.pdf

Jones, M. J. (2010). Accounting for the environment: Towards a theoretical perspective for environmental accounting and reporting. *Accounting Forum*, *34*(2), 123–138. doi:10.1016/j.accfor.2010.03.001

Jones, M. J., & Solomon, J. F. (2010). Social and environmental report assurance: Some interview evidence. *Accounting Forum*, *34*(1), 20–31. doi:10.1016/j.accfor.2009.11.002

Jorgensen, S., Mjos, A., & Pedersen, L. J. T. (2021). Sustainability reporting and approaches to materiality: Tensions and potential resolutions. *Sustainability Accounting, Management and Policy Journal*, 1-21. doi:10.1108/SAMPJ-01-2021-0009

Khan, S. (2020). Environmental accounting: One dimension of sustainable development. *Jaipuria International Journal of Management Research*, *6*(2), 38–41. doi:10.22552/jijmr/2020/v6/i2/205280

Khosroshahi, H., Azad, N., Jabbarzadeh, A., & Verma, M. (2021). Investigating the level and quality of the information in the environmental disclosure report of a corporation considering government intervention. *International Journal of Production Economics*, *235*, 1–16. doi:10.1016/j.ijpe.2021.108071

Laine, M., Jarvinen, J. T., Hyvönen, T., & Kantola, H. (2017). Ambiguity of financial environmental information: A case study of a Finnish energy company. *Accounting, Auditing & Accountability Journal, 30*(3), 593–619. doi:10.1108/AAAJ-02-2015-1961

Law Insider Dictionary. (2021). *Environmental Remediation Costs definition*. Retrieved from https://www.lawinsider.com/dictionary/environmental-remediation-costs

Leung, T. C. (2021). Environmental Accounting and Reporting Practices in Asian Countries. In *Routledge Handbook of Environmental Accounting*. Routledge. doi:10.4324/9780367152369-24

Liu, G., Yin, X., Pengue, W., Benetto, E., Huisingh, D., Schnitzer, H., Wang, Y., & Casazza, M. (2018). Environmental accounting: In between raw data and information use for management practices. *Journal of Cleaner Production, 197*, 1056–1068. doi:10.1016/j.jclepro.2018.06.194

Liu, Z., & Liu, M. (2021). Quality evaluation of enterprise environmental accounting information disclosure based on projection pursuit model. *Journal of Cleaner Production, 279*, 1–10. doi:10.1016/j.jclepro.2020.123679

MacLean, R., & Rebernak, R. (2007). Closing the credibility gap: The challenges of corporate responsibility reporting. *Environmental Quality Management, 16*(4), 1–6. doi:10.1002/tqem.20137

Maroun, W. (2019). Does external assurance contribute to higher quality integrated reports? *Journal of Accounting and Public Policy, 38*(4), 1–23. doi:10.1016/j.jaccpubpol.2019.06.002

Mata, C., Fialho, A., & Eugenio, T. (2018). A decade of environmental accounting reporting: What we know? *Journal of Cleaner Production, 198*, 1198–1209. doi:10.1016/j.jclepro.2018.07.087

Mathews, M. R. (1995). Social and environmental accounting: A practical demonstration of ethical concern? *Journal of Business Ethics, 14*(8), 663–671. doi:10.1007/BF00871347

Miklosik, A., & Evans, N. (2021). Environmental sustainability disclosures in annual reports of mining companies listed on the Australian Stock Exchange (ASX). *Heliyon, 7*(7), 1–11. doi:10.1016/j.heliyon.2021.e07505 PMID:34296012

Moneva, J. M., & Cuellar, B. (2009). The value relevance of financial and non-financial environmental reporting. *The Official Journal of the European Association of Environmental and Resource Economists, 44*(3), 441–456. doi:10.100710640-009-9294-4

Needles, B. E., Jr., Frigo, M. L., Powers, M., & Shigaev, A. (2016). Integrated reporting and sustainability reporting: An exploratory study of high performance companies. In M. J. Epstein, F. Verbeeten, & S. K. Widener (Eds.), Performance Measurement and Management Control: Contemporary Issues (pp. 41-81). Emerald Group Publishing Limited. doi:10.1108/S1479-351220160000031019

Nishitani, K., Haider, M. B., & Kokubu, K. (2020). Are third-party assurances preferable to third-party comments for promoting financial accountability in environmental reporting? *Journal of Cleaner Production, 248*, 1–13. doi:10.1016/j.jclepro.2019.119199

Official Journal of the European Union. (2014). *Directive 2014/95/EU of the European Parliament and of the Council of 22 October 2014 amending Directive 2013/34/EU as regards disclosure of non-financial and diversity information by certain large undertakings and groups*. Retrieved from https://eur-lex.europa.eu/legal-content/EN/TXT/PDF/?uri=CELEX:32014L0095&from=FI

Opferkuch, K., Caeiro, S., Salomone, R., & Ramos, T. B. (2021). Circular economy in corporate sustainability reporting: A review of organisational approaches. *Business Strategy and the Environment*, *30*(8), 4015–4036. doi:10.1002/bse.2854

Paternostro, S. (2020). Integrated Reporting and Social Disclosure: True Love or Forced Marriage? A Multidimensional Analysis of a Contested Concept. In L. Songini, A. Pistoni, P. Baret, & M. H. Kunc (Eds.), Non-Financial Disclosure and Integrated Reporting: Practices and Critical Issues (pp. 107-146). Emerald Publishing Limited. doi:10.1108/S1479-351220200000034008

Qian, W., Tilt, C., & Belal, A. (2021). Social and environmental accounting in developing countries: Contextual challenges and insights. *Accounting, Auditing & Accountability Journal*, *34*(5), 1021–1050. doi:10.1108/AAAJ-03-2021-5172

Radhouane, I., Nekhili, M., Nagati, H., & Pache, G. (2018a). Customer-related performance and the relevance of environmental reporting. *Journal of Cleaner Production*, *190*, 315–329. doi:10.1016/j.jclepro.2018.04.149

Radhouane, I., Nekhili, M., Nagati, H., & Pache, G. (2018b). The impact of corporate environmental reporting on customer-related performance and market value. *Management Decision*, *56*(7), 1630–1659. doi:10.1108/MD-03-2017-0272

Ribeiro, V. P., Aibar-Guzman, C., Aibar-Guzman, B., & Monteiro, S. (2016). Determinants of environmental accounting and reporting practices in Portuguese local entities. *Corporate Communications*, *21*(3), 352–370. doi:10.1108/CCIJ-11-2015-0071

Roberts, L., Hassan, A., Elamer, A., & Nandy, M. (2021). Biodiversity and extinction accounting for sustainable development: A systematic literature review and future research directions. *Business Strategy and the Environment*, *30*(1), 705–720. doi:10.1002/bse.2649

Rounaghi, M. M. (2019). Economic analysis of using green accounting and environmental accounting to identify environmental costs and sustainability indicators. *International Journal of Ethics and Systems*, *35*(4), 504–512. doi:10.1108/IJOES-03-2019-0056

Sadiq, M., & Tran, Q. H. (2020). Environmental Accounting Perception: Evidence from Nigerian Pharmaceutical Firms. *Systematic Reviews in Pharmacy*, *11*(1), 303–309. doi:10.5530rp.2020.1.39

Sarkar, S. H., & Ahmed, R. (2020). Corporate environmental accounting disclosure in Bangladesh. *International Journal of Applied Business and Economic Research*, *18*(1), 39–53.

Schaltegger, S., Etxeberria, I. A., & Ortas, E. (2017). Innovating corporate accounting and reporting for sustainability – Attributes and challenges. *Sustainable Development*, *25*(2), 113–122. doi:10.1002d.1666

Senn, J., & Giordano-Spring, S. (2020). The limits of environmental accounting disclosure: Enforcement of regulations, standards and interpretative strategies. *Accounting, Auditing & Accountability Journal, 33*(6), 1367–1393. doi:10.1108/AAAJ-04-2018-3461

Singh, A., Panackal, N., & Venkataramani, B. (2016). Environmental Accounting for Industrial Growth and Sustainable Development in India: An Overview and Theoretical Framework. In R. Dubey & A. Gunasekaran (Eds.), *Strategic Management of Sustainable Manufacturing Operations* (pp. 74–94). IGI Global. doi:10.4018/978-1-5225-0350-7.ch004

Skouloudis, A., Malesios, C., & Dimitrakopoulos, P. G. (2019). Corporate biodiversity accounting and reporting in mega-diverse countries: An examination of indicators disclosed in sustainability reports. *Ecological Indicators, 98*, 888–901. doi:10.1016/j.ecolind.2018.11.060

Solovida, G. T., & Latan, H. (2017). Linking environmental strategy to environmental performance: Mediation role of environmental management accounting. *Sustainability Accounting, Management and Policy Journal, 8*(5), 595–619. doi:10.1108/SAMPJ-08-2016-0046

Solum, S., Jacob, V. F., Levitt, M., & the Harvard Law School Forum on Corporate Governance. (2021). *The SEC's Upcoming Climate Disclosure Rules.* Retrieved from https://corpgov.law.harvard.edu/2021/09/01/the-secs-upcoming-climate-disclosure-rules/

Soriya, S. & Rastogi, P. (2021). A systematic literature review on integrated reporting from 2011 to 2020. *Journal of Financial Reporting and Accounting*, 1-22. doi:10.1108/JFRA-09-2020-0266

Speziale, M. (2019). Theoretical Perspectives on Purposes and Users of Integrated Reporting: A Literature Review. In S. O. Idowu & M. D. Baldo (Eds.), *Integrated Reporting: Antecedents and Perspectives for Organizations and Stakeholders* (pp. 13–60). Springer International Publishing Switzerland. doi:10.1007/978-3-030-01719-4_2

Stechemesser, K., & Guenther, E. (2012). Carbon accounting: A systematic literature review. *Journal of Cleaner Production, 36*, 17–38. doi:10.1016/j.jclepro.2012.02.021

Sustainability Accounting Standards Board. (n.d.). *Standards Overview.* Retrieved from https://www.sasb.org/standards/

Taygashinova, K., & Akhmetova, A. (2019). Accounting for environmental costs as an instrument of environmental controlling in the company. *Management of Environmental Quality, 30*(1), 87–97. doi:10.1108/MEQ-08-2017-0088

Tregidga, H. & Laine, M. (2021). On crisis and emergency: Is it time to rethink long-term environmental accounting? *Critical Perspectives on Accounting, 102311*, 1-11. doi:10.1016/j.cpa.2021.102311

United Nations Climate Change. (n.d.). *The Paris Agreement.* Retrieved from https://unfccc.int/process-and-meetings/the-paris-agreement/the-paris-agreement

United Nations Environment Programme Finance Initiative. (n.d.). *About Us.* Retrieved from https://www.unepfi.org/about/

U.S. Environmental Protection Agency. (1995). *An Introduction to Environmental Accounting As A Business Management Tool: Key Concepts and Terms*. Retrieved from https://www.epa.gov/sites/production/files/2014-01/documents/busmgt.pdf

Value Reporting Foundation. (n.d.). *Understanding the Value Reporting Foundation*. Retrieved from https://www.valuereportingfoundation.org/

Van Staden, C., Kern, T., McGuigan, N., & Wild, S. (2011). Social and Environmental Accounting and Accountability [Guest Editorial]. *Accounting Forum*, *35*(3), 127–129. doi:10.1016/j.accfor.2011.07.003

Vereckey, B. (2021). *5 forces remaking accounting*. Retrieved from https://mitsloan.mit.edu/ideas-made-to-matter/5-forces-remaking-accounting

Weidman, S. M., Welsh, C. N., & Bonino, L. N. (1994). Accounting for environmental remediation costs. *Business & Professional Ethics Journal*, *13*(1/2), 147–163. doi:10.5840/bpej1994131/23

Wellalage, N. H. & Kumar, V. (2021). Environmental performance and bank lending: Evidence from unlisted firms. *Business Strategy & the Environment*, 1-21. doi:10.1002/bse.2804

White, G. B. (2016). *Sustainability Reporting: Getting Started* (K. A. Merchant, Ed.; 2nd ed.). Business Expert Press.

World Economic Forum. (2022). *The Global Risks Report 2022*. Retrieved from https://www3.weforum.org/docs/WEF_The_Global_Risks_Report_2022.pdf

Zrnic, A., Starcevic, D. P., & Mijoc, I. (2020). Evaluating environmental accounting and reporting: The case of Croatian listed manufacturing companies. *Pravni Vjesnik (Journal of Law, Social Sciences and Humanities)*, *36*(1), 47-63. doi:10.25234/pv/8089

ADDITIONAL READING

Bang, Y., Lee, D. S., & Lim, S. (2019). Analysis of corporate CO2 and energy cost efficiency: The role of performance indicators and effective environmental reporting. *Energy Policy*, *133*, 1–10. doi:10.1016/j.enpol.2019.110897

Cho, C. H., Hageman, A. M., & Jerome, T. (2017). Eye-tracking experiments in social and environmental accounting research. *Social and Environmental Accountability Journal*, *37*(3), 155–173. doi:10.1080/0969160X.2017.1350588

Criado-Jiménez, I., Fernández-Chulián, M., Husillos-Carqués, F. J., & Larrinaga-González, C. (2008). Compliance with mandatory environmental reporting in financial statements: The case of Spain (2001-2003). *Journal of Business Ethics*, *79*(3), 245–262. doi:10.100710551-007-9375-7

Huang, C., & Kung, F. (2010). Drivers of environmental disclosure and stakeholder expectation: Evidence from Taiwan. *Journal of Business Ethics*, *96*(3), 435–451. doi:10.100710551-010-0476-3

Lehman, G., & Kuruppu, S. C. (2017). A framework for social and environmental accounting research. *Accounting Forum*, *41*(3), 139–146. doi:10.1016/j.accfor.2017.07.001

Moalla, M., Salhi, B., & Jarboui, A. (2021). An empirical investigation of factors influencing the environmental reporting quality: Evidence from France. *Social Responsibility Journal, 17*(7), 966–984. doi:10.1108/SRJ-02-2020-0065

Nicholls, J. A. (2020). Integrating financial, social and environmental accounting. *Sustainability Accounting. Management and Policy Journal, 11*(4), 745–769. doi:10.1108/SAMPJ-01-2019-0030

Parker, L. (2014). Constructing a research field: A reflection on the history of social and environmental Accounting. *Social and Environmental Accountability Journal, 34*(2), 87–92. doi:10.1080/0969160X.2014.938472

Qian, W., Hörisch, J., & Schaltegger, S. (2018). Environmental management accounting and its effects on carbon management and disclosure quality. *Journal of Cleaner Production, 174*, 1608–1619. doi:10.1016/j.jclepro.2017.11.092

Rezaee, Z., Alipour, M., Faraji, O., Ghanbari, M., & Jamshidinavid, B. (2021). Environmental disclosure quality and risk: The moderating effect of corporate governance. *Sustainability Accounting. Management and Policy Journal, 12*(4), 733–766. doi:10.1108/SAMPJ-10-2018-0269

KEY TERMS AND DEFINITIONS

Biodiversity (Biological Diversity): It refers to all different kinds of life on Earth, including all animals, plants, and micro-organisms, as well as the genetic information they contain and the ecosystems formed by them.

International Integrated Reporting Council (IIRC): IIRC is an organization including interest groups, such as companies, investors, standard setters, and regulators, that is formed in 2010 to promote integrated reporting by publishing a globally accepted framework.

International Organization of Securities Commissions (IOSCO): IOSCO is an international commission of global standard-setting for the regulators of securities markets.

Ozone Layer Erosion: The ozone layer is a region of Earth's stratosphere that protects human beings and other living things by absorbing the harmful ultraviolet radiation coming from the Sun. The erosion of the ozone layer is the gradual thinning of the layer due to manufactured chemicals, such as manufactured halocarbon refrigerants, propellants, solvents, and foam blowing agents.

US Securities and Exchange Commission (SEC): The Commission is a governmental body that is responsible for regulating securities and exchange markets in the US and protecting investors.

World Economic Forum: It is a not-for-profit international organization established in 1971 for public and private cooperation. The aim of this foundation is to promote entrepreneurship with high standards of governance and accountability.

Zoonotic Infectious Diseases: These are infectious diseases transmitted from animals to human beings caused by viruses, parasites, bacteria, and fungi. Zika virus, Ebola virus, SARS, and MERS have been classified as zoonotic pathogens which can cause different kinds of illnesses.

ENDNOTE

1 True and fair view concept is related to financial statements being free from material misstatements and faithfully representing the true financial position and performance of a company.

Chapter 2
Bibliometric Analysis of Social and Environmental Accounting Research and United Nations SDG Achievement

Graça Azevedo
https://orcid.org/0000-0002-6346-4035
University of Aveiro, Portugal & CICF, Indiana

Ana Fialho
CEFAGE, University of Évora, Portugal

Teresa Eugénio
https://orcid.org/0000-0003-1040-3164
CARME, Polytechnic of Leiria, Portugal

Maria da Conceição Tavares
University of Aveiro, Portugal

ABSTRACT

This chapter contributes to the knowledge and understanding of recent trends in social and environmental accounting with the Sustainable Development Goal of the UN Agenda 2030. This study has two specific objectives: 1) to analyse the scientific production that crosses the area of SEA with SDG achievement, quantifying the articles published by year and by title of origin, by citation and research network by author, and by journal impact and 2) to identify publications by author, affiliation, and country and by type, subject, and funder. It is important to understand how sustainability accounting research might contribute to responsible business management towards SDG. A bibliometric analysis of articles published in journals in the Scopus database was carried out in the period between 2016 and 2021. The results of this study describe the current state of research in SEA and SDG achievement and provide a framework that can guide researchers.

DOI: 10.4018/978-1-7998-9410-0.ch002

INTRODUCTION

Social, environmental and economic challenges are increasingly important and are on the global agenda because they affect plans to achieve sustainable development goals (SDG). The release of the 2030 Agenda by the United Nations in 2015 highlighted the relevance of these challenges and highlighted the role that all actors, governments, companies, and citizens have in the pursuit of the 17 SDG. Academic research has been following these issues and the growing scientific production of articles, studies, and reports are remarkable. Researchers in the area of accounting has also developed research in the field of social and environmental accounting and reporting, both from an internal and external perspective. Alongside this research, there has been, more timidly, some research regarding the involvement and contribution of companies in achieving the SDGs of the United Nations Agenda 2030.

Bebbington and Unerman (2018) highlight the holistic nature of the SDGs and question the tendency for companies to focus on a few SDGs that they believe are closer to their activity. This approach opens the possibility for academics to explore to what extent accounting and reporting can address these complexities and interconnections. Thus, it is hoped that the research tends to contribute to this direction.

This chapter survey the scientific production, published, from 2016 until the 1st half of 2021, in the Scopus database, that crosses the area of social and environmental accounting (SEA) with the SDG of the UN Agenda 2030 achievement. Research in the field of SEA has been growing over the last decades, reaching its peak in 2020, thus, we seek to understand if the SEA research has leveraged the development of research on SDGs.

Research based on the crossing of these two domains leads to very poor results, few references cross the SEA and the SDGs. Given the role of social and environmental accounting and reporting in companies, which is increasingly relevant given the global challenges that the SDGs aim to embrace, it is important to know what is being researched, the conclusions reached, and the contributions to the future development of research.

There are several studies in these areas that use bibliometric analysis (see Teixeira & Ribeiro, 2013, Farrukh et al., 2020, Fusco & Ricci, 2019, Petrushenko et al., 2020, Pizzi et al., 2020, Sweileh, 2020, Meseguer-Sánchez et al., 2021, Raman et al., 2021), but few have crossed these two topics (SEA and SDG)

Transnational companies, and the impact of their activity on the environment, should act as levers for achieving sustainable development. Prioritization and coordination across multiple sectors are necessary to achieve the SDGs (Adshead et al., 2020). Companies are key players in achieving the SDGs of the UN agenda.

Bebbington and Unerman (2020) conclude that accounting academics have not kept up with the relevance that SDGs have assumed globally. The research is scarce, and the pace of publication is slow despite some efforts that have been carried out for example in the call for papers for special issues of some journals in the area. In response to the challenge of their researchers, who seek to motivate accounting researchers to engage and contribute to the realization of the SDGs, we seek in this chapter to contribute by identifying possible avenues for this community engagement.

In particular, we intend to understand the evolution of research in the SDGs intersection with SEA research, in which many scientific fields have seen a large production of papers, which includes the pandemic period.

This study has two specific objectives:

1. Analyse the scientific production that crosses the area of SEA with SDG achievement, quantifying the articles published by year and by the title of origin; by citation and research network by the author; and by journal impact;
2. Identify publications by author, affiliation, and country; by type, subject, and funder.

To achieve these purposes, a bibliometric analysis of the literature published in the Scopus database from 2016 to the end of the 1st semester of 2021 is done. This analysis allows us to know the scientific production published in a recognized database and to analyze its contributions, conclusions, and directions for future research developments. The first article published with reference in the title to the "expression bibliometric" analysis is from 1972, indexed in the Scopus database. Since then, until today, this number has always been increasing, having exceeded 3000 references only in the year 2021. There is a growing trend to use this database for this type of analysis which allows one to explore, organize and evaluate a research topic of research, identifying the issues that have received the most attention and evaluating the characteristics and impact of published articles. It enables a better understanding of the topic of study, as well as identifying trends and gaps in academic research (Sáez-Martín et al. 2017, Baker et al. 2020; Rajan et al. 2020). Although this type of analysis cuts across all scientific areas, bibliometric studies predominate in the social sciences.

Thus, the results of this study describe the current state of research in SEA and SDG achievement and provide a framework that can guide researchers regarding the state of the art and future studies.

The remainder of the chapter is organized as follows: section two presents the research design; section three presents the results and analysis; and finally, section four concludes the chapter.

RESEARCH DESIGN

Sources

This study is based on a quantitative approach that takes as reference bibliometric indicators of scientific production according to their impact in the Scopus database. In this regard, bibliometric review studies use as main indicators the main articles, authors, journals, institutions, countries/regions, and keywords (Llanos-Herrera & Merigo, 2019).

Bibliometric techniques have been used mainly by information scientists to study the growth and distribution of scientific literature. Pritchard (1969) defined bibliometrics as the application of mathematical and statistical methods to books and other media (Lievrouw, 1989). Bibliometrics is a set of methods for measuring the academic impact of research publications and can easily be scaled from the micro-level (author, article, journal) to the macro-level (subject area, country, world) (Chung et al., 2015, Raman et al., 2021). Thus, to achieve the objectives of this work, a bibliometric analysis was developed from the Scopus database. The main objective of this methodology consists of identifying, organizing, and analyzing the main components within a specific research field (Lievrouw, 1989, Meseguer-Sánchez, et al., 2021).

This analysis allows us to know the scientific production published in a recognized database and to analyze its contributions, conclusions, and directions for future research developments. Since then, until today, the number of this kind of work has always been increasing, having exceeded 3000 references only in the year 2021. There is a growing trend to use this database for this type of analysis. Although this type of analysis cuts across all scientific areas, bibliometric studies predominate in the social sciences.

Period of Analysis

The period defined for the research in the Scopus database was between 2016 (after the UN Agenda 2030) and the first semester of 2021. With the choice of this time interval, it is intended that the research includes articles published from the year of publication of the United Nations Agenda 2030 up to the most recent time. In this period, it is possible to understand if the investigation increased in the first period of the pandemic, or if, on the contrary, a decrease in the number of studies is observed. The search was conducted in early July 2021.

Data Collection

For data collection, the following keywords were selected "Social and Environmental Accounting", limiting the search to the title of the article, abstract, and keywords, having obtained 3,685 document results. Next, the search was associated with "SDG Implementation", the search was based on the following terms:

(TITLE-ABS-KEY - social AND environmental AND accounting) AND (sdg AND implementation).

We then limited the bibliographic database exclusively to peer-reviewed scientific articles in scientific journals during 2016 and the first six months of 2021, which were classified in the item categories of Business, Management, and Accounting and Economics. With the search carried out, we found 16 papers which we will analyze in the next section.

BIBLIOMETRIC ANALYSIS ON SEA AND SDG RESEARCH

By Year and Source Title

The bibliometric analysis begins with the results obtained by year and by source title. As we can see, Figure 1 shows the evolution of documents collected between 2016 and the first half of 2021, after applying the parameters defined in the Scopus database, leading us to a total database of 16 documents. The almost constant growth of the number of publications is visible, except the year 2018-2019. After these first months, which we may consider as an adaptation to the new normality, we observe an increase in scientific production. We must however take into account that publications in journals classified in Scopus have long acceptance times, meaning that the decline observed in 2019, can be justified by other reasons that we were unable to ascertain. It is worth noting the growing interest in publishing in this area, the increasing number of publications in the last year (2021) is noteworthy, since it only includes one semester and even, so it is the year with the highest number of publications (see Figure 1).

From the 16 documents collected, we see from Figure 2 the journals to which correspond the major sources of publishing journals. Thus, we find that the Accounting Auditing and Accountability Journal has two published articles, while the journals Climate Policy, Critical Perspectives On Accounting, Earth S Future, and Ecological Economics have one article each.

Figure 1. Number of documents published, by year, during 2016 and 1st semester 2021
Source: Scopus, July 2021

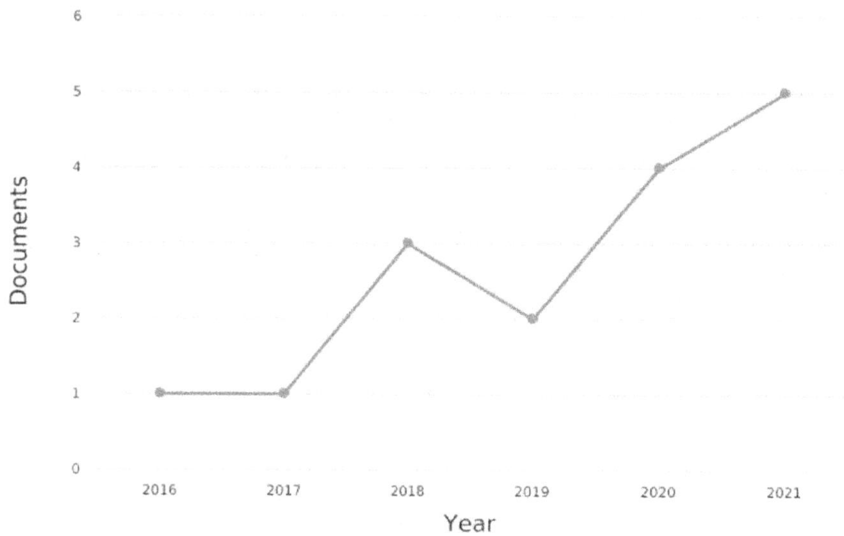

Figure 2. Documents published, by Source Title, during 2016 and 1st semester 2021
Source: Scopus, July 2021

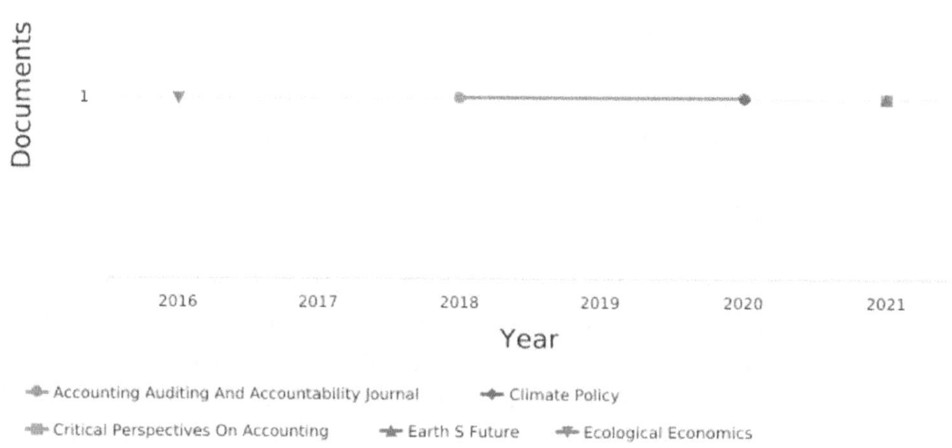

Raman et al. (2021) identified important connections between COVID-19 studies at the 17 SDGs, in particular on SDGs 3 (on health), 11 (on sustainable cities), and 13 (on climate action). Because the SDGs are universal there is a call for integrated implementation (Hülsmann & Ardakanian, 2018).

Citation and Research Network by Author by Year

The impact of research on the scientific community is one of the main indicators to consider and is assessed by the number of citations that a particular work has. Table 1 presents the articles in order of citation. The article that gathers the most citations was published in 2018 in Accounting, Auditing and Accountability Journal with 200 citations (Bebbington & Unerman, 2018) at the time of writing this chapter. The author analyzed the United Nations Sustainable Development Goals: An enabling role for accounting research. The second article that also stands out in terms of citations was published in 2017 and features 101 citations. This article was published in Sustainability and is about Systems thinking for life cycle sustainability assessment: A review of recent developments, applications, and future perspectives.

It is noteworthy that in fifth place appears an article published in 2020, with 16 citations. This article was published in the International Journal of Environmental Research and Public Health and examines the evolution of research on the socio-economic and environmental dimensions of university social responsibility.

We verified that of the 5 articles published in the first half of 2021, only one presents some citation (3 citations) and it was published in the Revista de Contabilidad - Spanish Accounting Review, with the theme Corporate social reporting and assurance: The state of the art.

From the analysis of Table 1, we conclude that the most recent articles have fewer citations, which is normal. However, we see that the themes analyzed are similar and the concerns with environmental, social, and sustainability issues have been growing and present a developing line of research.

Reconciling Table 1 with Table 2 we see that the citations are more concentrated in the last years, that is, in 2020 and 2021. Even the older articles have more citations in more recent years, which leads us to believe that, as previously mentioned, the theme is current and deserves more research.

When making a more detailed analysis of the collaborations, it is easy to see that most articles are written by more than one author. Thus, based on Table 3, we have 4 articles with five authors, 4 articles with four authors, 3 articles with three authors, 2 articles with two composite authors, and 3 articles written individually.

Continuing our analysis and taking into account the number of contributors to the articles as co-authors, we present in Table 4 the number of documents created by each, the total number of citations considering each author, the h-index of each author, and the number of co-authors that each author added in the explored and identified documents.

The author with the highest number of published papers, number of citations, h-index, and co-authors is Hall, Jim W. However, we found that this author is not the author of the most cited article, as we checked in Tables 1 and 2.

We also found that this author must be a reference author since he appears last in the publication and must work with junior researchers since his co-authors have a low number of citations and h-index.

It should also be noted that the authors with the lowest numbers in these components are Ivanović, Ivana Ravlić, Marszalek, Vali and Rösler, Branka all co-authors of the article 16 entitled: The analysis of the state of non-financial reporting in Croatia in 2017 and 2018.

Table 1. Cited articles during 2016 and 1st semester 2021

	Document title	Authors	Year	Source	Cited by
1	Achieving the United Nations Sustainable Development Goals: An enabling role for accounting research	Bebbington, J., Unerman, J.	2018	Accounting, Auditing and Accountability Journal 31(1), pp. 2-24	200
2	Systems thinking for life cycle sustainability assessment: A review of recent developments, applications, and future perspectives	Onat, N.C., Kucukvar, M., Halog, A., Cloutier, S.	2017	Sustainability (Switzerland) 9(5),706	101
3	The accounting push and the policy pull: Balancing environment and economic decisions	Vardon, M., Burnett, P., Dovers, S.	2016	Ecological Economics 124, pp. 145-152	32
4	Linking Environmental Management Accounting: A Reflection on (Missing) Links to Sustainability and Planetary Boundaries	Schaltegger, S.	2018	Social and Environmental Accountability Journal 38(1), pp. 19-29	31
5	Examining the research evolution on the socio-economic and environmental dimensions on university social responsibility	Meseguer-Sánchez, V., Abad-Segura, E., Belmonte-Ureña, L.J., Molina-Moreno, V.	2020	International Journal of Environmental Research and Public Health 17(13),4729, pp. 1-30	16
6	Ecological-economic sustainability of the Baltic cod fisheries under ocean warming and acidification	Voss, R., Quaas, M.F., Stiasny, M.H., (...), Reusch, T.B.H., Schmidt, J.O.	2019	Journal of Environmental Management 238, pp. 110-118	12
7	Sustainable development impacts of nationally determined contributions: assessing the case of mini-grids in Kenya	Dal Maso, M., Olsen, K.H., Dong, Y., Pedersen, M.B., Hauschild, M.Z.	2020	Climate Policy 20(7), pp. 815-831	6
8	Legal personality in Aotearoa New Zealand: an example of integrated thinking on sustainable development	Barrett, M., Watene, K., McNicholas, P.	2020	Accounting, Auditing and Accountability Journal 33(7), pp. 1705-1730	4
9	Application of set pair analysis in a comprehensive evaluation of water resource assets: A case study of Wuhan City, China	Zhang, J., Li, Y., Liu, C., (...), Jiang, L., Fu, J.	2019	Water (Switzerland) 11(8),1718	4
10	Key strategies to achieve the SDGs and consequences for monitoring resource use (Book Chapter)	Bringezu, S.	2018	Managing Water, Soil and Waste Resources to Achieve Sustainable Development Goals: Monitoring and Implementation of Integrated Resources Management pp. 11-34	4
11	Corporate social reporting and assurance: The state of the art I [Información social corporativa y aseguramiento: El estado de la cuestión]	García-Sánchez, I.-M.	2021	Revista de Contabilidad-Spanish Accounting Review 24(2), pp. 241-269	3
12	Benchmarking urban performance against absolute measures of sustainability – A review	Goodwin, K., Wiedmann, T., Chen, G., Teh, S.H.	2021	Journal of Cleaner Production 314,128020	0
13	Africa, from the past to the present: Moving the critical environmental accounting research on Africa forward (Book Chapter)	Denedo, M., Egbon, O.	2021	Environmental Accounting pp. 265-275	0
14	Infrastructure Strategies for Achieving the Global Development Agendas in Small Islands	Adshead, D., Román García, O., Thacker, S., Hall, J.W.	2021	Earth's Future 9(2),e2020EF001699	0
15	Sustainability at stake during COVID-19: Exploring the role of accounting in addressing environmental crises	Cho, C.H., Senn, J., Sobkowiak, M.	2021	Critical Perspectives on Accounting 102327	0
16	The analysis of the state of non-financial reporting in Croatia in 2017 and 2018 I [Analyse des stands der nichtfinanziellen berichterstattung in Kroatien in den jahren 2017 und 2018]	Omazić, M.A., Vukić, N.M., Ivanović, I.R., (...), Marszalek, V., Rösler, B.	2020	Socijalna Ekologija 29(2), pp. 207-226	0

Source: Own Development

Table 2. Cited articles, by year, during 2016 and 1st semester 2021

	Documents	Citations	<2017	2017	2018	2019	2020	2021	Subtotal	>2021	Total
	Total		**0**	**15**	**32**	**81**	**133**	**152**	**413**	**0**	**413**
15	Sustainability at stake during COVID-19: Exploring the role of accounting in addressing environmental crises	2021							0		0
11	Corporate social reporting and assurance: The state of the art ǀ [Información social corporativa y aseguramiento: El estado de la cuestión]	2021					1	2	3		3
12	Benchmarking urban performance against absolute measures of sustainability – A review	2021							0		0
13	Africa, from the past to the present: Moving the critical environmental accounting research on Africa forward (Book Chapter)	2021							0		0
14	Infrastructure Strategies for Achieving the Global Development Agendas in Small Islands	2021							0		0
8	Legal personality in Aotearoa New Zealand: an example of integrated thinking on sustainable development	2020					1	3	4		4
7	Sustainable development impacts of nationally determined contributions: assessing the case of mini-grids in Kenya	2020					2	4	6		6
5	Examining the research evolution on the socio-economic and environmental dimensions on university social responsibility	2020					3	13	16		16
16	The analysis of the state of non-financial reporting in Croatia in 2017 and 2018 ǀ [Analyse des stands der nichtfinanziellen berichterstattung in Kroatien in den jahren 2017 und 2018]	2020							0		0
9	Application of set pair analysis in a comprehensive evaluation of water resource assets: A case study of Wuhan City, China	2019				1		3	4		4
6	Ecological-economic sustainability of the Baltic cod fisheries under ocean warming and acidification	2019					6	6	12		12
10	Key strategies to achieve the SDGs and consequences for monitoring resource use (Book Chapter)	2018			2	1		1	4		4
4	Linking Environmental Management Accounting: A Reflection on (Missing) Links to Sustainability and Planetary Boundaries	2018			2	7	13	9	31		31
1	Achieving the United Nations Sustainable Development Goals: An enabling role for accounting research	2018			8	28	83	81	200		200
2	Systems thinking for life cycle sustainability assessment: A review of recent developments, applications, and future perspectives	2017		11	13	31	19	27	101		101
3	The accounting push and the policy pull: Balancing environment and economic decisions	2016		4	7	13	5	3	32		32

Source: Own Development

Table 3. Articles collaboration during 2016 and 1st semester 2021: number of authors

	Documents	Number of authors
1	Achieving the United Nations Sustainable Development Goals: An enabling role for accounting research	2
2	Systems thinking for life cycle sustainability assessment: A review of recent developments, applications, and future perspectives	4
3	The accounting push and the policy pull: Balancing environment and economic decisions	3
4	Linking Environmental Management Accounting: A Reflection on (Missing) Links to Sustainability and Planetary Boundaries	1
5	Examining the research evolution on the socio-economic and environmental dimensions on university social responsibility	4
6	Ecological-economic sustainability of the Baltic cod fisheries under ocean warming and acidification	5
7	Sustainable development impacts of nationally determined contributions: assessing the case of mini-grids in Kenya	5
8	Legal personality in Aotearoa New Zealand: an example of integrated thinking on sustainable development	3
9	Application of set pair analysis in a comprehensive evaluation of water resource assets: A case study of Wuhan City, China	5
10	Key strategies to achieve the SDGs and consequences for monitoring resource use (Book Chapter)	1
11	Corporate social reporting and assurance: The state of the art I [Información social corporativa y aseguramiento: El estado de la cuestión]	1
12	Benchmarking urban performance against absolute measures of sustainability – A review	4
13	Africa, from the past to the present: Moving the critical environmental accounting research on Africa forward (Book Chapter)	2
14	Infrastructure Strategies for Achieving the Global Development Agendas in Small Islands	4
15	Sustainability at stake during COVID-19: Exploring the role of accounting in addressing environmental crises	3
16	The analysis of the state of non-financial reporting in Croatia in 2017 and 2018 I [Analyse des stands der nichtfinanziellen berichterstattung in Kroatien in den jahren 2017 und 2018]	5

Source: Own Development

By Journal Impacts

To finish this section, we perform the bibliometric analysis of journal impacts. In Table 5 we present some facts about the journals where the 16 articles found in the Scopus database were published. The articles were almost all published in different journals. The only journal that published two articles of those analyzed was the Accounting, Auditing and Accountability Journal (AAAJ).

In what concerns the impact of the journal with the most citations are *Sustainability*, where the second most cited article was published. The journal where the most cited article was published is AAAJ which does not stand out in terms of citations. It is also verified that the journal that presents the higher CiteScore in 2020 is *Earth's Future*. These differences can be justified by the fact that the open-access journal is the most cited journal as is the case of *Sustainability* and the *Earth's Future*. The issue of open access is increasingly controversial and debated in the scientific community, defended by some and called into question by others, and with no end in sight yet.

Table 4. Research activity and facts for the authors of the articles published during 2016 and 1st semester 2021

	Author	Documents by author	Total citation	h-index	Co-authors
1	Bebbington, Jan	85	4407	30	82
	Unerman, Jeffrey	56	3783	30	24
2	Cihat Onat, Nuri	56	1715	24	84
	Küçükvar, Murat	86	2856	34	90
	Halog, Anthony Basco	68	1483	18	75
	Cloutier, Scott A.	22	478	9	54
3	Vardon, Michael J.	32	597	14	118
	Burnett, Peter	4	38	3	6
	Dovers, Stephen	93	3062	26	133
4	Schaltegger, Stefan C.	124	8043	44	82
5	Meseguer-Sánchez, Víctor	6	30	3	10
	Abad-Segura, Emilio	36	336	12	29
	Belmonte-Ureña, Luis J.	49	616	16	54
	Molina-Moreno, Valentín	30	425	13	41
6	Voss, Rüdiger	69	1699	23	205
	Quaas, Martin Friedrich	99	2654	26	285
	Stiasny, Martina H.	12	208	8	112
	Reusch, Thorsten	188	10444	56	668
	Schmidt, Jörn Oliver	61	1102	20	364
7	Dal Maso, Mirko	2	12	2	6
	Olsen, Karen Holm	14	480	6	23
	Dong, Yan	21	338	9	78
	Pedersen, Mathilde Brix	6	106	5	8
	Hauschild, Michael Zwicky	35	426	12	138
8	Barrett, Mereana	3	28	2	24
	Watene, Krushil	12	88	5	24
	McNicholas, Patty	9	438	6	11
9	Zhang, Jingdong	65	833	15	84
	Li, Yanan	14	50	5	33
	Liu, Chaoyang	36	398	10	69
	Jiang, Luping	5	13	3	15
	Fu, Jiatian	4	8	2	14
10	Bringezu, Stefan	77	2060	23	155
11	García-Sánchez, Isabel María	176	5553	38	65
12	Goodwin, Kylie	1	0	0	3
	Wiedmann, Thomas Oliver	122	9877	47	288

continues on following page

Table 4. Continued

	Author	Documents by author	Total citation	h-index	Co-authors
	Chen, Guangwu	20	470	11	52
	Teh, Soo Huey	9	158	5	23
13	Denedo, Mercy	6	46	3	25
	Egbon, Osamuyimen	4	17	2	4
14	Adshead, Daniel	6	160	5	27
	Román García, Orlando	1	0	0	3
	Thacker, Scott	21	372	11	41
	Hall, Jim W.	299	10718	54	582
15	Cho, Charles H.	62	3429	23	79
	Senn, Juliette	4	18	2	4
	Sobkowiak, Madlen	3	10	1	4
16	Omazić, Mislav Ante	12	9	1	24
	Vukić, Nikolina Markota	2	1	1	6
	Ivanović, Ivana Ravlić	1	0	0	5
	Marszalek, Vali	1	0	0	5
	Rösler, Branka	1	0	0	5

Source: Own Development

By Author, Affiliation, and Country

In this subsection, the analysis continues with the identification and characterization of the reference authors in the field of Research on Social and Environmental Accounting and SDG implementation. However, the results do not point to the existence of reference authors since only authors with 1 published article and no more were found. Thus, the authors found were Abad-Segura, E., Adshead, D., Barrett, M., Bebbington, J., Belmonte-Ureña, L.J., Bringezu, S., Burnett, P., Chen, G., Cho, C.H., and Cigula, K. (Figure 3).

Regarding the affiliation of these authors, Figure 4 shows that each author belongs to a different University, i.e., we have United Nations Office for Project Services UNOPS, Área de Economía Financiera y Contabilidad Accounting, University of Essex, Arizona State University, Universität Leipzig, University of Zagreb, The Australian National University, Danmarks Tekniske Universitet, Christian-Albrechts-Universität zu Kiel, and Universidad de Almería. This shows that this subject is recent and of interest in different parts of the world.

When analyzing the research done by country or territory in the area of Research on Social and Environmental Accounting and SDG implementation, we find that much more research is needed in general terms due to the scarcity of research on the topic. Furthermore, a comparative analysis of Research on Social and Environmental Accounting and SDG implementation, within and between regions is needed to observe whether the results vary with country heterogeneity or whether this heterogeneity does not influence the results.

Table 5. Facts related to the Journals of the articles published during 2016 and 1st semester 2021

	Review	CiteScore 2020	SJR 2020	SNIP 2020	Citations 2017-2020	Documents 2017-2020
1	Accounting, Auditing and Accountability Journal	6.0	1.741	1.888	2.097	351
2	Sustainability (Switzerland)	3.9	0.612	1.242	97.894	24.902
3	Ecological Economics	9.1	1.917	1.920	11.361	1.255
4	Social and Environmental Accountability Journal	2.0	0.361	0.517	148	75
5	International Journal of Environmental Research and Public Health	3.4	0.747	1.356	65.479	19.020
6	Journal of Environmental Management	9.8	1.441	1.888	51.386	5.227
7	Climate Policy	7.6	1.764	1.512	2.920	384
9	Water (Switzerland)	3.7	0.718	1.179	33.299	8.887
10	Managing Water, Soil and Waste Resources to Achieve Sustainable Development Goals… (Book Chapter)					
11	Revista de Contabilidad-Spanish Accounting Review	2.7	0.378	0.972	209	77
12	Journal of Cleaner Production	13.1	1.937	2.475	203.300	15.505
13	Routledge Handbook of Environmental Accounting (Book Chapter)					
14	Earth's Future	11.3	2.641	2.211	4.279	379
15	Critical Perspectives on Accounting	6.6	2.042	2.261	975	147
16	Socijalna Ekologija	0.4	0.162	0.449	23	53

Source: Own Development

Figure 5 presents the documents by country or territory during the 2016 and 1st semesters of 2021. So, we see that research on this topic is being focused on Australia and the United Kingdom leading the top ten, followed by Germany, China, Denmark, Spain, and the United States. At the bottom of the top ten are Canada, Croatia, and France.

This would mean that much more research is needed in countries within Europe, where there is still a dearth of research on the topic.

By Type, Subject, and Funding Sponsor

The analysis of the type of work, the subject where the work is located, and the source of funding also contributes to a better understanding of the research being published. We should point out that by this stage we are considering all types of documents published under the keyword of Research on Social and Environmental Accounting and SDG implementation for the period of analysis, whether full articles or not, as will be described in Figure 6 shows us that the most published documents during this period were primarily in the form of scientific articles (62.5%), reviews accounted for only 18.8%, while book chapters and short surveys accounted for 12.5% and 6.3%, respectively.

Figure 3. Documents published, by author, during 2016 and 1st semester 2021
Source: Scopus, July 2021

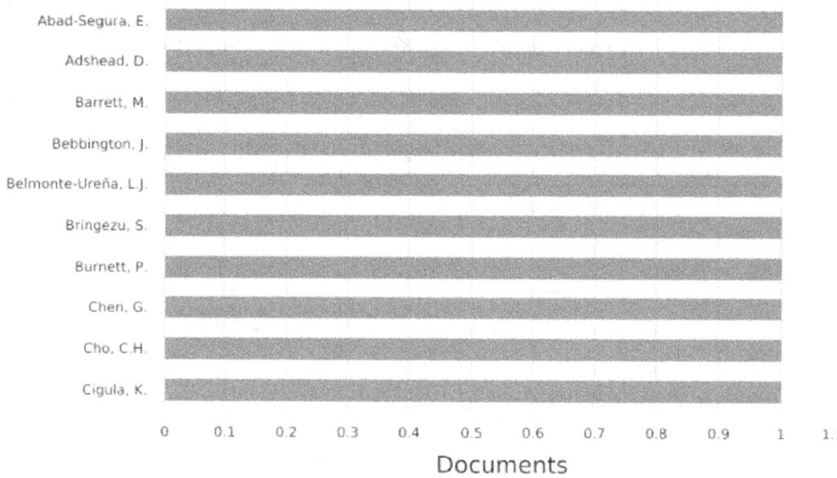

Figure 4. Documents published by affiliation during 2016 and 1st semester 2021
Source: Scopus, July 2021

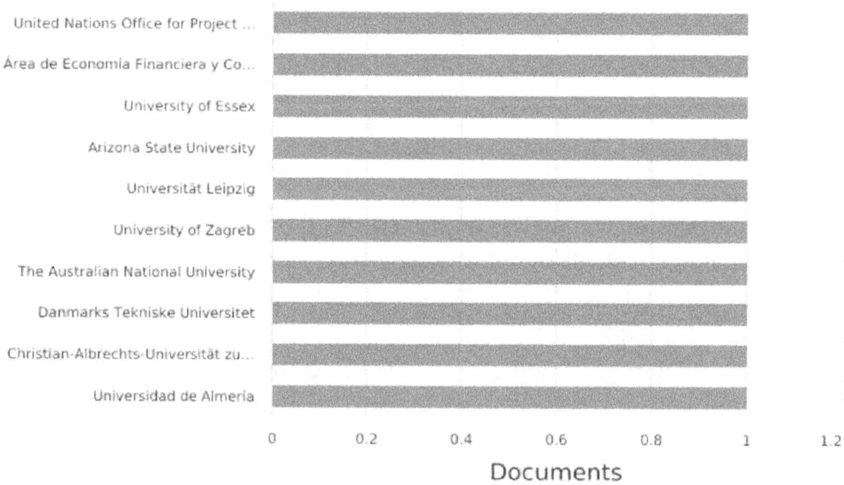

Regarding the main thematic areas of publication, we observe in Figure 7 that the majority (28.2%) is within the Environmental Science item area, followed by Business, Management and Accounting (17.9%), Economics, Econometrics, and Finance 12.8%, and Social Sciences with 10.3%.

However, it appears that Research on Social and Environmental Accounting and SDG implementation is being explored in other areas of knowledge, such as Earth and Planetary Sciences, Agricultural and Biological Sciences, Energy, Engineering, Biochemistry, Genetics, and Molecular Biology, and Decision Sciences.

Figure 5. Documents by country or territory during 2016 and 1st semester 2021
Source: Scopus, July 2021

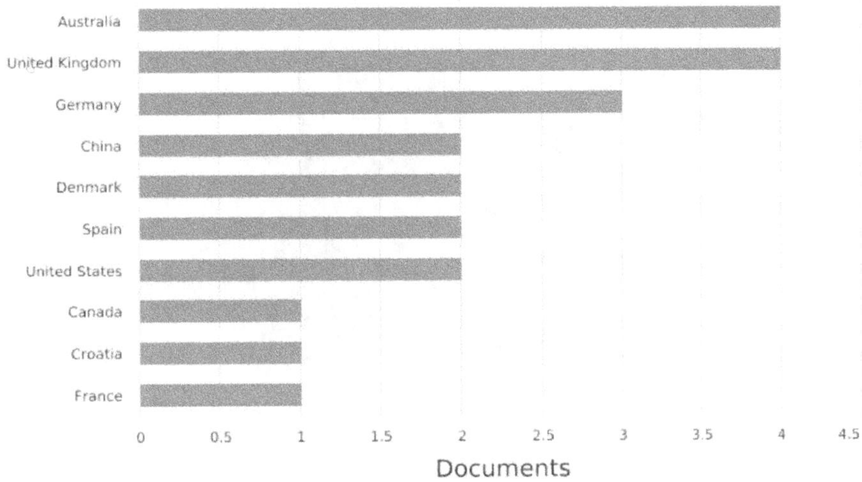

Figure 6. Documents published, by type, during 2016 and 1st semester 2021
Source: Scopus, July 2021

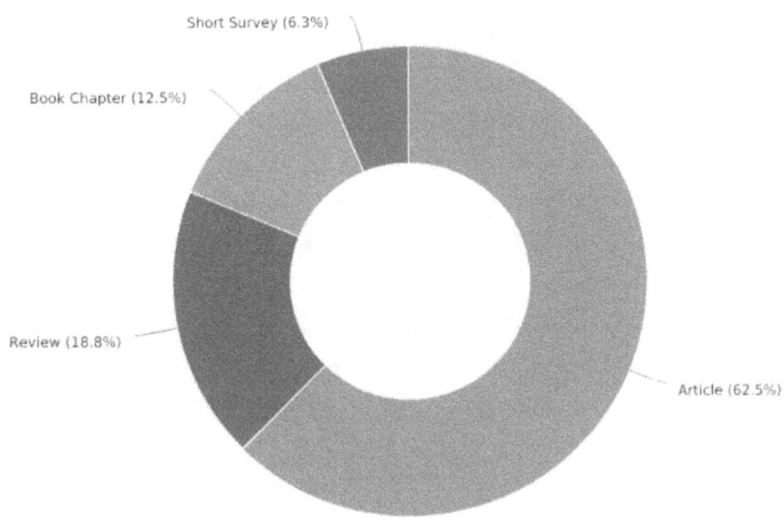

As for the sponsor of the funding of these documents, we can see from Figure 8 that we have one funder per document, equally split between the Australian Research Council, Bundesministerium für Bildung und Forschung, China Postdoctoral Science Foundation, Department of Science and Technology, Hubei Provincial People's Government, Engineering and Physical Sciences Research Council, Horizon 2020 Framework Programme, Hubei Technological Innovation Special Fund, Ministry of Education of the People's Republic of China, National Natural Science Foundation of China, and Norges Forskningsråd (Figure 8).

Figure 7. Documents by Subject Area published during 2016 and 1st semester 2021
Source: Scopus, July 2021

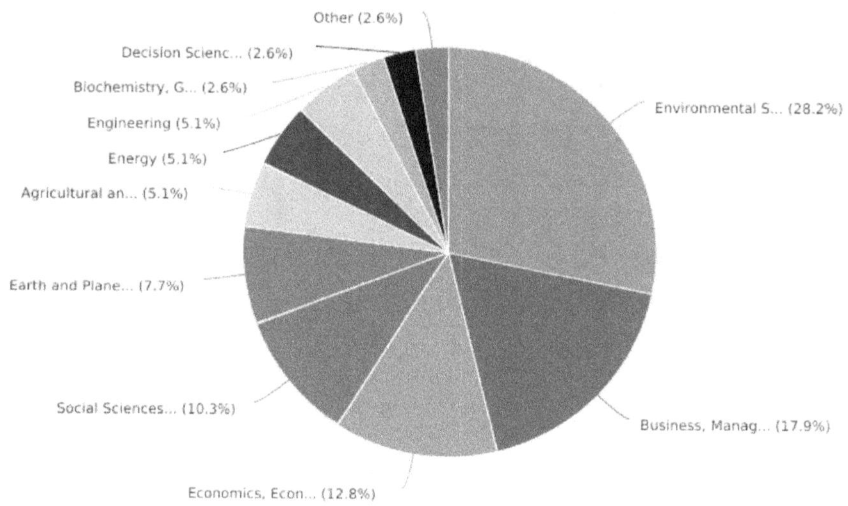

Figure 8. Documents published, by founding sponsor, during 2016 and 1st semester 2021
Source: Scopus, July 2021

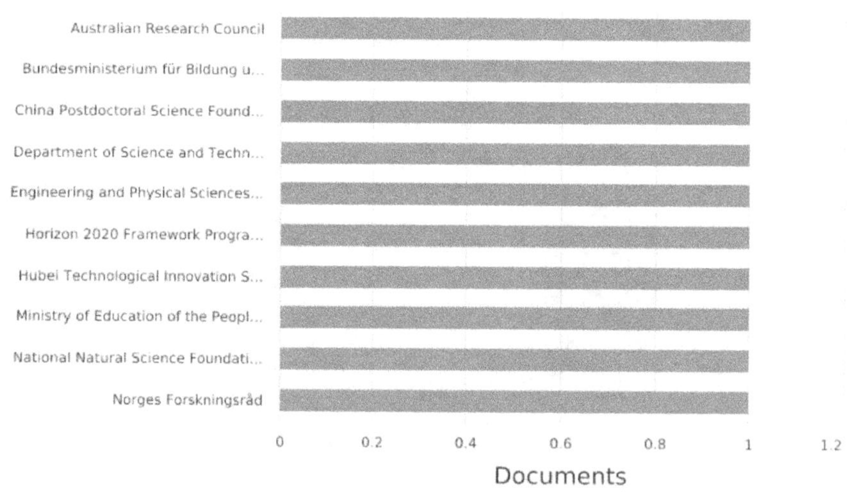

CONCLUSION

From the analysis of the results obtained with the bibliometric review, it is possible to draw some conclusions about the research evolution in the field that crosses the SEA and the SDGs achievement. The scope of the work, the approach and methodology used, the main conclusions and contributions are important to identify the current trends and recent developments for the future of these research areas and at their intersection. Since SEA research could contribute to the achievement of the SDGs it is important to develop and deepen it. Analyzing the scientific production there is a growing interest in publishing at the intersection between SEA and SDGs. This evidence can result from different reasons, the actual global challenges or the health crisis, which forced to be confined, contributing to greater investment in research by academics in accounting and related areas, can be two examples.

Bebbington and Unerman published a theoretical paper, in 2018, in Accounting, Auditing and Accountability Journal, that gathers the most citations are. the authors point out the potential of the SDGs in accounting research. With the UN 2030 agenda a path has been opened which cannot be ignored, there are multiple opportunities offered to the academic community for the affirmation of research in the fields of accounting and sustainability. The journal with the most citations is Sustainability and the journal that presents the higher CiteScore in 2020 is Earth's Future, demonstrating the relevance and timeliness of the research topics to global challenges that affect the lives of communities and society. By identifying publications by author, affiliation, and country, it is verified that, despite the absence of reference authors, the study theme under analysis is of interest to researchers in different parts of the world. It is in Europe where there is the greatest scarcity of research in the area. In identifying publications by type, subject and funder, there is a supremacy of scientific articles and in the area of environmental sciences, followed by accounting and related areas. There is also a funder associated with the publication.

This research provides an overview of the state of the art in the field that crosses the SEA and SGD achievement research. The bibliometric analysis developed demonstrates the little scientific production in this field and highlights its importance. Despite the growth, the intersection of SEA and SGD is still shy and with a small number of case studies. It is important to raise awareness among institutions, government, practitioners, and academics about the importance of developing more research that contributes to the SDG achievement.

The literature review is limited to publications in journals indexed to the Scopus database. Thus, for future research, we suggest the inclusion of other databases, for a more in-depth study of the state of the art in the intersection of SGD achievement and SEA and reporting.

ACKNOWLEDGMENT

This research is financed by National Funds of the FCT—Portuguese Foundation for Science and Technology within the projects UIDB/04928/2020; UIDB/04007/2020, and under the Scientific Employment Stimulus - Institutional Call CEECINST/00051/2018.

REFERENCES

Adshead, D., Román García, O., Thacker, S., & Hall, J.W. (2021). Infrastructure Strategies for Achieving the Global Development Agendas in Small Islands. *Earth's Future, 9*(2), e2020EF001699.

Baker, H. K., Pandey, N., Kumar, S., & Haldar, A. (2020). A bibliometric analysis of board diversity: Current status, development, and future research directions. *Journal of Business Research, 108*, 232–246. doi:10.1016/j.jbusres.2019.11.025

Barrett, M., Watene, K., & McNicholas, P. (2020). Legal personality in Aotearoa New Zealand: An example of integrated thinking on sustainable development. *Accounting, Auditing & Accountability Journal, 33*(7), 1705–1730. doi:10.1108/AAAJ-01-2019-3819

Bebbington, J., & Unerman, J. (2018). Achieving the United Nations Sustainable Development Goals: An enabling role for accounting research. *Accounting, Auditing & Accountability Journal, 31*(1), 2–24. doi:10.1108/AAAJ-05-2017-2929

Bebbington, J., & Unerman, J. (2020). Advancing research into accounting and the UN sustainable development goals. *Accounting, Auditing & Accountability Journal, 33*(7), 1657–1670. doi:10.1108/AAAJ-05-2020-4556

Bringezu, S. (2018). Key strategies to achieve the SDGs and consequences for monitoring resource use. *Managing Water, Soil and Waste Resources to Achieve Sustainable Development Goals: Monitoring and Implementation of Integrated Resources Management*, 11-34.

Cho, C. H., Senn, J., & Sobkowiak, M. (2021). Sustainability at stake during COVID-19: Exploring the role of accounting in addressing environmental crises. *Critical Perspectives on Accounting*, 102327.

Chung, J., & Tsay, M. Y. (2017). A bibliometric analysis of the literature on open access in Scopus. *Qualitative and Quantitative Methods in Libraries, 4*(4), 821–841.

Dal Maso, M., Olsen, K. H., Dong, Y., Pedersen, M. B., & Hauschild, M. Z. (2020). Sustainable development impacts of nationally determined contributions: Assessing the case of mini-grids in Kenya. *Climate Policy, 20*(7), 815–831. doi:10.1080/14693062.2019.1644987

Denedo, M., & Egbon, O. (2021). Africa, from the past to the present: Moving the critical environmental accounting research on Africa forward. *Environmental Accounting*, 265-275.

Farrukh, M., Meng, F., Raza, A., & Tahir, M. S. (2020). Twenty-seven years of Sustainable Development Journal: A bibliometric analysis. *Sustainable Development, 28*(6), 1725–1737. doi:10.1002d.2120

Fusco, F., & Ricci, P. (2019). Qual é o estoque da situação? Uma análise bibliométrica sobre a pesquisa em contabilidade socioambiental no setor público. *Revista Internacional de Gestão do Setor Público*, 1-38.

García-Sánchez, I.-M. (2021). Corporate social reporting and assurance: The state of the art [Información social corporativa y aseguramiento: El estado de la cuestión]. *Revista de Contabilidad-Spanish Accounting Review, 24*(2), 241–269. doi:10.6018/rcsar.409441

Goodwin, K., Wiedmann, T., Chen, G., & Teh, S. H. (2021). Benchmarking urban performance against absolute measures of sustainability – A review. *Journal of Cleaner Production, 314*, 128020. doi:10.1016/j.jclepro.2021.128020

Hülsmann, S., & Ardakanian, R. (Eds.). (2018). *Managing water, soil, and waste resources to achieve sustainable development goals: monitoring and implementation of integrated resources management*. Springer. doi:10.1007/978-3-319-75163-4

Lievrouw, L. A. (1989). The invisible college reconsidered: Bibliometrics and the development of scientific communication theory. *Communication Research, 16*(5), 615–628. doi:10.1177/009365089016005004

Llanos-Herrera, G., & Merigo, J. M. (2018). Overview of brand personality research with bibliometric indicators. *Kybernetes, 48*(3), 546–569. doi:10.1108/K-02-2018-0051

Meseguer-Sánchez, V., Abad-Segura, E., Belmonte-Ureña, L.J., & Molina-Moreno, V. (2020). Examining the research evolution on the socio-economic and environmental dimensions of university social responsibility. *International Journal of Environmental Research and Public Health, 17*(13), 1-30.

Meseguer-Sánchez, V., Gálvez-Sánchez, F. J., López-Martínez, G., & Molina-Moreno, V. (2021). Corporate Social Responsibility and Sustainability. A Bibliometric Analysis of Their Interrelations. *Sustainability, 13*(4), 1636. doi:10.3390u13041636

Omazić, M.A., Vukić, N.M., & Ivanović, I.R., Marszalek, V., & Rösler, B. (2020). The analysis of the state of non-financial reporting in Croatia in 2017 and 2018 [Analyse des stands der nichtfinanziellen berichterstattung in Kroatien in den jahren 2017 und 2018]. *Socijalna Ekologija, 29*(2), 207.

Onat, N. C., Kucukvar, M., Halog, A., & Cloutier, S. (2017). Systems thinking for life cycle sustainability assessment: A review of recent developments, applications, and future perspectives. *Sustainability, 9*(5), 706. doi:10.3390u9050706

Petrushenko, Y., Vadym, A., Vorontsova, A., & Ponomarenko, O. (2020). Sustainable development goals as a tool for strategic planning in communities: A bibliometric analysis of research. In *E3S Web of Conferences* (Vol. 202, p. 03005). EDP Sciences. doi:10.1051/e3sconf/202020203005

Pizzi, S., Caputo, A., Corvino, A., & Venturelli, A. (2020). Management research and the UN sustainable development goals (SDGs): A bibliometric investigation and systematic review. *Journal of Cleaner Production, 276*, 124033. doi:10.1016/j.jclepro.2020.124033

Rajan, R., Dhir, S., & Sushil. (2020). Alliance termination research: A bibliometric review and research agenda. *Journal of Strategy and Management, 13*(3), 351–375. doi:10.1108/JSMA-10-2019-0184

Raman, R., Vinuesa, R., & Nedungadi, P. (2021). Bibliometric Analysis of SARS, MERS, and COVID-19 Studies from India and Connection to Sustainable Development Goals. *Sustainability, 13*(14), 7555. doi:10.3390u13147555

Sáez-Martín, A., López-Hernandez, A. M., & Caba-Perez, C. (2017). Access to public information: A scientometric study of legal versus voluntary transparency in the public sector. *Scientometrics, 113*(3), 1697–1720. doi:10.100711192-017-2541-5

Schaltegger, S. (2018). Linking Environmental Management Accounting: A Reflection on (Missing) Links to Sustainability and Planetary Boundaries. *Social and Environmental Accountability Journal, 38*(1), 19–29. doi:10.1080/0969160X.2017.1395351

Sweileh, W. M. (2020). Bibliometric analysis of scientific publications on "sustainable development goals" with emphasis on "good health and well-being" goal (2015–2019). *Globalization and Health, 16*(1), 1–13. doi:10.118612992-020-00602-2 PMID:32723366

Teixeira, L. M. S., & Ribeiro, M. S. (2014). Estudo bibliométrico sobre as características da contabilidade ambiental em periódicos nacionais e internacionais [Bibliometric study about environmental accounting caracteristics in nacional and international journals]. *Revista de Gestão Social e Ambiental, 8*(1), 20. doi:10.24857/rgsa.v8i1.810

Vardon, M., Burnett, P., & Dovers, S. (2016). The accounting push and the policy pull Balancing environment and economic decisions. *Ecological Economics, 124*, 145–152. doi:10.1016/j.ecolecon.2016.01.021

Voss, R., Quaas, M. F., Stiasny, M. H., Hänsel, M., Stecher Justiniano Pinto, G. A., Lehmann, A., Reusch, T. B. H., & Schmidt, J. O. (2019). Ecological-economic sustainability of the Baltic cod fisheries under ocean warming and acidification. *Journal of Environmental Management, 238*, 110–118. doi:10.1016/j.jenvman.2019.02.105 PMID:30849595

Zhang, J., Li, Y., Liu, C., Qu, Z., Li, F., Yang, Z., Jiang, L., & Fu, J. (2019). Application of set pair analysis in a comprehensive evaluation of water resource assets: A case study of Wuhan City, China. *Water (Basel), 11*(8), 1718. doi:10.3390/w11081718

KEY TERMS AND DEFINITIONS

Bibliometric Analysis: Is the attempt to quantitatively assess the academic quality of journals or authors by statistical methods such as citation rates.

Sustainable Development Goals (SDG): Are the blueprint to achieve a better and more sustainable future for all. They address the global challenges we face, including poverty, inequality, climate change, environmental degradation, peace, and justice. The 17 SDG are integrated - they recognize that action in one area will affect outcomes in others and that development must balance social, economic, and environmental sustainability. Countries have committed to prioritizing progress for those who are furthest behind. The SDG are designed to end poverty, hunger, AIDS, and discrimination against women and girls. The creativity, know-how, technology, and financial resources from all of society are necessary to achieve the SDG in every context.

United Nations (UN): Is an intergovernmental organization aiming to maintain international peace and security, develop friendly relations among nations, achieve international cooperation, and be a center for harmonizing the actions of nations. It is the world's largest and most familiar international organization. The UN is headquartered on international territory in New York City and has other main offices in Geneva, Nairobi, Vienna, and The Hague.

Chapter 3
Stranded Assets in the Context of the Low Carbon Economy:
A Bibliometric Study

Juliano Almeida Faria
Federal University of Sergipe, Brazil

Sonia Maria Da Silva Gomes
Faculty of Accounting Sciences, Federal University of Bahia, Brazil

André Luis Rocha Souza
Federal Institute of Education, Science and Tecnology of da Bahia, Brazil

Viviane Helena Torinelli
Foudation Dom Cabral, Brazil

ABSTRACT

The generation of stranded assets affects the largest oil, gas, and coal companies in the world; the entire resulting structural chain; and also carbon-intensive industries. Thus, this chapter aimed to understand the state of the art of the stranded assets theme produced in the scientific world. The investigation showed that the Paris Agreement contributed to increase the amount of research on stranded assets with a growth of 91.09%, when compared to the pre-agreement period. Although the number of recent researches is significant, the theme is concentrated in two prominent authors who collaborate with each other and have a strong influence on the other works identified in the other clusters, serving as references in works of the sample extracted from the Scopus database. Ben Caldecott and Adrien Vogt-Schilb were identified as prominent on the topic investigated due to the significant number of citations in works around the world.

DOI: 10.4018/978-1-7998-9410-0.ch003

INTRODUCTION

In recent decades, climate change has been increasingly perceived by people and most of its causes are related to intense anthropic actions on natural resources (Caldecott, 2017). Society's responses in terms of mechanisms available for the reuse of natural resources have not yet managed to be sufficient to enable the maintenance of a harmonious interaction with the planet. This harmonization is a challenge for everyone who lives on earth, as the lack of it impacts the economy, biodiversity and the relationship between man and his habitat. Nevertheless, the context of the global pandemic, caused by the virus Sars-CoV-2 (COVID19), causing significant losses to society, came as a catalyst for listening to research that demonstrates how climate risks can affect turnover in the world (WEF, 2021).

The World Economic Forum's The Global Risk Report 2021 lists the greatest threats to global economies. According to this report, social and environmental aspects are the causes of the top five risks in terms of probability and of seven out of ten risks in terms of global economic impact. They lead the risks related to extreme weather conditions, failures in mitigation, adaptation to climate change and infectious diseases (WEF, 2021).

In the same direction, the report of the European Risk Council, signed by Gros et al (2016), already warned that keeping global warming below 2°C requires substantial reductions in global greenhouse gas (GHG) emissions in the coming decades . To reduce these emissions, economies must reduce their carbon intensity. Given current technology, this implies a decisive shift in the energy sector, with a strong impact on fossil fuels and related physical capital (Gros et al, 2016; Pfeiffer et al 2018), which converges with the Paris Agreement, signed in 2015 by 195 countries (including Brazil), which contemplated the goal of maintaining global warming at 2°C compared to the pre-industrial era (IPCC, 2018; PFEIFFER et al, 2018; BOS, GUPTA, 2019; ZAKLAN, WACHSMUTH, DUSCHA, 2021).

In this way, not only these resources, but also all investments made around the main function of fossil fuel extraction, may be compromised (Caldecott 2017; SEI et al, 2020; Wilson and Caldecott, 2021). Examples include drilling platforms, seismic vessels, processing terminals, oil pipelines and oil tankers (GPIF, 2019; Carbon Tracker Initiative, 2021). On the other hand, over the next decade, the energy sector is expected to invest around 7.2 trillion dollars in power plants and grids around the world, much of it in coal and gas plants that emit CO2 (Pfeiffer et al 2018). Although the growth of emissions commitments has slowed in recent years, generators currently in operation still commit to emissions (PFEIFFER et al 2018; VAN DER PLOEG and REZAI, 2020).

This context requires a change in the energy generation paradigm. With regard to the use of fossil fuels, the Paris Agreement requires that 30% of current oil reserves, 50% of gas reserves and 80% of coal reserves are no longer consumed, becoming retained or English assets. stranded assets (GPIF, 2019; BOS E GUPTA, 2019; PAPATERRA, 2019; FIBRAS, 2020; SEI ET AL, 2020; SEN AND VON SCHICKFUS, 2020).

According to Caldecott, Howarth and McSharry (2013, p. 13), "stranded assets are assets that will suffer from unforeseen events or premature reductions, devaluations or conversion into liabilities". For the International Energy Agency - IEA (2013), stranded assets are those investments that have already been made, but that at some point before the end of their economic life (as assumed at the investment decision point), they are no longer able to obtain an economic return. Deloitte (2016), on the other hand, considers that they are assets that have become obsolete or in default, but which must be recorded in the Balance Sheet as a loss of profit.

In this vein, it is highlighted that the aforementioned concept is still fragmented (Shimbar, 2021) denoting the need for a better understanding of the scope through scientific research. The existence of stranded assets has been presented by Perez (1985) and Caldecott, 2017) from a perspective inherent to the process of changing economic paradigms. Perez (1985) and Caldecott (2017) cite the era of steam, technology and oil to exemplify some of the paradigms already experienced by world society. At this time, the current paradigm shift is recognized, from the oil era to the low carbon era, which is characterized by the mass production of energy from renewable sources, that is, with low consumption of fossil sources characterized by their finiteness and high potential for GHG generation (VAN DER PLOEG AND REZAI, 2020; SHIMBAR, 2021).

Along these lines, it is noteworthy that research on stranded assets has a relatively restricted scope, with a sectorial and geographic imbalance evident in the literature. Furthermore, stranded assets are mostly discussed with little connection to working in other research domains; stranded assets research focuses primarily on calculating the accumulated amount of assets that can be stranded over the long term; there may be a disconnect between value destruction and the value creation aspects of the transition to a low carbon economy; and there appears to be a paucity of empirical studies to answer many questions about the market implications of stranded assets, particularly in the energy sector (Shimbar, 2021). Therefore, more advanced studies are needed to develop a deeper understanding of the macro and microeconomic impacts of stranded assets, a promising path for future research.

Thus, this research aimed to understand the state of the art of stranded assets produced in the scientific world. Specifically, it seeks to identify prominent authors in the study of stranded assets with articles published in international journals; know the main occurrences of research concentration related to the stranded assets theme; and identify trends in driving research on the impacts of stranded assets in the context of the low-carbon economy.

This research is justified by the importance that the subject has been acquiring, especially since the Paris Agreement; it is also an area with a wide profusion of meanings; fragmented literature on stranded assets (Shimbar 2021) and the low volume of research on the topic (Bos and Gupta, 2019; Ansari and Holz, 2020). Thus, it is necessary to identify and gather the main articles published in journals around the world, authors, respective research networks and trends on stranded assets. With this, it will be possible to inform the reader which researches have been found so far, describing opportunities to expand existing knowledge, that is, as drivers for future research.

The results presented allow us to know the profile of the authors of the topic, the citations and references used and the clarification of the most significant groups of themes and approaches, reaching new horizons of research, in addition to providing insights into future investigations and the volume and characteristics of research published (Oliveira et al, 2019; Simões and Souza, 2020). Thus, the research can contribute to compressing all those interested in a clearer understanding of the impacts of stranded assets in the current and initial low-carbon era, including helping international policy makers able to avoid further actions that strengthen GHG mitigation. After all, it is assumed that a small number of researches on a recent and promising subject may mean that the subject needs to be further studied given its potential to contribute to the entire planet.

The research is structured in this introduction, which addresses the context and macro aspects related to the stranded assets theme. Next, the theoretical and conceptual aspects on the subject are presented, through a literature review. Therefore, there are the methodological procedures adopted to achieve the chapter's objectives. In the fourth section, there are the research results and, finally, in the fifth and last section, the final considerations.

LITERATURE REVIEW

The Paris Agreement and the Low Carbon Era

At the climate summit held in April 2021, global leaders signaled actions with short, medium and long-term impacts for the entire planet, including the confirmation of actions in order to direct world economies towards compliance with the Paris Agreement, signed in 2015. However, the intention of the leaders opens the opportunity for several questions about the next investments supported by a structure of carbon generation. In this phase of transition to the low carbon era, it may not be as fast and abrupt as the burning of fossil fuels may exceed the carbon budget signed in the aforementioned agreement (VAN DER PLOEG, 2020).

To follow a consistent 1.5°C path, the world will need to decrease fossil fuel production by about 6% a year between 2020 and 2030. Instead, however, countries are planning and projecting an average annual increase of 2%, which by 2030 would result in more than double the production compatible with the 1.5° C limit, including encouraging intensive use of carbon in the post-Covid-19 pandemic period (SEI ET AL, 2020; WILSON AND CALDECOTT, 2021).

This scenario raises questions about the feasibility of complying with the Paris Agreement and, on the other hand, sheds light on the need to reduce the volume of GHG emissions by 2050. Gros and Schoenmaker (2016) argue that a transition to the era Low carbon needs to be slow and careful, otherwise the recognition of sudden obsolescence caused by a rapid change in environmental policy would be inevitable, and could precipitate a radical and not fully anticipated reassessment of carbon intensive assets, including fossil fuel and other reserves active.

The current structure of production of goods and services is strongly based on carbon emissions, increasing the challenge for the transition from a decarbonized economy, especially with regard to the impacts of the generation of stranded assets. In essence, the transition into other times has also generated stranded assets (Peter 1985; Caldecott, 2017). However, this time, it adds the fact that the volume of world production is more significant than in previous transitions, making it unprecedented. Thus, the consequences of this transition, whether accelerated or not, for this and future generations, are still being studied. With financing policies, the financial market can avoid a disruptive and disorderly energy transition and macroeconomic shocks, by directing capital towards climate-friendly technologies (SEN E VON SCHICKFUS, 2020).

The threshold set for global warming implies a tighter carbon budget, which could at some future date, as yet unknown, make fossil fuel extraction more voracious and accelerate global warming (Van Der Ploeg, 2020). However, this position compromises compliance with the Paris Agreement is under threat, given that, although the growth of emission reduction commitments has advanced in recent years, current emission levels still point to emissions above levels compatible with the scenario average of 1.5°C-2°C (PFEIFFER ET AL, 2018).

On the other hand, if the 2030 threshold is a level less stringent than cost-effective, its path needs to become heavily accelerated after 2030 (Zaklan, Wachsmuth and Duscha, 2021). This is not only more expensive, but it can also lead to greater impact on the generation of stranded assets due to the long investment cycles inherent in the industrial sector. This is another argument for setting a strict 2030 threshold, thus providing a clear sign of long-term investment. In contrast, setting the 2030 limit in excess of cost (stranded assets) versus benefit ($CO2$ reduction), for example, leading to a linear limit trajectory, can result in high investments in premature technologies (Zaklan, Wachsmuth and Duscha, 2021), also impacting the cost of transitioning to a low-carbon economy. An adequate transition values the social cost of transition, but can lead to premature withdrawal of existing polluting capacities and significant private costs in the form of lost assets (ROZENBERG, VOGT-SCHILB & HALLEGATTE, 2020).

Without concerted global action, individual fossil fuel producers may simply try to extract as much as they can in the short term. The global community has a role to play in ensuring cooperation to avoid the green paradox, addressing the issue of carbon leakage and ensuring that developing countries can pursue a sustainable development path (Bos and Gupta, 2019). Sharing the remaining carbon budget in a cost-effective and equitable way by allocating country-specific extraction limits can be a key to preventing world temperatures from rising above the 1.5-2°C limit.

In fact, keeping global warming below 2°C will require substantial reductions in global GHG emissions over the next few decades. To reduce emissions, economies must reduce their carbon intensity at a pace not yet agreed upon among researchers. Inevitably, the transition to a low-carbon economy negatively affects the asset value and valuation of carbon-intensive industrial companies, especially the fossil fuel sector (SHIMBAR, 2021).

The end result is that each future scenario includes climate-related financial risk, although the level and form of underlying uncertainty varies (Sen and Von Schickfus, 2020). A high carbon scenario would generate considerable financial risk from uncertain extreme events and adverse trends. A low-carbon path would moderate such climate hazards, but would produce transitional financial risk from the possible adoption of new climate policies and technologies, including stranded assets risks (Rudebusch,2021). The current environment is conducive (and necessary) to the development of more studies in order to shed light on the issues that guide the future uncertainty inherent to the shift to the low carbon era, its main driver so far, the Paris Agreement.

Concepts and Impacts of Stranded Assets

The concepts of stranded assets are still in the process of consolidation given their complexity and novelty, especially when discussed in the context of the current transition to the low carbon era (Shimbar, 2021). The breadth of stranded assets concepts reveals the academic effort in the search for understanding the theme that is used in different contexts and, given its explanatory potential, for understanding the impacts on the global economy. In the context of the assessment and accounting of stranded assets, accounting professionals have international accounting standards (to which Brazil is a signatory), the English International Accounting Standards (IAS), which guide the treatment in the face of the identification of unproductive assets in the entities' equity.

For this purpose, IAS16 and IAS36 are used as guiding instruments on how to account for the identified non-productive assets and how to assess the assets' capacity to become idle, respectively. In Brazil, NBCTG01 (R4) establishes procedures to ensure that its assets are accounted for at an amount that does not exceed their recoverable amounts. If this is the case, the asset is characterized as subject to loss recognition. The Standard requires an entity to recognize adjustment for impairment losses and establishes appropriate disclosures (CFC, 2021).

In the context of energy production, the IEA (2013) defines stranded assets as those investments that have already been made, but which, at some point before the end of their economic life (as assumed at the investment decision point), are no longer capable to obtain an economic return. The Carbon Tracker Initiative (2020) also uses this definition of economic loss and adds that they are also a result of changes in the market and regulatory environment associated with the transition to a low carbon economy. The stranding of climate-related assets still refers to the depreciation of assets - such as resource reserves, infrastructure or industries - resulting from unforeseen changes, such as the tightening of climate policies (ANSARI E HOLZ, 2020).

The Generation Foundation (2013, p. 21) defines a stranded asset "as an asset that loses economic value well ahead of its expected useful life, whether as a result of changes in legislation, regulations, market forces, disruptive innovation, social norms or environmental shocks." Caldecott, Howarth, and McSharry (2013) proposed a definition to cover all of these different definitions, in which they present stranded assets as assets that suffered from unforeseen events or premature reductions, devaluations or conversion into liabilities. stranded assets is a term that indicates physical and financial assets that suffer unexpected loss of value with abrupt and premature depreciation, becoming practically non-negotiable (CAMPOS, 2021).

Among the existing assets, most likely to become stranded assets, the assets that are at greatest risk of becoming stranded during the energy transition to the low-carbon era, as society seeks to limit global warming below 2°C (Shimbar) deserve to be highlighted., 2021). The fossil fuel sector is more prone to the emergence of stranded assets and should receive more attention from national and international policy makers (Ansari and Holz, 2020). There are already examples of coal mines, coal and gas plants and other hydrocarbon reserves that will be trapped by the low carbon transition (GROS et al, 2016; PFEIFFER et al, 2018, GPIF, 2019; PAPATERRA, 2019; ANSARI and HOLZ, 2020, FIBRAS, 2020; SEI et al, 2020; SEN and VON SCHICKFUS, 2020; CAMPOS, 2021).

Assets in fossil fuel industries are at risk of becoming stranded and losing market value due to unexpected advances in renewable technology and governments stepping up climate policies in light of Paris commitments to limit global warming to 1.5 or 2° Ç. Stranded assets arise due to uncertainty about the future timing of these two types of events and substantial costs of adjusting intertemporal and cross-sector investment (Van Der Ploeg and Rezai, 2020). It should be added that not only the raw material may be retained, but also all assets linked to the extraction, production, processing and distribution structure, such as drilling platforms, vessels, processing terminals, pipelines, oil tankers, up to the end as service stations. sale of fuel.

Companies that earn significant revenues from fossil fuel-related activities, rely on fossil fuel reserves for their market valuations, or invest heavily in fossil fuel-related activities (such as exploration), run the risk of becoming stranded assets (GPIF, 2019). Decarbonizing the energy sector to limit the temperature rise well below 2°C will result in lost assets and replacement of the capital stock before the end of its technical useful life, resulting in US$927 billion of stranded assets in the energy sector global by 2050 (SAYGIN ET AL, 2019).

According to Van Der Ploeg and Rezai (2020), asset stranding primarily affects the top 20 oil, gas and coal companies, responsible for at least a third of global warming since 1965, and also affects carbon-intensive industries, such as steel, aluminum, cement, plastics and greenhouse horticulture. Sen and Von Schickfus (2020) believe that additional research in similar contexts can help to generalize these results, including to identify the important factors in the formation of expectations regarding climate-related risks.

As such, stranded assets encompass a variety of different factors, including: economic stranding - due to a change in relative costs/prices; physical stranding - due to distance / flood / drought; and regulatory backlog – due to a change in policy legislation (Carbon Tracker Initiative, 2020). On the other hand, Caldecott, Howarth and McSharry (2013) proposed a typology for different risks related to the environment that could cause stranded assets, as shown in Table 1.

Table 1. Typology of risks causing stranded assets

Type	Risk Class	Exemples
Physical	Environmental Challenges	Climate change, degradation of natural capital.
	Transformations in natural and fossil resources	Shale gas abundance, phosphate shortage.
Social	New government regulations	Carbon pricing, air pollution regulation.
	Clean technology cost reduction	Photovoltaic solar energy, wind energy, electric vehicles.
	Evolutions in social norms and consumer behavior	Campaign on fossil fuel divestment / product certification.
	legal changes	Fiduciary Obligations or Publicity or Information Disclosure Requirements.

Source: Direct Translation from Caldecott, Howarth e McSharry (2013, page 09).

The risks causing stranded assets apply in different ways to all countries around the globe and, although they are more intense in developed competitive markets, while leaving other countries behind can only lead to (more) mechanisms for the imprisonment of fossil resources, inflicting ripple effects across the economy and undermining long-term sustainable development (Bos and Gupta, 2019). In this context, Papaterra (2019) states that Brazil, as an important social actor in the world oil and natural gas sector, sooner or later will have to face the problem of the risk of setting aside its oil and natural gas resources. Lagging countries need to be sensible when deciding which resources to develop to avoid carbon lock-in and whether the phasing in of renewable energy could avoid the creation of stranded assets (BOS and GUPTA, 2019).

One of the paths is associated with the financial market, given that, in order to align capital flows with the objectives of the Paris Agreement, financial institutions must decarbonise primary market transactions, as they continue to provide new capital for the real economy which can create carbon lock-in and the risk of stranded assets (Wilson and Caldecott, 2021). According to Saygin et al, (2019) delaying actions to mitigate climate change until 2030 doubles the value of stranded assets and countries should consider the characteristics of the asset's age profile in their environmental policy. Early action and prevention of investments in new carbon-intensive assets can minimize the risks of stranded assets.

According to Van Der Ploeg and Rezai (2020) and Sen and Von Schickfus, (2020) a disorderly transition to a carbon-free economy will lead to the generation of stranded assets and a significant increase in lawsuits from those who are directly or indirectly impacted by the restriction of use of these assets. Furthermore, the authors add that institutional investors should be aware of these financial risks. In the meantime, Bos and Gupta (2019) signaled the need to develop more studies on stranded assets because with the progressive introduction of renewable energies inherent to the low carbon era, it is necessary to decide which resources to develop to avoid blocking energy. carbon in order to avoid the creation of stranded assets.

METHOD

In this investigation, an eminently bibliographical study was carried out, as it was proposed to identify and discuss the topic in evidence, seeking the content produced as a basis. The main articles published in the Scopus international database were identified by means of bibliometric mapping using articles published from 1989 to July 2021, when this research was closed. This technique allows the researcher to map the state of the art in a given area through the characterization of bibliometric parameters (Oliveira et al, 2019; Simões and Souza, 2020) based on a co-occurrence matrix. In this way, the distance between the nodes of the bibliographic network is observed on a map, which represents the existing relationship between the researches.

To carry out this study, the free software VOSviewer (Visualization of Similarities Viewer) developed by Van Eck and Waltman (2010) was used. Recent researches such as García-Machado (2018), Palludeto and Felipini (2019), de Oliveira et al. (2019), Burton, Kumar and Pandey (2020) and Simões and Souza (2020) used this software whose main characteristic is the analysis and visualization of a large set of bibliographic data considering a distance-based approach (based on the approximation relationship between surveys). Using this software, the analyzes shown in Table 2 were performed using bibliometric mapping.

Table 2. Procedures for meeting the respective specific objectives of the work

Collection/analysis instrument	Specific objetives
(i) Bibliometric Network of Co-authorship – Observes how research is distributed around the world and the existing ties of publication.	1 - identify prominent authors in the study of stranded assets with articles published in international journals;
(ii) Citation Analysis – Identifies the research most cited by the articles in each sample period cited by the articles in the period under analysis.	
(iii) Analysis of Reference Quotations – Observes the relationships between different publications, as the greater the number of times two articles are cited by a third party, the stronger the relationship of quotation between the two cited works.	2 - know the main occurrences of research concentration related to the stranded assets theme.
(iv) Co-occurrence Map – Identifies the main search terms used in the articles by year of publication, indicating history and possible future research trends.	3 - identify trends in directing research on the impacts of stranded assets

Source: Adapted from Souza e Simões (2020).

The sample of this research is formed by 294 articles published in English, obtained from the Scopus database, meeting the criterion of containing in its content the words "stranded assets". It is important to note that the selection parameters described do not include the epistemological classification of the analyzed articles.

ANALYSIS

Next, Table 3 shows the number of articles per period, the evolution of the growth rate of publications, as well as the mean and standard deviation.

More specifically, Figure 1 shows the evolution of the number of publications per year, with publications up to July 2021 (23 publications) already surpassing the historical average, which is 10.65 and approaching the average post-Paris Accord (2015), which is 28.33.

The Paris Agreement (2015) represented a milestone in the development of research on stranded assets, creating an upward convergence point of scientific production, which is why, in this initial stage, the sample period was divided into two: pre-Agreement of Paris and post-Paris Agreement. Thus, the analysis was directed to the post-Paris Agreement period.

Table 3. Initial description of the sample, by period

Period	Number of papers	Nominal Growth	Media	Standard Deviation	Average Growth
Pre 2015	101	-	5,05	2,93	-
Post 2015	193	91,09%	28,33	11,38	461,06%

Source: Research data

Figure 1. Number of publications
Source: Research data

Through the construction of the bibliometric network of co-authorship and citation analysis, it was possible to meet the specific objective 1, from which it was possible to identify the prominent authors in the study of stranded assets with articles published in international journals, as shown in Figure 2.

Figure 2. Prominent authors in the study of stranded assets
Source: Research data

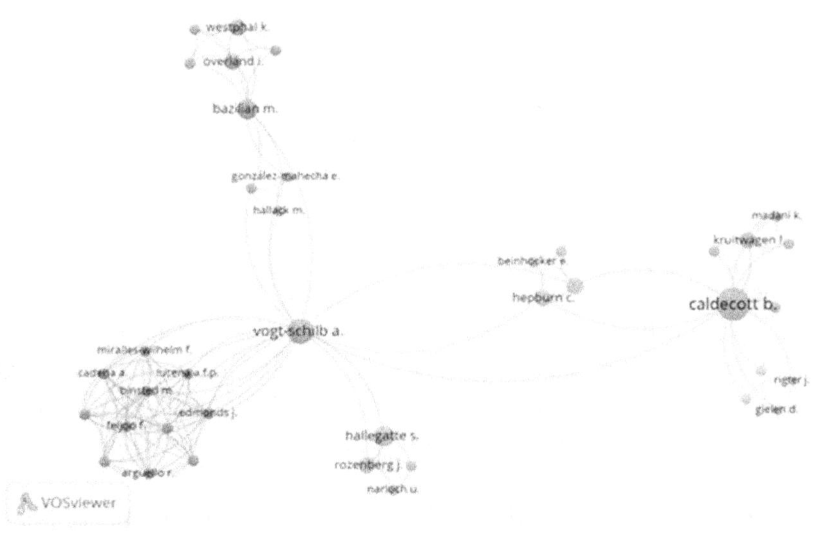

In this graphical representation, each circle represents an author and the larger the circle size, the greater its influence on the research collaboration network. In turn, the distance and width of the link reflect the degree of collaboration between the authors. It is possible to observe the collaboration between the influential authors of the network, Ben Caldecott and Adrien Vogt-Schilb, determined by the number of publications they had together. Ben Caldecott is a British researcher and specialist in sustainable finance, and serves as the founding director of the Sustainable Finance Program at Oxford University's School of Business and Environment. Adrien Vogt-Schilb is a senior economist, specialist in climate change for the climate change division of the Inter-American Development Bank.

The analysis of citations identifies the research most cited by articles in the post-Paris agreement sample period, as well as the research cited and published in the analyzed period. By criterion of graphic exposure, Figure 3 shows only articles that received at least two citations in the sample.

Figure 3. Most cited authors in the sample
Source: Research data

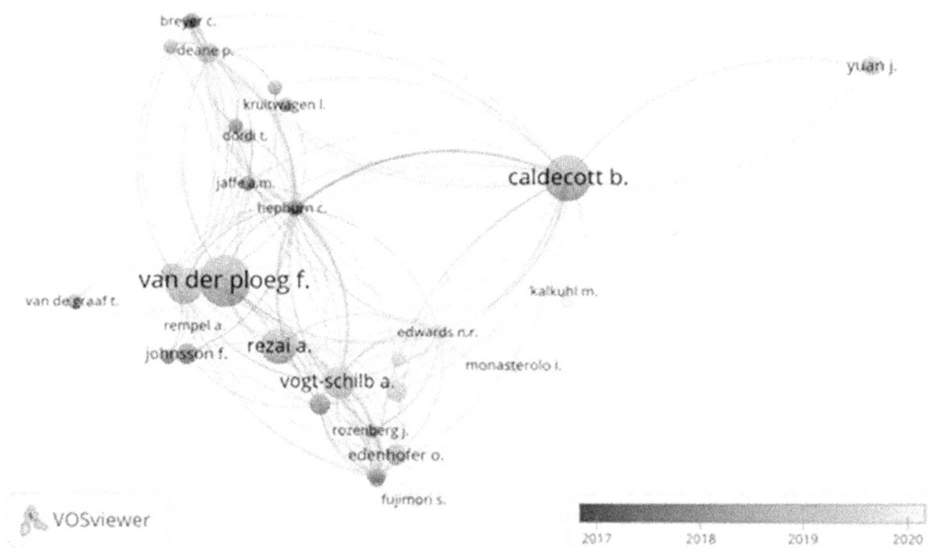

Figure 3 shows the age perception of the cited works, given that throughout the most recent researches it has been citing the authors Ben Caldecott and, more recently, Adrien Vogt-Schilb. Of note is the author Frederick van der Ploeg, who is Professor of Economics and Research Director at the Oxford Center for the Analysis of Resource Rich Economies, University of Oxford and Adjunct Professor of Economics, University of Amsterdam, having also served as Co-Investigator, Advanced Grant to Combat Climate Change.

Furthermore, it is also possible to verify authors with research that, despite being published more recently (in yellow in Figure 3), already receive at least two citations, namely: Arthur Rempel, University of Amsterdam; Matthias Kalkuhl, Professor of Climate Change, Development and Economic Growth at the Faculty of Economic and Social Sciences at the University of Potsdam; Irene Monasterolo, Assistant

Professor of Climate Economics and Finance at the University of Vienna; Shinichiro Fujimori, Kyoto University Associate Professor (Higher School of Engineering); and Ali Shimbar, Macquarie Business School, Macquarie University, Sydney, Australia.

This result suggests that the stranded assets theme has motivated authors in different parts of the world to develop research highlighting the worldwide concern about its scope. However, there is still a low concentration of this research, substantiating the recent approach to the topic, considered recent in academia, reinforcing the need for continuity in the development of new research. At the same time, it reinforces the expectation of a continuous growth in the volume of publications, while more authors can create scientific collaboration ties in favor of a better understanding of the impacts of stranded assets in the current shift to a low-carbon economy.

To meet specific objective 2 - to know the main occurrences of concentration of research related to the stranded assets theme - the cocitation analysis was used. A co-citation occurs when two articles are cited by a third party and thus, the greater the number of times two articles are cited by a third party, the stronger the co-citation relationship between the two cited works (Palludeto and Felipini, 2019) . It is noteworthy that the co-citation is aimed at identifying the references used by the articles in the sample. However, as several of the cited articles were published before the period of analysis of this research, the cited articles tend not to be part of the sample period (Simões and Souza, 2020). In the cocitation network map (Figure 4), each circle represents an article and the larger the circle size, the greater the amount of cocitations that occurred. In the analysis of quotations, the sample articles are grouped into clusters, considering the minimum number of 02 citations.

Figure 4. Cocitation analysis
Source: Research Data

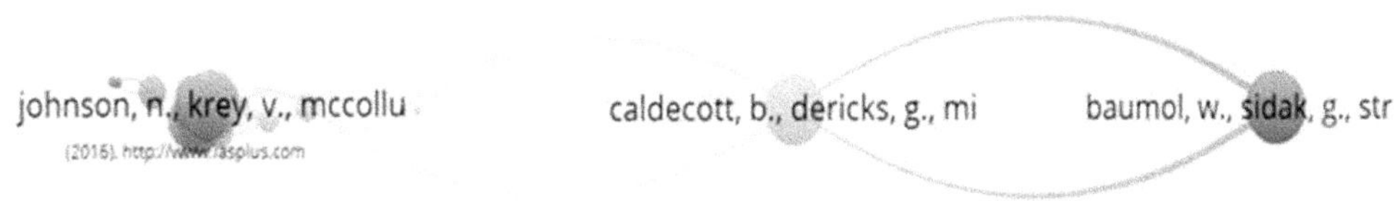

As a result of the sample, the works with the highest number of references in references reveal the influence of these works in the discussion of the stranded assets theme, since they are the most strongly cited in the following works.

The research by Baumol and Sidak (1995), observed in the analysis of cocitations (Figure 4), approached the discussion from the perspective of stranded costs, for the electricity sector. According to the authors, these costs represent expenses incurred by a utility company in the past when fulfilling its obligation to serve all customers in the area in which it held an exclusive franchise, granted to it under the traditional regulatory regime. The authors concluded that the role of regulatory bodies is fundamental in protecting consumers and in evaluating energy companies' assets, highlighting the conduct of the transition to avoid generation and stranded costs (retained costs).

The research by Johnson et al (2014) (Figure 4) sought to better understand the relationship between short-term climate policy and stranded coal capacity (assuming a long-term goal of limiting warming to 2 °C) and explores strategies to reduce idle capacity. The authors mainly suggest that short-term strengthening of climate policy (ie, reducing the global greenhouse gas emission target by 2030) in general reduces stranded coal capacity and its costs. This can occur by reducing electricity demand through improvements in energy intensity and/or by keeping existing plants operating through useful life extensions.

Finally, research by Caldecott, Dericks, and Mitchell (2015) (Figure 4) analyzed the world's most carbon-inefficient and highly polluting global energy stock. According to the authors, their research results could encourage companies to: i) publicly confirm their exposure and the proportion of their total portfolio generation that is subcritical, ii) disclose which proportion is most at risk, for example, in terms of intensity carbon, air pollution and water stress, iii) disclose how much of your pipeline is subcritical and how this may change portfolio risk exposure, and iv) describe the strategies employed at an asset level and in a portfolio to minimize carbon intensity and manage harmful contributions to air pollution and water stress.

It can be concluded that the articles presented have a strong relationship with the other clusters, as well as being used by other researches together, serving as references in the works of the sample. The work of Baumol and Sidak (1995), as one of the oldest, may suggest that the researches moved to a stage more focused on finding solutions that would contribute to improving the performance of energy companies to advance in discussions for other companies, especially from the Paris Agreement (2015).

The Co-occurrence Map resource was used to identify trends in directing research on the impacts of stranded assets, the third objective of this research. This map identifies the main search terms used in the articles by year of publication, shown in Figure 5. This feature allows identifying the frequency of co-occurrence of terms in the main fields of the sample, presenting them by year of publication, allowing interpretation of the associated theme flow to stranded assets over the period studied, in this case, post-Paris Agreement.

Figure 5. Map of co-occurrence by groups

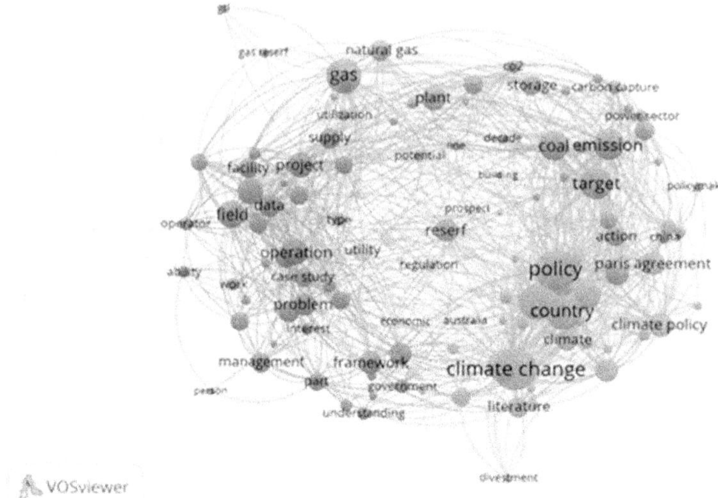

Figure 6. Map of co-occurrence by period
Source: Survey data

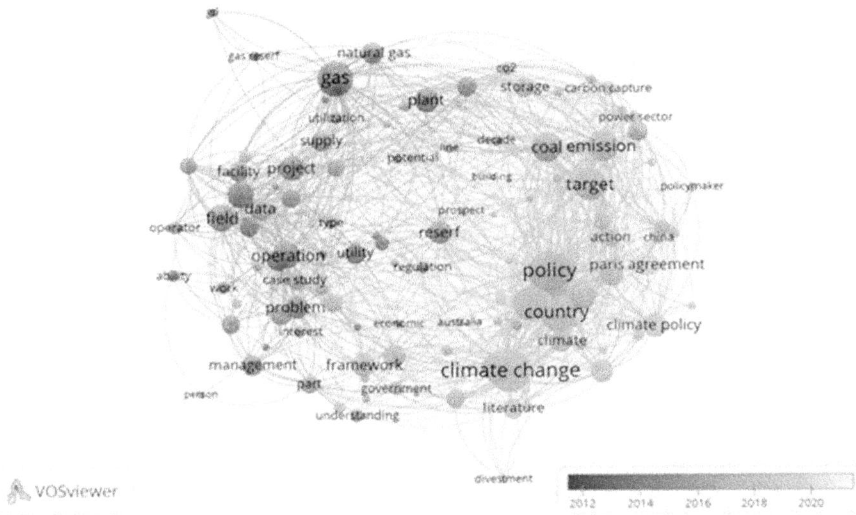

118 terms mentioned at least seven times were found in the sample, representing a gross total of 2,137 occurrences in the sample. From this base, VOSViewer allows the presentation by groups (clusters), revealing greater proximity of terms depending on the size of the circle which is represented, in addition to its connection through lines. It can be seen that the volume of citations of more recent articles is growing, which means that there has been a greater demand for the topic in recent years. From Figure 6, the eight main terms found in the sample were selected, based on the highest percentage of citations in the text map. Although these items represent 6.78% of the total of 118 occurrences, the number of these occurrences repeated in the sample is equivalent to 14.09% of the total (Table 4).

Table 4. Quantitative of main terms

Termo	%	Termo	%	Termo	%	Termo	%
Climate change	2,29	*Country*	1,92	*Coal*	1,59	*Climate policy*	1,31
Policy	2,25	*Transition*	1,87	*Investor*	1,59	*Framework*	1,26
Total							**14,09**

Source: Survey data

Directing the look at these terms, that is, the most frequent in recent research, the temporal interconnection of these terms with others is evidenced, which broadens the perspective of future research trends (Figure 7).

Figure 7. Main terms and intertemporal connections
Source: Survey data

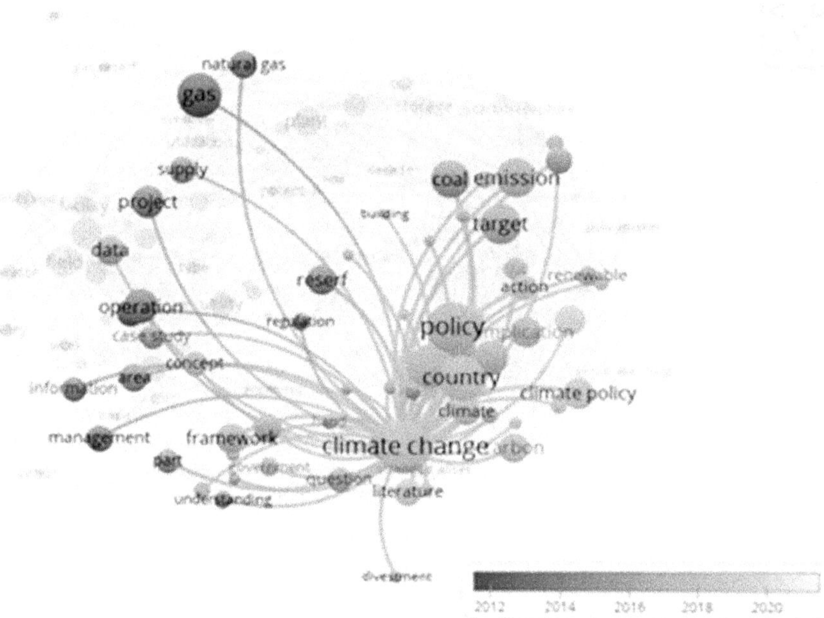

In addition, the association of the term coal (coal) appears to be stable in terms of the volume of citations, however, it maintains an intense relationship with the gas cluster and terms such as target, emission and climate change in a more recent cluster, confirming the impact of this discussion, especially in fossil fuel extractive companies (Van Der Ploeg and Rezai (2020). This term is also associated with the term investor. This is because, as has been predicted in the literature (Sen and Von Schickfus, 2020), it reflects the concern of investors in companies with intense generation of GHG in their operations, including coal extractive. This type of industry is characterized by high investments with respective durability (long useful life) which can be strongly compromised in view of the goals of each signatory country of the Paris Agreement (GPIF, 2019; SAYGIN et al, 2019).

Finally, it is worth emphasizing the positioning of the term framework both in terms of time and location between clusters linked to technical operational terms (pre-Paris Agreement research) and more recent terms (post-Paris Agreement) as inherent to the climate change cluster . This can mean a continuous flow process of scientific investigations, on stranded assets supported by the discussion of a framework with a more pragmatic and conservative approach for a more structured perspective on current climate policies and proposals to fulfill the goals of the Paris Agreement.

CONCLUSION

This research aimed to contribute to the understanding of the profile of international research on stranded assets in the context of the low carbon era. Specifically, we sought to identify prominent authors in the study of stranded assets with articles published in international journals; know the main occurrences of research concentration related to the stranded assets theme; identify trends in research driving the impacts of stranded assets.

It was identified that authors/researchers Ben Caldecott and Adrien Vogt-Schilb were identified as prominent on the instigated topic, supported by the significant number of citations in works around the world. The collaboration between them can be seen, as well as the most cited were Ben Caldecott, Adrien Vogt-Schilb and Frederick van der Ploeg. It is worth highlighting the finding of recent works and citations from authors namely: Arthur Rempel, Matthias Kalkuhl, Irene Monasterolo, Shinichiro Fujimori and Ali Shimbar. With regard to cocitations, the research by Baumol and Sidak (1995), Johnson et al (2014) and Caldecott, Dericks and Mitchell (2015) stood out.

The investigation showed that the Paris Agreement contributed to increase the amount of research on stranded assets with a growth of 91.09%, when compared to the pre-Agreement period. In addition, the volume of citations of more recent articles increased, denoting an effort to deepen the theme and its impact on the global economy. Finally, the most common terms (14.09% among all terms) and their respective associations were surveyed in a structured network demonstrating an intertemporal connection between them, supported by the timeline of pre- and post-Paris Agreement publications. Note the direction for research related to climate policy, framework, renewable energy among others in a main cluster of Climate Change. It can be said that the researches are in the initial phase of building the knowledge and gaps identified on the stranded assets topic regarding the context of transition to a low-carbon economy.

Despite the consensus on the financial risk, in the oil and gas sector, arising from the global efforts towards a low carbon economy, there are several investors today betting on these companies in search of higher dividends. For these investors, global efforts towards a low-carbon economy discourage new investments in the sector, reducing capex and causing a high level of cash balance in some companies in the sector. Furthermore, with the reduction of investments in the sector, demand may outstrip supply for an indefinite time horizon, pushing commodity prices up, improving cash and profits.

Unlike investors who are adopting the strategy of being the first to leave this sector, those who remain invested in it foresee a time horizon in which the oil and gas industry can still generate a lot of return for its shareholders. Therefore, the discussion of stranded assets has a significant time factor to be noted. For impact investors, leaving now is justified by minimizing negative socio-environmental externalities, regardless of any losses in the potential future return of discarded investments. However, for the vast majority of investment managers, who work with mandates from investors who do not intend to give up on competitive market returns, the main question revolves around identifying the ideal withdrawal time, demanding the intensity of studies that they are able to assist investors in this transition period.

This research is limited in itself to the choice of method, especially in the chosen database and in the technique of analyzing bibliometric data. For future research, it is suggested to deepen the study of the techniques used by the main authors, as well as to identify collaborative research networks on stranded assets in order to measure the synergy directed to the theme at this historic moment characterized by the shift to a low-carbon economy and uncertainty in relation to the pandemic context.

REFERENCES

Ansari, D., & Holz, F. (2020). Dawud; HOLZ, Franziska. Between stranded assets and green transformation: Fossil-fuel-producing developing countries towards 2055. *World Development*, *130*, 104947. doi:10.1016/j.worlddev.2020.104947

Bos, K., & Gupta, J. (2019). Stranded assets and stranded resources: Implications for climate change mitigation and global sustainable development. *Energy Research & Social Science, 56*, 1-15.

Burton, B., Kumar, S., & Pandey, N. (2020). Twenty-five years of The European Journal of Finance (EJF): A retrospective analysis. *European Journal of Finance, 26*(18), 1–25. doi:10.1080/1351847X.2020.1754873

Caldecott. (2016). *Stranded assets: A climate risk challenge*. IDB.

Caldecott, B. (2017). Introduction to special issue: Stranded assets and the environment. *Journal of Sustainable Finance & Investment, 7*(1), 1–13. doi:10.1080/20430795.2016.1266748

Caldecott, B., Howarth, N., & McSharry, P. (2013). *Stranded assets in Agriculture: Protecting Value from Environment-Related Risks*. Smith School of Enterprise and the Environment, University of Oxford.

Campos, R. I. (2021). *Mudanças climáticas e estabilidade financeira: incertezas, riscos e o papel dos bancos centrais* (Master's dissertation). Programa de PósGraduação em Economia da Faculdade de Economia, Administração, Contabilidade e Gestão de Políticas Públicas da Universidade de Brasília.

Carbon Tracker Initiative. (2020). *Fault Lines: How diverging oil and gas company strategies link to stranded asset risk*. Author.

Carbon Tracker Initiative. (2021). *The burning need to cut oil dependence in the energy transition*. Author.

Deloitte. (2016). *IAS 16 — Property, Plant and Equipment*. Disponível em: https://www.iasplus.com/en/standards/ias/ias16

FIBRAS. (2020). *Finanças Sustentáveis: Um Panorama*. Projeto Finanças Brasileiras Sustentáveis.

García-Machado, J. J. (2018). The latest streams in finance research: An updated bibliometric mapping based on co-occurrence data. In *Forum Scientiae Economia* (pp. 7–25). Wydawnictwo Naukowe Akademii WSB.

Generation Foundation. (2013). *Stranded Carbon Assets*. Disponível em: http://genfound.org/media/pdf-generationfoundation-stranded-carbon-assets-v1.pdf

GPIF. (2019). *Climate Related Portfolio Risk Assessment. Trucost Analysis Supporting GPIF's Disclosure in line with TCFD Recommendations*. GPIF.

Gros, D., & Schoenmaker, D. (2016). *Too late, too sudden: Transition to a low-carbon economy and systemic risk*. Reports of the Advisory Scientific Committee. European Systemic Risk Board.

IEA. (2013). Redrawing The Energy Climate Map. World Energy Outlook Special Report.

Intergovernmental Panel on Climate Change. (2018). *Global warming of 1,5oC: Summary for policymakers*. IPCC.

Oliveira. (2019). Bibliometric method for mapping the state-of-the-art and identifying research gaps and trends in literature: An essential instrument to support the development of scientific projects. In Scientometrics Recent Advances. IntechOpen.

Palludeto, A. W. A., & Felipini, A. R. (2019). Panorama da literatura sobre a financeirização (1992-2017): Uma abordagem bibliométrica. *Economia e Sociedade, 28*(2), 313–337. doi:10.1590/1982-3533.2019v28n2art02

Papaterra. (2019). *Stranded assets e o petróleo no Brasil. Trabalho de Conclusão de Curso - Monografia da Escola Superior de Guerra. Curso de Altos Estudos de Política e Estratégia (CAEPE).* Escola Superior de Guerra.

Pfeiffer, A., Hepburn, C., Vogt-Schilb, A., & Caldecott, B. (2018). Committed emissions from existing and planned power plants and asset stranding required to meet the Paris Agreement. Alexander Pfeiffer, Cameron Hepburn, Adrien Vogt-Schilb and Ben Caldecott. *Environmental Research Letters, 13*(5), 054019. doi:10.1088/1748-9326/aabc5f

Reports of the Advisory Scientific Committee No 6. (2016). *Too late, too sudden: Transition to a low-carbon economy and systemic risk.* ESRB Advisory Scientific Committee.

Rozenberg, J., Vogt-Schilb, A., & Hallegatte, S. (2020). Instrument choice and stranded assets in the transition to clean capital. *Journal of Environmental Economics and Management, 100*, 102183. doi:10.1016/j.jeem.2018.10.005

Rudebusch. (2021). Climate Change Is a Source of Financial Risk. FRBSF Economic Letter, 3, 1-6.

Saygin. (2019). Power sector asset stranding effects of climate policies. Energy Sources, Part B: Economics, Planning, and Policy, 14(4), 99-124.

SEI, IISD, ODI, E3G, & UNEP. (2020). *The Production Gap Report: 2020 Special Report.* Authors.

Sen, S., & von Schickfus, M.-T. (2020). Climate policy, *stranded assets*, and investors' expectations. *Journal of Environmental Economics and Management, 100*, 102277. doi:10.1016/j.jeem.2019.102277

Shimbar, A. (2021). Environment-related stranded assets: An agenda for research into value destruction within carbon-intensive sectors in response to environmental concerns. *Renewable & Sustainable Energy Reviews, 144*, 111010. doi:10.1016/j.rser.2021.111010

Simões. (2020, September/December). Panorama da Literatura sobre a Governança Corporativa: Uma Análise Bibliométrica das Bases Teóricas e Abordagens mais Utilizadas em Artigos Revista de Gestão. *Finanças e Contabilidade, 10*(3), 62–82.

Tavares. (2019). *Política energética em um contexto de transição: a construção de um regime de baixo carbono* (Doctoral thesis). Universidade Federal do Rio de Janeiro, Instituto de Economia, Programa de Pós-Graduação em Economia da Indústria e Tecnologia.

University of Cambridge Institute for Sustainability Leadership. (2020). *Bank 2030: Accelerating the transition to a low carbon economy.* CISL.

Van Der Ploeg, . (2020). Race to burn the last ton of carbon and the risk of stranded assets. *European Journal of Political Economy, 64*, 101915. doi:10.1016/j.ejpoleco.2020.101915

Van Eck, N., & Waltman, L. (2010). Software survey: VOSviewer, a computer program for bibliometric mapping. *Scientometrics, 84*(2), 523–538. doi:10.100711192-009-0146-3 PMID:20585380

WEF. (n.d.). *World Economic Forum Global Risks. Perception Survey 2019-2020*. Author.

Wilson, C., & Caldecott, B. (2021). *Breaking the Bond: Primary Markets and Carbon-Intensive Financing*. University of Oxford Smith School of Enterprise and the Environment Working Paper 21-05.

Zaklan, A., Wachsmuth, J., & Duscha, V. (2021, July 3). The EU ETS to 2030 and beyond: Adjusting the cap in light of the 1.5°C target and current energy policies. *Climate Policy*, *21*(6), 778–791. Advance online publication. doi:10.1080/14693062.2021.1878999

Chapter 4
Supply Chain Pressure on Environmental Management Systems Adoption:
A Systematic Literature Review

Moritz Keilhauer
Nova School of Business and Economics, Portugal

Marta Almeida
https://orcid.org/0000-0001-7081-4964
Nova School of Business and Economics, Portugal

ABSTRACT

As climate change is becoming more visible, individuals, governments, and firms around the globe increasingly try to mitigate their ecological impact. For companies, the implementation of (certified) environmental management systems (e.g., ISO 14001 or EMAS) can improve their environmental performance. Through a systematic literature review, the past and current research regarding the effects of supply chain pressure on the adoption of EMS is analysed and synthesized. In total, 59 papers are assessed in detail, revealing several research gaps. These include the lack of research on EMAS and the role of female CEOs.

INTRODUCTION

Climate change is one of the most pressing topics of the current generation. The negative impact affects both people and nature alike (Zscheischler and Seneviratne 2017). The main reason for human-made climate change is increasing Green House Gases (GHG), particularly carbon dioxide (CO_2), which raise temperatures through the greenhouse effect (Masson-Delmotte et al. 2018; Oreskes 2004). Thus, the current social, political, scientific and economic consent is to find ways to mitigate this impact. Proposed

DOI: 10.4018/978-1-7998-9410-0.ch004

actions include appeals to individuals and, more importantly, stricter regulations and legislation with the intended effect to incentivise businesses to operate more sustainably, e.g., through carbon taxes.

One way companies can improve their environmental performance is through the use of Environmental Management Systems (EMS) (Iraldo, Testa, and Frey 2009). EMS can aid in this regard as they set guidelines on how companies can effectively and continuously achieve environmental objectives and targets by monitoring their operations and impact on the environment (Sturm and Upasena 1998). According to Melnyk, Sroufe, and Calantone (2003, p. 332), *an EMS involves the formal system and database which integrates procedures and processes for the training of personnel, monitoring, summarising and reporting of specialised environmental performance information to internal and external stakeholders of the firm*. Therefore, one can consider EMS as an effective and widespread tool in reducing pollution (Morrow and Rondinelli 2002; Marouen Amiri et al. 2015).

However, the implementation of EMS systems is often expensive from a financial and human resources perspective (Salim 2018). Reasons to implement EMS include companies' greenwashing, intrinsic motivation, and exerted pressure by the supply chain. A supply chain includes *all the activities in delivering a product from raw material through to the customer, including sourcing raw materials and parts, manufacturing and assembly, warehousing and inventory tracking, order entry and order management, distribution across all channels, delivery to the customer, and the information systems necessary to monitor all of these activities* (Lummus et al. 2001, p.428). In particular, customers often pressure their suppliers to adopt an EMS. In this context, customers often have more power than their suppliers and can force them to adhere to their requirements (Kaminski 2013). Thus, supply chain pressure is an essential factor in the diffusion of EMS.

This chapter provides an overview on the impact that supply chain pressure has on the adoption of EMS by large firms and small and medium enterprises (SMEs) through a systematic literature review on the topic. The research questions read as follows:

1. What trends exist regarding the number of publications on this topic over time?
2. Which countries and industries are the studies based on?
3. Are the research methods primarily quantitative, theoretical or case study based?
4. What are the main findings from current research?

To answer the research questions, this study analyses and synthesises existing literature exploring International Organization for Standardization (ISO) 14001, Eco-Management and Audit Scheme (EMAS), and EMS in general in the context of supply chain pressure. The results of this literature review are helpful to inform future research as they will reflect the current state of affairs, guide further research avenues, and reflect current developments. This is valuable for researchers, as well as practitioners, as important insights are summarised, which can act as guidelines.

This chapter develops as follows; next we provide a contextual background to the development of EMS from an historical perspective as well as providing an overview of the most prevalent EMS. Then, we explain the research design of this systematic literature review. This is followed by a discussion of the main findings and finally we present the concluding remarks and ideas for future research.

CONTEXTUAL BACKGROUND

EMS

An EMS is the umbrella term for a management tool or framework that can take different forms. Firms can either implement EMS following existing guidelines, which can be certified (e.g., ISO 14001 or EMAS certified), or develop them independently (Zilahy 2017). In both scenarios, the implementation is voluntary. Nevertheless, customers often demand an EMS from their suppliers as a condition for conducting business with them (Elbarkouky and Abdelazeem 2013).

Firms use EMS to improve their environmental performance through practices that follow the Plan-Do-Check-Act (PDCA) cycle. The PDCA cycle helps firms to implement and improve their changes efficiently and continuously (Zilahy 2017). Thus, it is essential to improve an EMS after its set up continuously. Another essential aspect is to ensure that employees have environmental awareness and receive training in this regard (Yiridoe and Marett 2004). According to (Smith 1993), an EMS is also ideal for identifying, among others, environmental effects caused by the firm through its normal operations or incidents.

History of EMS

The history of EMS began in the early 1990s with the introduction of the British Standard (BS) 7750, a now-retired standard. Later EMS such as the ISO 14001 built upon this. The need for such environmental standards arose because governments increasingly required environmental protection practices that firms needed to fulfil. To cope with the ever-rising costs and complexities of fulfilling these requirements, firms started to incorporate their environmental management practices into broader and defined systems (Zilahy 2017).

In 1992, 175 countries signed the Rio Declaration at the United Nations "Conference on Environment and Development". This declaration had the intended purpose of *Working towards international agreements which respect the interests of all and protect the integrity of the global environmental and developmental system* (General Assembly 1992, p. 1). Arising from this, ISO saw the need for a global EMS standard and released ISO 14001 in 1993 (Kaminski 2013). Since then, EMS popularity has increased worldwide, leading to the introduction of additional EMS standards, for example, EMAS. The following sub-sections give a short overview of the most important past and current EMS.

BS 7750

BS 7750, introduced in 1992, was the first certifiable EMS. This standard laid out the groundwork which other EMS standards would later use as a blueprint (Kollman and Prakash 2002). The specifications included the necessity *to form a company environmental policy and appoint a member of the board to be responsible for environmental matters [and recommended] that an environmental awareness committee should be set up with representatives of the whole work force to disseminate information and to bring forward suggestions for improvement* (Smith 1993, p. 278). However, the former was not mandatory (Smith 1993).

BS 7750 was officially terminated in September 1996 and superseded by ISO 14001, an international standard. At this point, approximately 200 British firms were BS 7750 certified. A transition period existed until April 1997 where firms that fulfilled ISO 14001's requirements could transfer to the new standard (Holt 1998).

ISO 14001

ISO used BS 7750 as a template to develop ISO 14001. Consequently, the standard itself sets the requirements for an EMS, as BS 7750 did. However, according to Delmas (2002) and Prajogo et al. (2012), ISO 14001 has the added benefit that SMEs can increase their international trade as the EMS guidelines are globally accepted.

The ISO organisation recognised the need for an international EMS standard mainly based on two factors. First, the earlier mentioned Rio Declaration, and second, the 1986 General Agreement on Tariffs and Trade (GATT) Uruguay round. The GATT's purpose, now the World Trade Organization (WTO), was to liberate trade by eliminating trade barriers. A total of 123 nations took part in the Uruguay round. The creation of ISO 14001 helped mitigate these trade barriers by establishing a common supranational standard that firms could use to signal their adherence to a minimum level of environmental action and protection (Van der Veldt 1997).

Introduced in 1996, the ISO 14001 guidelines include five key principle sections. These fundamental principles are policies on what the firm must do to be ISO 14001 compliant and include: an environmental policy; planning; implementation and operation; checking and corrective action; and review. For example, the firm has to set objectives and targets, document their EMS, train their employees, and commit to continuous improvement and pollution prevention (Van der Veldt 1997).

ISO 14001 is continuously updated. The last update was in 2015 with the revised ISO 14001:2015. This version includes supply chain issues and requires firms to shift towards improving their EMS and their actual environmental performance. Firms can use the standard itself without certification (ISO 2015). Certification may be advisable as it signals the proper implementation of ISO 14001 by the firm to customers and business partners (Van der Veldt 1997; Orcos and Palomas 2019). This certification, however, cannot be done by ISO itself. An audit must be performed by a third-party certification body instead (ISO, n.d.). In 2019 ISO has also launched a flexible approach to phased implementation of ISO 14001, with a special attention to help SMEs in improving their environmental performance, the ISO 14005:2019.

In 2019, 312,580 firms in 177 countries were ISO 14001 certified, not least because the standard is internationally recognized and available to virtually all types of firms and organisations (ISO 2021).

EMAS

In July 1993, the European Commission introduced EMAS, becoming operational two years later. While EMAS builds on the EU-regulation (EC) No 1221/2009, organisations and firms' implementation is voluntary as with ISO 14001. Initially, the standard was limited to the industrial sector and European countries. However, both limitations were lifted in later revisions in 2001 and 2011, respectively (European Commission, n.d.).

In general, EMAS covers all aspects of ISO 14001 and extends these with further and stricter requirements. Compared to ISO 14001, EMAS is primarily concerned with results and environmental performance instead of systems and processes. In addition to ISO 14001 requirements, EMAS demands that certified firms regularly publish environmental statements, let their employees participate directly, e.g. in the preparation of environmental statements, continuously improve their environmental performance and not just EMS, and ensure their compliance with local law (Albelda 2011; Boiral and Henri 2012; European Commission 2018).

Furthermore, one of the most significant differences to ISO 14001 is the certification process itself. ISO 14001 firms get certified by certification bodies that comply with ISO's Committee on Conformity Assessment (CASCO) standards and attain accreditation through national bodies. For EMAS, the local government itself accredits certification bodies (European Commission 2011; ISO, n.d.).

Additionally, EMAS has procedures for helping SMEs become certified and maintain their certification, thus overcoming some of the barriers faced by SMEs in relation to the implementation of EMS. Barriers include but are not limited to size-proportionate fees, the requirement to publish the environmental statement only every second year instead of every year, and an entire certification cycle of four instead of three years (European Commission, n.d.).

EMAS Global was developed in 2010, enabling EMAS certification to firms that operate entirely or to some extent outside of Europe. The same standards and processes as for companies operating entirely in Europe are applicable except that they need to get certified through an European country that offers this service (European Commission, n.d.). In September 2020, 12,751 sites and 3,838 organisations were certified (European Commission 2020). Despite the availability of EMAS global, due to the historical geographical focus of EMAS certification, it has a lower rate of implementation compared to ISO 14001.

The next section explains the steps taken in this research.

RESEARCH DESIGN

This chapter adopts a systematic literature review to study the supply chain pressure in the adoption of EMS. A systematic literature review involves the *systematic accumulation, analysis and reflective interpretation of the full body of relevant empirical evidence related to a question* (Rousseau et al. 2008, p. 475). The idea is to use transparent and replicable procedures to minimize errors. This chapter adds to the existing research literature given that, to the best of the authors knowledge, no systematic literature review has been published to date on this topic. Nevertheless, it is a pressing topic, as supply chain pressure may increase the adoption rate of EMS and thus help the environment. In this sense, this chapter helps to identify the scientific contributions in relation to EMS adoption and supply chain pressure and lead to the effective use of evidence.

In order to address the research questions, the aim of this systematic literature review is to find and analyze all relevant studies on the topic within two major academic databases: Web of Science and Scopus. These are well-established, multi-disciplinary research platforms containing a broad and up-to-date variety of peer-reviewed journals. To effectively search selected databases for scientific papers and articles in the context of this chapter, relevant search terms were defined, as detailed in Table 1:

Table 1. Research methodology (Source: own illustration)

Database	Search Terms	Results
Scopus	TITLE-ABS-KEY (({iso 14001} OR {environmental management system} OR {emas} OR {eco management and audit scheme} OR {bs 7750}) AND (supply* OR {gscm})) AND (LIMIT-TO (DOCTYPE, "ar")) AND (LIMIT-TO (LANGUAGE, "English"))	220
Web of Science	(("iso 14001" OR "environmental management system" OR "emas" OR "eco management and audit scheme" OR "bs 7750") AND (supply* OR "gscm"))	232

Here, the parentheses were used to limit the search to the exact result with this term. For both searches, the asterisk was used as a wildcard to include terms such as "supply chain" or "supply pressure". The search was further limited to articles published in English. No limitations were made regarding the publication year, where the search included all publications until the end of 2021.

The selection of articles for this study observed the following steps. First, the articles that were duplicated between both databases were removed, leading to a total of 351 articles. The following step was to eliminate studies that do not meet the topic of the research with a title screening, in which 28 articles were further eliminated. Examples include clinical research papers focusing on "encephalo-myo-arterio-synangiosis", which is also abbreviated as EMAS, and blood supply reconstruction. The last step was to read abstracts of the 323 articles left to assess their relevance. For example, papers which used ISO 14001 certified companies as a data basis but did not explore the connection to the supply chain were considered not relevant. Similarly, papers which looked at corporate social responsibility more broadly, mentioning ISO 14001 or EMAS certification and supply chain, but did not relate these two, were also considered not relevant to this analysis. This last step resulted in 59 articles relevant to the research, which were fully analysed in the context of the systematic literature review. This review involved the creation of an MS Excel database with the record of the information obtained from the selected articles. The information recorded included the articles' general information, citation and specific information such as country researched, organization's size, industry, research method used and type of EMS. The selection process was structured as summarized in Figure 1:

Figure 1. Research results (Source: own creation)

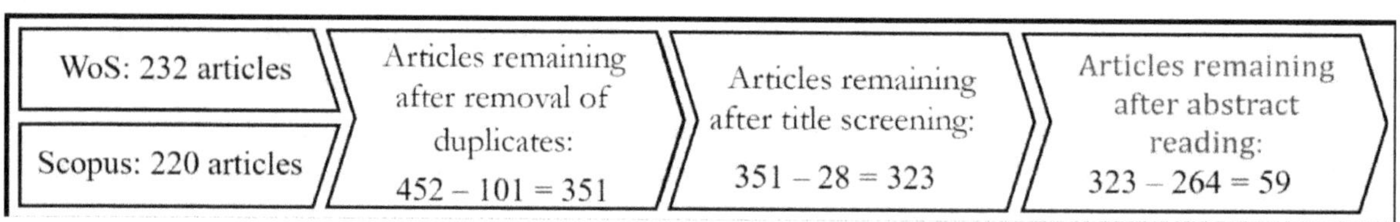

The findings from the systematic literature review were summarized in a narrative synthesis which are presented in the following section.

FINDINGS

This section shows a descriptive overview of research trends in this analysis according the four research questions.

RQ 1: What trends exist regarding the number of publications on this topic over time?

This research identified 59 articles published in a total of 40 different journals. Four journals published more than three articles; seven in the "Journal of Cleaner Production", five in "Corporate Social Responsibility and Environmental Management", four in "Business Strategy and Environment" and three in "Sustainability". Equally diverse is the number of authors, 156 different authors contributed to the 59 articles. Only 11 authors wrote or contributed to more than one paper. The authors with most papers written were Zimon, with four papers, Chiarini, Nawrocka, Sarkis, Sroufe and Madzik, with three papers each.

Figure 2 highlights an increasing attention by researchers to EMS in the context of supply chains. While the number of publications was low or even zero during the early and mid-2000s, publications increased considerably in more recent years. With the number of publications over the last five years counting to more than half the whole number of publications. Most articles in a single year were published in 2020 with nine articles. This research trend is not limited to EMS and supply chains but reflects the generally increasing attention that EMS receives and the increasing rate of ISO 14001 certifications (Salim et al. 2018).

Figure 2. Publications per year (Source: own illustration)

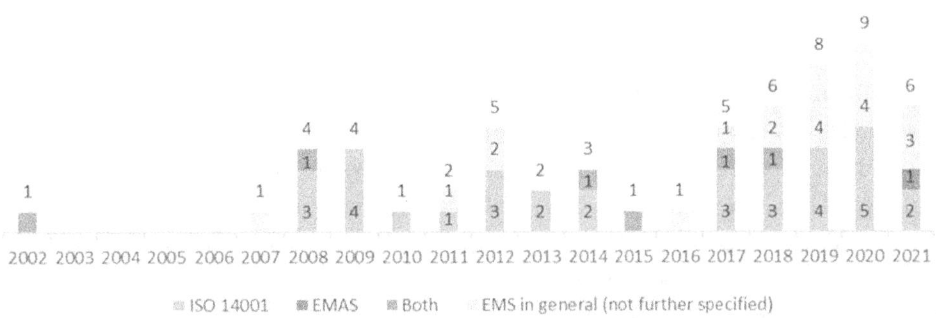

The majority of the papers included in this literature review analyse ISO 14001 rather than EMAS, suggesting that academic coverage is scarce. 56% consider only ISO 14001, 2% only EMAS, 10% both, and 32% consider EMS in general without further specifying them. However, Daddi et al. (2018) found that EMAS certifications were decreasing between 2010 and 2015, mainly due to *the lack of financial and human resources, the lack of market and stakeholder recognition, and the unclear added value of EMAS* (p. 2358). Nevertheless, a study using more than 400 European EMAS registered companies was published in 2021, signalling that EMAS might start getting research attention. However, against expectations, EMAS Global was not mentioned once. This may be due to the relatively short existence or that EMAS Global is not yet in the focus of researchers and practitioners.

RQ 2: Which countries and industries are the studies based on?

Figure 3 shows the countries examined in the papers. The term "Multiple" is used for papers that study more than three countries, while "N/A" refers to papers that do not clarify the dataset's region or if it is purely theoretical.

Figure 3. Examined countries (Source: own illustration)

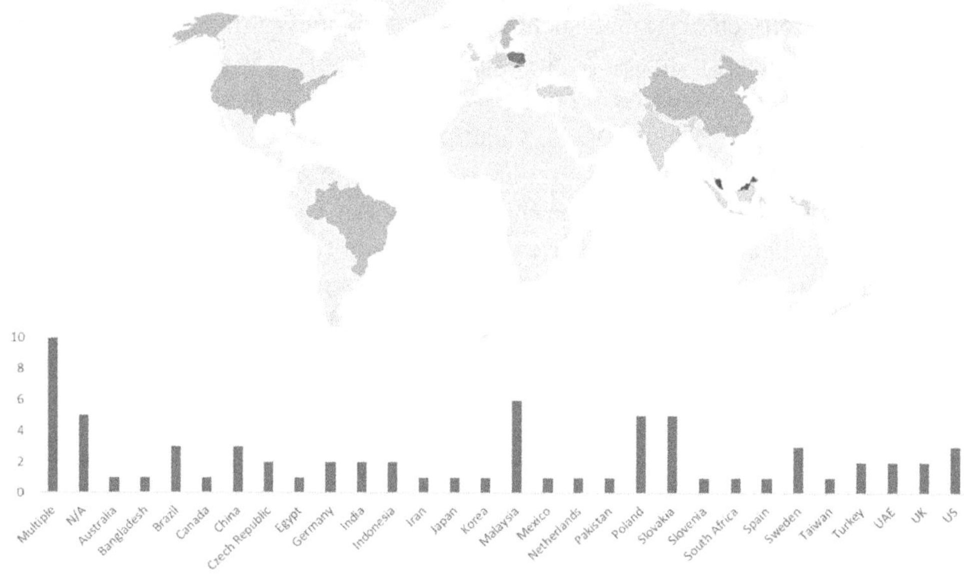

Interestingly, the number of papers investigating EMS in the United States (U.S.) is relatively high, with three papers. This is surprising, as EMS adoption is generally relatively low in the U.S. compared to other developed countries (Delmas 2000; Lee et al. 2017). Malaysia stands out as the country counting the highest number of published research. This concentration might be a result of the 2020 environmental agenda fostered in this Asian country (Jamal et al. 2020). It is also noticeable that few European countries are represented, even though Europe is often perceived as a leader in environmental protection (Oberthür and Kelly 2008; Sommer and Hain 2017), as exemplified by the European Green Deal. However, research on European countries often compares multiple countries. If more than three countries are analysed, the article is counted in the "Multiple" category. Most of these articles in the "Multiple" category use international datasets of various countries with the objective to generalize rather than compare different countries or regions. Thus, the number of examined countries does not equal the number of analysed articles. It is also noticeable that South America is only represented by research in Brazil. Otherwise, most world regions are represented except Russia and Africa, omitting Egypt and South Africa.

Figure 4 depicts that the majority of papers analysed large firms, 33% conducted research solely on SMEs and 30% on large firms and SMEs in conjunction.

Figure 4. Analysed firm size (Source: own illustration)

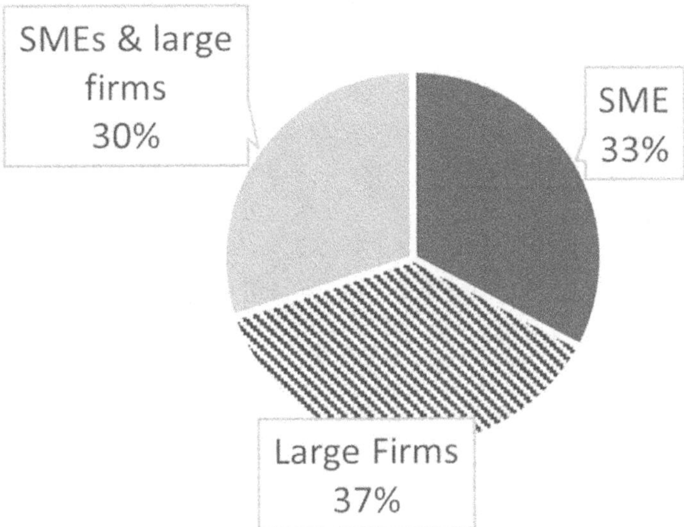

As for the industries studied, according to the Global Industry Classification Standard, 13 different industries (or, when not possible, the broader industry group) were identified. However, authors did not always specify the industries that were researched; instead, their analyses often refer generally to the manufacturing sector. Here, it is notable that the automobiles and components industry is most frequently analysed; possibly because of the extensive supply chain in this industry. Figure 5 lists industries that were investigated more than once.

Figure 5. Industries that appear in more than one research paper (Source: own illustration)

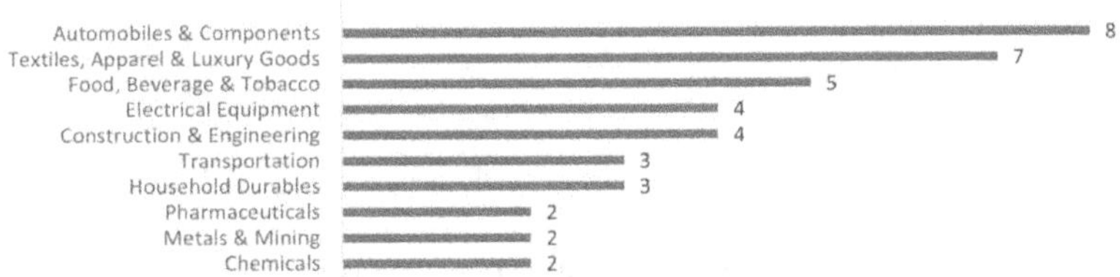

When analysing company size together with the most represented industries, the results show that while papers with the most represented industries include both SMEs and large firms in conjunction, it is possible to identify that large firms are of metals and mining, household durables, transportation, construction and automobile industries. SMEs include the industries of electrical equipment, food and beverages and textile. Figure 6 depicts the distribution between industries and firm size.

Figure 7 shows the number of industries studied in each country. The analysis shows that for the majority of the countries studied, the studies focused on one industry.

Figure 6. Industries and company size (Source: own illustration)

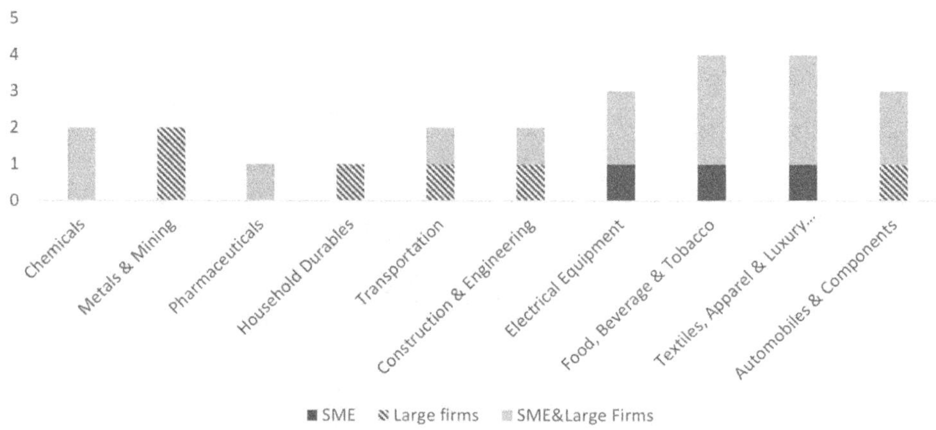

Figure 7. Industries and countries (Source: own illustration)

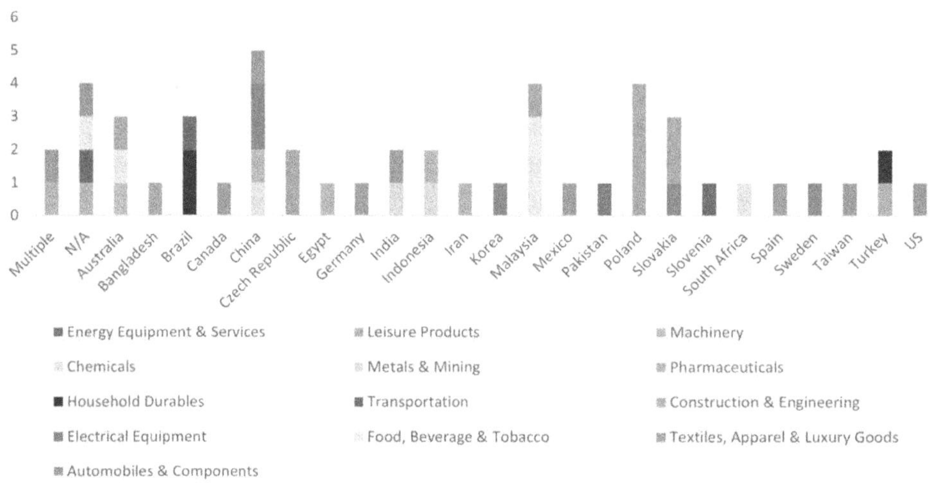

The comparison between Figures 3 and 7 show that those countries which were most represented do not always coincide with those for which more industries were studied. An example is Poland, which was featured on four studies on a single industry, curiously the textile industry is not counted among Poland's main exports. On the contrary, the single study on Australia focused three different industries. This suggests that most studies are focused on single countries and industries, which might both provide richer insights into those, but at the same time demonstrate limitations in terms of conclusions and generalizations. In addition, it suggests that there is scope for further studies to explore different industries in different countries to advance further the knowledge on the impact of supply chain pressure on EMS adoption.

RQ 3: Are the research methods primarily quantitative, theoretical or case study based?

The analysed papers were predominantly quantitative as shown in Figure 8. 47 papers employed quantitative methods, e.g., regressions. Ten employed case studies and 21 papers were in part or entirely theoretical. Notably, papers with a theoretical contribution increased in recent years. Since 2016, 17 papers were theoretical. In addition, a trend towards statistical decisions models is discernible.

Figure 8. Employed study type (Source: own illustration)

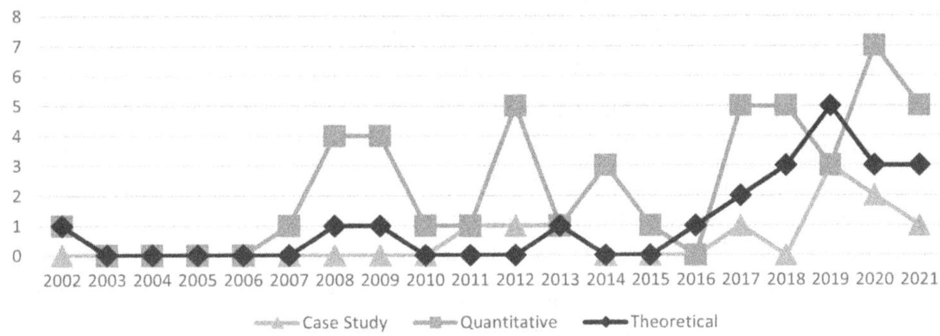

RQ 4: What are the main findings of the studies?

To answer RQ 4 with the main findings, this section identifies the main themes found on the studies. It starts with a brief overview of the adoption rate of certified EMS, followed by an explanation of the main findings in relation to the EMS and suppliers' selection. It then covers the findings that relate EMS adoption with sustainable and green SCM, the drivers and barriers to EMS adoption and outlooks on EMS adoption related to regions, industries and firm size and age.

Adoption of EMS

According to Bajec, Tuljak-Suban, and Krmac (2015) ISO 14001 was adopted by 31.25% of Slovenian logistic companies, whereas EMAS was adopted by merely 6.25%. Another study by Beske, Koplin, and Seuring (2008) reveals that 78% of surveyed firms operating in the German automotive industry are ISO-14001 certified. Moreover, 51% of responding firms require their suppliers to be ISO-certified. The significantly smaller share of EMAS (67%) certified companies, compared to ISO 14001, is partly due to the restriction to the European Union (EU) and the international business operations of most suppliers(Beske, Koplin, and Seuring 2008).

EMS and Supplier's Selection

Gupta, Soni, and Kumar (2019) illustrate society's shift towards demanding more sustainable behaviour from firms. While traditionally, firms were primarily concerned with maximising profits, they are often legally required to incorporate environmental aspects in their practices. However, they often experience difficulties selecting a green supplier because of the lack of transparency and comparison impeding their selection process. The authors find that EMS is among the essential criteria in green supplier selection in India's automotive sector. Other scholars support this view and argue that companies prefer suppliers concerned with environmental issues and hence, facilitate EMS (e.g., Shojaei and Bolvardizadeh 2020; dos Santos, Godoy, and Campos 2019; Jiang et al. 2018). Thus, supplier selection plays a crucial role in gaining a competitive advantage and can be facilitated through EMS (e.g., Ecer 2020; Mishra et al. 2019; Akcan and Taş 2019; Aiyub et al. 2009).

Jiang et al. (2018) argue further that EMS needs to be more strongly considered by automotive manufacturers when choosing green suppliers. Suppliers should strive to improve the performance of their EMS to serve manufacturers' needs. Furthermore, EMSs are essential for international expansion as overseas buyers expect their suppliers, in this case, Indian steel firms, to comply with environmental standards (i.e., ISO 14001 certification) (Prasad et al. 2018). Hence, EMS certification is often a key criterion in sustainable supply chain management (SSCM). Wu, Santoso, and Roan (2017) found that in the Indonesian coal industry firms that implement a globally standardised and certifiable EMS can better prove their sustainability credentials (Wu, Santoso, and Roan 2017). This suggests that globally standardised and certifiable EMS are key to organizations to demonstrate to their clients their sustainability practices.

The urge to enforce sustainable supply chain solutions is accelerating in the Polish textile industry (Zimon 2020). To define best practices, assessing suppliers to ISO 14001 standards are beneficial. Al-Refaie, Al-Tahat, and Lepkova (2020) show at the example of logistic, pharmaceutical, and food companies that ISO 14001 implementation can indirectly encourage all stakeholders to follow more sustainable practices. According to Chiarini (2012), ISO 14001 is essential to include suppliers in an enterprise's EMS. Based on two case studies of one Slovakian and one Chinese electronic manufacturing firm, Medina-Serrano et al. (2019) show that endorsing a supplier's code of conduct and its observance along the supply chain is necessary for achieving sustainable procurement.

Furthermore, academic literature shows a preference for ISO-certified suppliers. When selecting a green supplier, ISO 14001 is often a prerequisite for manufacturing businesses across various sectors (Wiengarten, Pagell, and Fynes 2013). Although ISO 14001 is rarely an obligatory requirement, those suppliers are often favoured by Swedish manufacturing firms as other supplier audits demand many resources (Nawrocka, Brorson, and Lindhqvist 2009). The ISO 14001 certificate's value as evidence of environmental sustainability is a combined effect of the supplier's environmental goals, the customer's advancement of supply chain practices, and the certifying bodies' objectives (Nawrocka, Brorson, and Lindhqvist 2009). González, Sarkis, and Adenso-Díaz (2008) find a positive relationship for Spanish automotive firms between a certified EMS and the environmental requirements these firms set for their suppliers.

ISO certification is crucial for firms exporting their products internationally (e.g., Jiang et al. (2018)). Scur and Barbosa (2017) find that all multinational corporations in their sample of home appliance manufacturers in Brazil are ISO 14001 certified. Often customers are based abroad, hindering the supervision of companies' ecological procedures. As such, customers favour international accreditation (i.e., ISO 14001) to guarantee their suppliers' adequate environmental performance (Scur and Barbosa 2017).

Contrarily, Chiarini (2017) asserts that supply chain managers of European manufacturing firms operating in the oil & gas, automotive & motorcycle, and construction & mining sectors declare a weak belief in ISO 14001 certification. Chiarini (2017) notes that some managers question whether ISO 14001 improves environmental outcomes. Other managers express that they met companies with an ISO 14001 certified EMS *that did not even respect environmental laws and regulations* (Chiarini 2017, p.107). ISO

14001 has signalling power that the certified firm is adhering to specific standards, yet the certification alone cannot be trusted entirely as the rigorousness of certification bodies differs (Nawrocka 2008b).

EMS Adoption and Sustainable and Green Supply Chain Management

While the adoption of certified EMS signals organizations environmental practices, Zimon and Madzik's (2019) results do not support that ISO 14001 is crucial for improving SSCM. Zimon and Madzik (2019) find that despite the survey's majority observed a positive effect of ISO 14001 implementation on enhanced environmental measures in the supply chain, achieving sustainable supply chains is not explicitly apparent. The survey consists of 28 Polish and Swedish textile firms' managers, of which 51% of respondents regard ISO 14001 as helpful for SSCM, while 23% could not identify a positive effect.

Another essential aspect is that firms may choose to include supply chain partners, particularly suppliers and customers, in EMS. This leads to more considerable sustainable outcomes than narrowing down to single firms (e.g., Prajogo, Tang, and Lai (2014)). Prajogo, Tang, and Lai (2014, p. 565) assert a positive relationship between EMS diffusion and *green products, green processes, and green supply chain management* (GSCM) for Australian firms in various sectors. Other authors, e.g., Hajikhani, What, and Idris (2012), Zimon and Madzik (2019) and Rasit et al. (2019) support this view. Firms that incorporate ISO 14001 address environmental issues along the entire value chain. Together with suppliers, these standards can foster pro-environmental responses among stakeholders. ISO 14001 can be a valuable starting point for the overall SSCM framework, encouraging that as many members in the supply chain of textile firms as possible align with its policies (Zimon 2020). Other evidence suggests that ISO 14001 certified manufacturing enterprises in Japan evaluate their supplier's green performance 40% more often compared to non-certified firms and demand 50% more often that their suppliers implement ecological procedures (Arimura, Darnall, and Katayama 2011).

Surprisingly, the nexus between ISO 14001 and GSCM practices has received scarce coverage. This is partly because ISO 14001 does not require entities to assess their supply chain decisions' environmental outcomes. However, ISO 14001 certified entities might be capable of adopting GSCM at a low cost. That indicates that ISO 14001 certified firms are more inclined to employ GSCM efforts to encourage suppliers to mitigate their environmental effects. If this is the case, the environmental benefits of ISO 14001could be far more substantial than previously thought. Furthermore, Arimura, Darnall, and Katayama (2011) discover that plants headquartered overseas and plants of public companies are less apt to evaluate supplier's green performance (Arimura, Darnall, and Katayama 2011).

According to Razak et al. (2016), entities in the public sector (e.g., universities) can oversee the green performance of supply chains. This is due to their powerful position arising from their spend in the private sector in which they could encourage improvement efforts to achieve the highest overall benefit. This can be a notable evolution in ecological management attempts since it mitigates an individualistic and self-served attitude (Razak et al. 2016).

Based on Arimura's, Daranall's, and Katayama's (2011) observations from Canada, France, Germany, Hungary, Japan, Norway, and the U.S., government incentives lead to a 7% increase in the likelihood that ISO 14001-certified firms will evaluate suppliers' environmental performance and increase by 8% the likelihood that ISO 14001 users will expect certain environmental practices from their suppliers. This suggests that incentivising green behaviour is an effective response to environmental issues.

A study conducted by Bremmers et al. (2007) finds a significant effect of stakeholder groups on EMS adoption in the Dutch agri-food sector. They suggest that commercial primary stakeholders (i.e., governments, customers) hold a more salient position in EMSs implementation than commercial secondary stakeholders (i.e., environmental organisations).

Interestingly, although ISO 14001 certification is not mandatory, Scur and Barbosa (2017) found that significant differences exist between home appliance firms that possess ISO 14001 certification and those that do not when considering GSCM. An observable connection exists between a certified company and the requirements placed on suppliers concerning environmental issues. ISO 14001-certified firms have on average the highest green purchasing scores relative to non-certified firms (Scur and Barbosa 2017).

Scholars assess opposing views on potential drivers and barriers that suppliers face in their EMS implementation (e.g., Zhu, Cordeiro, and Sarkis 2012; Elbarkouky and Abdelazeem 2013; Salim and Padfield 2017; Scur and Barbosa 2017; Zimon 2020). The following two sub-sections analyse the drivers and barriers of EMS adoption.

Drivers of EMS Adoption

An important driver for EMS adoption is market competitiveness since many customers demand ISO 14001 certification (Elbarkouky and Abdelazeem 2013). In a study conducted in China by Zhu, Tian, and Sarkis (2012), results show that while low awareness of green protection dominates, market competitiveness of the supply chain is a crucial driver for ISO 14001 certification. Respondents of a survey, conducted by Lee and Kim (2009), reported that marketing risks (49%), brand risks (42%), and customer demand (56%) are essential factors in adopting corporate social responsibility (CSR) standards, including ISO 14001. The survey's respondents were Korean electronic manufactures.

Kollman and Prakash (2002) found that network effects, domestically and globally, play a prominent role in the adoption of ISO 14001 and EMAS, as well as domestic governmental promotion. However, in the latter's absence, global market pressure or pressure from supranational policymakers may compensate for this. Kollman and Prakash (2002) reached this conclusion by examining the U.S., Germany, and the UK. Compared to the UK and Germany, the U.S. has a low adoption rate due to the absence of promotion and network effects.

Moreover, international supply chain pressure plays a vital role. Zhu, Tian, and Sarkis (2012) find that Chinese firms that are expanding internationally face trade barriers (e.g., EU's Restrictions of Hazardous Substances Directive (RoHS)). Chinese suppliers need to be ISO 14001 certified or design EMSs by ISO 14001 standards for many international corporations. The majority of ISO 14001 certifications occurred to satisfy international pressure and environmental standards. They find that supply chain pressures, regulations, and environmentally conscious business opportunities are the main drivers for implementing GSCM practices. Finally, many manufacturing firms face a wide range of international environmental and social legislation (e.g., RoHS directives) (Zhu, Tian, and Sarkis 2012).

Interestingly, institutional pressure varies internationally and domestically. Zhu, Cordeiro, and Sarkis (2012) conclude that only international institutional pressure positively affects green practices such as ISO 14001 by Chinese firms. Inside China, economic and legal pressure, which would usually compel the companies to act ecologically, is lacking. This shows the importance of putting international pressure on Chinese manufacturers to add to the pressures imposed on them through domestic regulation, competition, and costumers. Furthermore, the authors show that foreign ownership or joint venturing positively affects green practices for Chinese manufacturers. An explanation is their more considerable environmental concern, resources, and expertise, which their Chinese counterparts can leverage (Zhu, Cordeiro, and Sarkis 2012).

Other drivers include legal requirements and reputation. Lee and Kim (2009) find that the most significant driver for introducing CSR standards, including ISO 14001, are legal requirements. 67% of respondents stated that they introduced CSR standards because local regulations compelled it (Lee and Kim 2009). Concerning reputation, Malaysian food & beverage firms regard ISO 14001 to improve their image and reputation and ultimately generate competitive advantage (Salim and Padfield 2017).

Barriers of EMS Adoption

In terms of barriers, scholars argue that high implementation costs are the main barrier to adapt EMS (e.g., Zhu, Tian, and Sarkis 2012; Salim and Padfield 2017; Scur and Barbosa 2017; Zimon 2020). ISO 14001 entails high costs due to the implementation of environmental-friendly concepts (e.g., energy-saving methods) (Zimon 2020) and high investment costs (Salim and Padfield 2017). Initial certification costs for ISO 14001 range between $24,000 to $128,000, and yearly maintenance costs amounts from $5,000 to $10,000 (Salim and Padfield 2017; Curkovic and Sroufe 2011). High costs affect product prices, and customers are less willing to pay higher prices (Zimon and Madzik 2019). Thus, firms are less interested in undertaking eco-friendly activities. Similarly, Scur and Barbosa (2017) state that implementation costs are high because domestic consumers in Brazil prioritise price in their purchase decision. Although they are aware of sustainability, they tend to choose the product with the lowest price if they have the same quality (Scur and Barbosa 2017).

Salim and Padfield (2017) found that in Malaysia, the government can help to overcome this barrier by taking an active role, introducing stricter ecological legislation and tax subsidies. In their survey, conducted with 42 Malaysian food & beverage firms, only 50% of respondents state that they are well-informed with the standard's benefits and registration. Particularly in developing countries, considering the current political and regulatory setting, firms are uncertain whether ISO 14001 implementation enhances their firm performance given high registration costs and unsure benefits (Salim and Padfield 2017).

Additionally, authors argue that some countries and industries lack pressure concerning GSCM (e.g., Scur and Barbosa (2017)). Barriers include the absence of regulations, governmental support, and social pressure. Salim and Padfield (2017) support this argument with the example of Malaysia. SMEs dominate the food and beverage industry, targeting mainly the domestic and non-ecological-sensitive export markets. Combined with a low level of ecological awareness in society, low pressure from the demand side has led to a small number of ISO 14001 adoptions in Malaysia (Salim and Padfield 2017). However, the results of their study show that approximately 67% of large firms and 56% of SMEs are willing to adapt their business to ISO 14001 standards. In Malaysian industries, low demand-side and supply-side pressures to implement ISO 14001 standards exist. Governments may consider providing both demand-side pressure and supply-side incentives, especially for SMEs, as large companies are more inclined to adopt ISO 14001 quickly given their export requirements. This will enhance the ability of SMEs to endorse ISO 14001, potentially raising their involvement in the export market in the future. The domestic Malaysian market is expected to request adjustments as a result of the emerging trend towards sustainable goods (Salim and Padfield 2017).

To overcome the barriers above, organisations may cooperate with different stakeholders, i.e., companies, society, governments, universities, consultants, green land developers, and non-governmental organisations (Elbarkouky and Abdelazeem 2013). Zhu, Tian, and Sarkis (2012) argue that firms can imitate the procedures from firms that have already adapted GSCM practices, and as such, they can learn from their experience and reduce costs.

Regional Outlook on EMS Adoption

Customer preferences and pressure in eco-conscious countries, namely Finland, Japan, Germany, and Denmark, are, to a greater extent promoting domestic and foreign suppliers to implement ISO 14001. Existing environmental laws may act as a possible driver for these requirements. Customer pressure exists as these countries are among the global leaders in environmental policy. As such, many organisations in eco-conscious countries are reluctant to conduct business with suppliers that do not fulfil the customer's firms environmental requirements, evidenced by the adoption rate of ISO14001 in supplier markets (Nishitani 2010).

Furthermore, for firms operating in eco-conscious markets, ISO 14001 implementation is more accessible due to their, on average, better financial performance. Since they are already ISO 14001 certified, they demand their suppliers to follow suit in the international supply chain as the supplier's EMS quality reflects theirs (Nishitani 2010). According to Beske, Koplin, and Seuring (2008), the requirement to implement quality and environmental standards is not difficult for German or (Western) European suppliers to the automotive industry, as these standards are widespread. Surprisingly, Nawrocka (2008b) concludes that in Sweden, while ISO 14001 is used in assessing suppliers, it is not a required condition and ultimately not a decisive factor for the electronic manufacturers interviewed by Nawrocka (2008a). Yet, Swedish companies exerted pressure on Polish firms to adopt ISO 14001. Without the pressure, they would not implement an EMS and certify it according to interviews with 12 Polish manufacturing firms (Nawrocka 2008b).

As companies in cross-border supply chains work intensively together, developed countries spread ISO 14001 to exporting countries, especially in the Far East (To and Lee 2014). When globalisation accelerated, emerging countries (e.g., Brazil, Russia, India) began to adopt ISO 14001 countrywide to enhance their competitiveness. However, Australia, Thailand, and Brazil suffered a notable decline since 2006. Intriguingly, Brazil and Thailand were perceived as pioneers in adopting the standard. Some events may have happened in these countries (e.g., pollution harms triggered by popular ISO 14001-certified corporations) that led organisations to doubt the effectiveness of the standard. The correlation analysis of the quantity of ISO 14001 certificates worldwide and the volume of exports reveals a strong association of ISO 14001 certification with cross-border trade. The results exhibit that the strength of the relation declines as ISO 14001 implementation reaches a more mature stage (To and Lee 2014).

In a different manner, González, Sarkis, and Adenso-Díaz (2008) find that the internationalisation of Spanish automotive firms, evidenced by import and export rates of these companies, has no significant impact on supply chain pressure to implement EMS. However, they note that the outcome might depend on the severity of environmental policies (e.g., severe in the EU and U.S.; not severe in Mexico and China).

Industry Outlook on EMS Adoption

While EMS is gaining importance and awareness, diffusion differs across industries. Delmas and Montiel (2009) conclude that suppliers with highly specialised assets within the automotive industry are more likely to react to customer pressure and less reluctant to adapt ISO-14001 certification. Also, Islam et al. (2018) find that ISO 14001 certification points to high importance in the leather industry. However, no firm in their sample is certified. Their study shows that participating in ISO 14001, selecting green suppliers, and working together with customers towards better environmental performance is among the most critical aspects in this industry. Therefore, managers may adhere to ISO 14001 and other EMS to respond to environmental pressures (Islam et al. 2018).

Moreover, the development of an ecological supply chain differs across manufacturing and service sectors (Chiarini 2014). Respondents in the manufacturing sector place higher importance on auditing the supplier's premises for enhanced ecological outcomes than respondents in the service sector. The latter perceive auditing their suppliers as too time-consuming and thus, do not conduct audits. In terms of costs, Nawrocka (2008b) concludes that audited firms did not believe auditing was too costly compared to ISO 14001 itself.

Firm Size and Age Outlook on EMS Adoption

Several studies assess whether firm-level factors such as firm size and age impact EMS implementation (e.g., González, Sarkis, and Adenso-Díaz 2008; Urban and Govender 2012; Graafland 2018; Younis and Sundarakani 2019). Their assertion is vital because although SMEs generate only small savings, their overall impact is significant as many SMEs exist. In Europe, SMEs contribute to 64% of industrial pollution (Graafland 2018).

Graafland (2018) reveals that ISO 14001 certification has no direct impact on environmental performance. However, he notices a significant indirect effect. Certification encourages involvement in external networks (e.g., supply chain, training institutes, or local organisations), positively impacting their members' environmental performance. This is particularly important for SMEs because they often lack the resources and knowledge to respond to ecological problems. Thus, they depend on the expertise of external parties to implement necessary processes. ISO 14001 can help in this regard as the certification enables SMEs to collaborate with other network partners. SMEs lacking technical capacity and legal & business expertise can overcome this by cooperating with suppliers or customers (Machado, Vivaldini and Oliveira, 2020). Therefore, supply chain management and the practices adopted throughout the supply chain influence environmental results (Graafland 2018).

González, Sarkis, and Adenso-Díaz (2008) find that environmental pressure on automotive suppliers grows with the customer's firm size. Bigger firms can exert stronger influence on their suppliers compared to smaller firms. Larger firms tend to have more influence and resources to request such practices. The authors stress out that regardless of firm size, firms benefit from implementing ISO 14001 or EMAS. Contrarily, Curkovic and Sroufe (2011) found that customers in the automotive sector force ISO 14001 requirements on their suppliers but do not provide any further guidance. However, small suppliers voiced their need for this, e.g., in the form of industry networks (Curkovic and Sroufe 2011).

Other scholars find that large South African chemical firms and multinational enterprises that are exporting products are more engaged in EMS practices because they need to fulfil global sustainability goals (Urban and Govender 2012). Furthermore, they signal to their stakeholders (customers, governments, and NGOs) that they are environmental-conscious.

Regarding firm age, Delmas and Montiel (2009) find that younger automotive suppliers are more likely to react to customer pressure of introducing ISO 14001 certification. One possibility could be that the image of ISO 14001 is more valuable to firms striving to become established in the industry. It is difficult for customers to assess the quality of younger suppliers as they lack a proven track record. ISO 14001 implementation would advocate that the firm is acting environmentally conscious, providing insights of quality (Delmas and Montiel 2009).

The next section discusses the findings of this systematic literature review, presents its limitations and provides suggestions for future research.

CONCLUSION

The first finding illustrates firms' shift towards more sustainable practices. Traditionally, firms selected the supplier with the lowest costs but increasingly firms prefer to select suppliers that are environmentally conscious and implement EMS in their business model. By including the best interests of various stakeholder groups, EMS serves as an excellent tool to combat agency costs.

When investigating the adoption rate of ISO 14001 and EMAS, the findings suggest that ISO 14001 is more frequent than EMAS (Beske, Koplin, and Seuring 2008; Bajec, Tuljak-Suban, and Krmac 2015). A potential reason could be EMAS limitation to European countries up to 2009 followed by the economic crisis (Daddi et al. 2021). Furthermore, interviews conducted by Daddi et al. (2018) show that EMAS has stricter requirements than ISO 14001 and is less known among relevant stakeholders. EMAS certification involves stricter requirements and perhaps more effort. Nawrocka (2008b) criticises that although ISO 14001 certification proves that firms comply with specific standards, certification alone is not entirely trustworthy because of their certification bodies. As an example, interview experts claim that Polish certification bodies differ in their quality and strictness. In contrast to ISO 14001, EMAS certifications bodies are more regulated by the government and thus, perceived as more trustworthy. As trust plays a distinctive role in business relations, a managerial recommendation would be for firms to consider EMAS certification instead of ISO 14001. Future research could explore whether companies take for granted the trustworthiness of the EMS certification from their suppliers, or if (and how) they assess suppliers' environmental capabilities. In addition, future research could survey companies to explore whether they differentiate suppliers' assessment with ISO 14001 and EMAS, and how they do so. Another fruitful area for further research would be to explore the absence of supply chain pressure for EMAS Global.

While environmental awareness is also rising in developing countries, some authors point toward the social aspect of sustainability (e.g., Beske, Koplin, and Seuring 2008; Lee and Kim 2009). So far, EMS standards are neglecting this aspect. Consumers increasingly expect to see sustainable behaviour of firms in prominent areas such as the environmental impacts of product design and manufacturing processes and considering workers' rights and firm governance. While both ISO 14001 and EMAS push organizations towards a life-cycle perspective on environmental practices, to be recognised as a genuinely sustainable and encouraging firm, firms could choose to endorse social values as well. Future research could explore current practices on broader sustainability and develop a framework to incorporate both environmental and social practices that could be adopted by standard setters.

Concerning drivers and barriers of ISO 14001 certification, most of the authors agree that the main barrier associated with certification is its high cost (e.g., Zhu, Tian, and Sarkis 2012; Zhu, Cordeiro, and Sarkis 2012; Elbarkouky and Abdelazeem 2013; Salim and Padfield 2017; Scur and Barbosa 2017; Zimon 2020). To become certified, it costs firms resources, time, and money. This is particularly challenging for SMEs, younger firms, firms with specialised products, or firms in developing countries (e.g., Delmas and Montiel 2009; Zhu, Tian, and Sarkis 2012, Graafland 2018), as they lack the required resources. However, certification appears to be often a prerequisite for them as they usually do not have a disclosed track record of their adherence to environmental laws and environmental efforts beyond that. This poses a threat for small and young firms as they could become excluded from the market due to new entry barriers. Simply reducing the certification costs, however, may not be a suitable solution to this problem. First, in the case of ISO 14001, ISO does not directly influence the costs. Second, a race to the bottom cost-wise may affect the certification body's quality and could potentially further harm the already questioned reputation of ISO 14001. What could alleviate this problem is the support of small suppliers' customers who require EMS certification with knowledge and assist them with its implementation. From this, knowledge networks could emerge from which small firms and large firms could profit. Future research adopting interventionist approach could promote such knowledge networks by bringing together companies, researchers, students and certification bodies. Moreover, the necessary mechanisms to develop a support and learning network for firms could be elucidated by further research.

Furthermore, authors state that low awareness and no pressure exist in some countries, so firms are reluctant to become certified (e.g., Scur and Barbosa (2017)). However, Salim and Padfield (2017) note that in Malaysia, this is likely to change in the future, which is evidenced by the high number of studies published in this country following its' 2020 environmental agenda. Also, the pressure might increase as United Nations aim to achieve 17 sustainable development goals (SGDs) by 2030. The implementation of EMS would be beneficial in this regard.

While academia discussed firm-level characteristics such as age and size widely, existing studies neglected characteristics of the CEO. Yet, the gender of top executives influences a firm's success (Hambrick 2007). As women put a greater emphasis on helping people instead of financial matters (Betz, O'Connell, and Shepard 1989), it may be the case that female CEOs are more likely to implement EMS compared to their male counterparts. Future research could address the firm's diversity influence on the adoption of EMS.

While this systematic literature review uses research papers indexed by two scientific publication repositories, some relevant papers may be missing. Other databases, such as Google Scholar and papers in non-English languages, could be included in future studies to get an even more complete and more representative picture of the current scientific status-quo regarding EMS and supply chain pressure.

Furthermore, otherwise relevant papers may be excluded as the used search terms may not encompass all possible search terms combinations. A more comprehensive search may bring up other relevant papers. However, this study also shows that interviews may be an exciting source of insights. This means that interviewees also mention relevant points relating to the link between EMS and supply chain in interviews that are not primarily addressing this topic. Thus, these papers may be missed in the analysis as their focus lies elsewhere. It is therefore difficult to circumvent this problem completely. A possible solution could be to rely on the known databases, search them with keywords and follow citations within research papers.

Additionally, current research topics offer ample opportunities for expansion with new research questions, which will help to understand better the connection between EMS and supply chain pressure in full. These research opportunities include the actual testing of the recently developed theoretical models for supplier selection, how governments can increase their role in promoting EMS certification, for example, by using the globally agreed-upon SDGs 2030, and a complete analysis of countries and sectors, as these seem to differ significantly in their demand for EMS certification. Another fertile ground for further research would be to explore how the supply chain disruption arising from the Covid-19 pandemic fostered or hindered the pressure for EMS.

Repeating existing research could be worthwhile to analyse if social and business perception towards environmental issues has changed over time, thus influencing how firms chose and assess their suppliers. Finally, recently developed theoretical decision models on how to select a supplier based on EMS, e.g., by Akcan and Taş (2019), Ecer (2020), and Shojaei and Bolvardizadeh (2020), may be applied to a variety of real-world cases.

ACKNOWLEDGMENT

This work was funded by Fundação para a Ciência e a Tecnologia (UID/ ECO/00124/2019, UIDB/00124/2020, UIDP/00124/2020 and Social Sciences DataLab - PINFRA/22209/2016), POR Lisboa and POR Norte (Social Sciences DataLab, PINFRA/22209/2016).

REFERENCES

Aiyub, K., Arifin, K., Awang, A., & Jahi, J. M. (2009). Environmental Performance in Small and Medium Sized Enterprises (SMEs) Certified to ISO 14001 in the United Kingdom. *International Business Management, 3*(1), 7–14.

Akcan, S., & Taş, M. A. (2019). Green Supplier Evaluation with SWARA-TOPSIS Integrated Method to Reduce Ecological Risk Factors. *Environmental Monitoring and Assessment, 191*(12), 736. doi:10.100710661-019-7884-3 PMID:31707512

Al-Refaie, A., Al-Tahat, M., & Lepkova, N. (2020). Modelling Relationships Between Agility, Lean, Resilient, Green Practices in Cold Supply Chains Using ISM Approach. *Technological and Economic Development of Economy, 26*(4), 675–694. doi:10.3846/tede.2020.12866

Albelda, E. (2011). The Role of Management Accounting Practices as Facilitators of the Environmental Management: Evidence from EMAS Organisations. *Sustainability Accounting, Management and Policy Journal, 2*(1), 76–100. doi:10.1108/20408021111162137

Arimura, T. H., Darnall, N., & Katayama, H. (2011). Is ISO 14001 a Gateway to More Advanced Voluntary Action? The Case of Green Supply Chain Management. *Journal of Environmental Economics and Management, 61*(2), 170–182. doi:10.1016/j.jeem.2010.11.003

Bajec, P., Tuljak-Suban, D., & Krmac, E. (2015). Do ISO Standards Favour Logistics Provider Efficiency, Competitiveness and Sustainability? A Slovenian Perspective. *International Journal of Logistics Management, 26*(2), 275–295. doi:10.1108/IJLM-01-2013-0006

Beske, P., Koplin, J., & Seuring, S. (2008). The Use of Environmental and Social Standards by German First-Tier Suppliers of the Volkswagen AG. *Corporate Social Responsibility and Environmental Management, 15*(2), 63–75. doi:10.1002/csr.136

Betz, M., O'Connell, L., & Shepard, J. M. (1989). Gender Differences in Proclivity for Unethical Behavior. *Journal of Business Ethics, 8*(5), 321–324. doi:10.1007/BF00381722

Boiral, O., & Henri, J. F. (2012). Modelling the Impact of ISO 14001 on Environmental Performance: A Comparative Approach. *Journal of Environmental Management, 99*(May), 84–97. doi:10.1016/j.jenvman.2012.01.007 PMID:22326757

Bremmers, H., Omta, O., Kemp, R., & Haverkamp, D. J. (2007). Do Stakeholder Groups Influence Environmental Management System Development in the Dutch Agri-Food Sector? *Business Strategy and the Environment*, *16*(3), 214–231. doi:10.1002/bse.480

Chiarini, A. (2012). Designing an Environmental Sustainable Supply Chain through ISO 14001 Standard. *Management of Environmental Quality*, *24*(1), 16–33. doi:10.1108/14777831311291113

Chiarini, A. (2014). Strategies for Developing an Environmentally Sustainable Supply Chain: Differences Between Manufacturing and Service Sectors: Strategies for Developing an Environmentally Sustainable Supply Chain. *Business Strategy and the Environment*, *23*(7), 493–504. doi:10.1002/bse.1799

Chiarini, A. (2017). Environmental Policies for Evaluating Suppliers' Performance Based on GRI Indicators: A Multi-Regression Model for Suppliers' Environmental Performance. *Business Strategy and the Environment*, *26*(1), 98–111. doi:10.1002/bse.1907

Curkovic, S., & Sroufe, R. (2011). Using ISO 14001 to Promote a Sustainable Supply Chain Strategy. *Business Strategy and the Environment*, *20*(2), 71–93. doi:10.1002/bse.671

Daddi, T., Giacomo, M. R., Frey, M., & Iraldo, F. (2018). Analysing the Causes of Environmental Management and Audit Scheme (EMAS) Decrease in Europe. *Journal of Environmental Planning and Management*, *61*(13), 2358–2377. doi:10.1080/09640568.2017.1395316

Daddi, T., Heras-Saizarbitoria, I., Marrucci, L., Rizzi, F., & Testa, F. (2021). The effects of green supply chain management capability on the internalisation of environmental management systems and organisation performance. *Corporate Social Responsibility and Environmental Management*, *28*(4), 1241–125328. doi:10.1002/csr.2144

Delmas, M. A. (2000). Barriers and Incentives to the Adoption of ISO 14001 by Firms in the United States. *Duke Environmental Law and Policy Forum*, *11*(1), 1–38.

Delmas, M. A. (2002). The Diffusion of Environmental Management Standards in Europe and in the United States: An Institutional Perspective. *Policy Sciences*, *35*(1), 91–119. doi:10.1023/A:1016108804453

Delmas, M. A., & Montiel, I. (2009). Greening the Supply Chain: When Is Customer Pressure Effective? *Journal of Economics & Management Strategy*, *18*(1), 171–201. doi:10.1111/j.1530-9134.2009.00211.x

Ecer, F. (2020, Feb.). Multi-Criteria Decision Making for Green Supplier Selection Using Interval Type-2 Fuzzy AHP: A Case Study of a Home Appliance Manufacturer. *Operational Research*.

Elbarkouky, M. M. G., & Abdelazeem, G. (2013). A Green Supply Chain Assessment for Construction Projects in Developing Countries. *WIT Transactions on Ecology and the Environment*, *179*, 1331–1341. doi:10.2495/SC131132

European Commission. (2011). *EMAS - Factsheet.* Retrieved October 2020, from https://ec.europa.eu/environment/emas/pdf/factsheets/EMASiso14001_high.pdf

European Commission. (2018). *Case Study – Involving Employees in Implementing EMAS.* Retrieved October 2020, from https://ec.europa.eu/environment/emas/pdf/other/Case%20study_Involving%20employees.pdf

European Commission. (2020). *Official Statistics of the European EMAS Helpdesk - Organisations and Sites per Country (September 2020).* Retrieved October 2020, from https://ec.europa.eu/environment/emas/pdf/statistics/EMASStatistics_September%202020_final.pdf

European Commission. (n.d.a). *Policy - EMAS Regulation.* Retrieved October 2020, from https://ec.europa.eu/environment/emas/emas_publications/policy_en.htm

European Commission. (n.d.b). *SME Corner.* Retrieved October 2020, from https://ec.europa.eu/environment/emas/join_emas/what_if_i_am_an_sme_en.htm

General Assembly. (1992). Report of the United Nations Conference on Environment and Development. In *Proceedings of the Conference* (Vol. 2). United Nations. Retrieved October 2020, from https://www.un.org/en/development/desa/population/migration/generalassembly/docs/globalcompact/A_CONF.151_26_Vol.I_Declaration.pdf

González, P., Sarkis, J., & Adenso-Díaz, B. (2008). Environmental Management System Certification and Its Influence on Corporate Practices: Evidence from the Automotive Industry. *International Journal of Operations & Production Management, 28*(11), 1021–1041. doi:10.1108/01443570810910179

Graafland, J. J. (2018). Ecological Impacts of the ISO14001 Certification of Small and Medium Sized Enterprises in Europe and the Mediating Role of Networks. *Journal of Cleaner Production, 174*(February), 273–282. doi:10.1016/j.jclepro.2017.10.322

Gupta, S., Soni, U., & Kumar, G. (2019). Green Supplier Selection Using Multi-Criterion Decision Making under Fuzzy Environment: A Case Study in Automotive Industry. *Computers & Industrial Engineering, 136*(October), 663–680. doi:10.1016/j.cie.2019.07.038

Hambrick, D. C. (2007). Upper Echelons Theory: An Update. *Academy of Management Review, 32*(2), 334–343. doi:10.5465/amr.2007.24345254

Holt, D. (1998). The Perceived Benefits of an Environmental Management Standard. *Business Process Management Journal, 4*(3), 204–213. doi:10.1108/14637159810224313

Iraldo, F., Testa, F., & Frey, M. (2009). Is an Environmental Management System Able to Influence Environmental and Competitive Performance? The Case of the Eco-Management and Audit Scheme (EMAS) in the European Union. *Journal of Cleaner Production, 17*(16), 1444–1452. doi:10.1016/j.jclepro.2009.05.013

Islam, M. S., Tseng, M. L., Karia, N., & Lee, C. H. (2018). Assessing Green Supply Chain Practices in Bangladesh Using Fuzzy Importance and Performance Approach. *Resources, Conservation and Recycling, 131*(April), 134–145. doi:10.1016/j.resconrec.2017.12.015

ISO. (2015). *Introduction to ISO 14001:2015.* Retrieved October 2020, from https://www.iso.org/files/live/sites/isoorg/files/store/en/PUB100371.pdf

ISO. (2021). *1. ISO Survey 2019 Results - Number of Certificates and Sites per Country and the Number of Sector Overall.* Retrieved October 2020, from https://isotc.iso.org/livelink/livelink?func=ll&objId=21412354&objAction=Open&nexturl=%2Flivelink%2Flivelink%3Ffunc%3Dll%26objId%3D18808772%26objAction%3Dbrowse%26viewType%3D1

ISO. (n.d.). *Certification*. Retrieved October 2020, from https://www.iso.org/certification.html

Jamal, N. M., Wong, W. K., Saat, M. M., & Othman, A. (2020). A Relationship Analysis between Green Supply Chain Practices, Environmental Management Accounting and Performance. *Journal of Sustainability Science and Management, 15*(3), 113–133.

Jiang, P., Hu, Y. C., Yen, G. F., & Tsao, S. J. (2018). Green Supplier Selection for Sustainable Development of the Automotive Industry Using Grey Decision-Making. *Sustainable Development, 26*(6), 890–903. doi:10.1002d.1860

Kaminski, I. (2013). *Road from Rio: The Past, the Present and the Future of ISO 14001*. Retrieved October 2020, from https://www.bsigroup.com/LocalFiles/en-GB/iso-14001/whitepapers/BSI-ISO-14001-article-The-ENDS-Report-Road-from-Rio-UK-EN.pdf

Kollman, K., & Prakash, A. (2002). EMS-Based Environmental Regimes as Club Goods: Examining Variations in Firm-Level Adoption of ISO 14001 and EMAS in U.K., U.S. and Germany. *Policy Sciences, 35*(1), 43–67. doi:10.1023/A:1016071810725

Lee, K. H., & Kim, J. W. (2009). Current Status of CSR in the Realm of Supply Management: The Case of the Korean Electronics Industry. Supply Chain Management: An International Journal, 14(2), 138–48.

Lee, S. M., Noh, Y., Choi, D., & Rha, J. S. (2017). Environmental Policy Performances for Sustainable Development: From the Perspective of ISO 14001 Certification: The Financial Effects of ISO 14001. *Corporate Social Responsibility and Environmental Management, 24*(2), 108–120. doi:10.1002/csr.1395

Machado, M. C., Vivaldini, M., & de Oliveira, O. J. (2020). Production and supply-chain as the basis for SMEs' environmental management development: A systematic literature review. *Journal of Cleaner Production, 273*, 123141. doi:10.1016/j.jclepro.2020.123141

Marouen, A. (2015). Environmental Management System: Environmental Impacts and Productivity. *International Journal of Business and Management, 10*(11), 107. doi:10.5539/ijbm.v10n11p107

Masson-Delmotte, V., Zhai, P., Pörtner, H. O., Roberts, D., Skea, J., Shukla, P. R., & Pirani, A. (2018). Global Warming of 1.5°C. An IPCC Special Report on the Impacts of Global Warming of 1.5°C above Pre-Industrial Levels and Related Global Greenhouse Gas Emission Pathways, in the Context of Strengthening the Global Response to the Threat of Climate Change. *Sustainable Development, and Efforts to Eradicate Poverty.*

Medina-Serrano, R., González-Ramírez, M., Gascó-Gascó, J. L., & Llopis-Taverner, J. (2019, December). Prácticas de Evaluación de Proveedores Sostenibles a Través de La Cadena de Suministro. *Dirección y Organización*, (69), 13–26. doi:10.37610/dyo.v0i69.558

Melnyk, S. A., Sroufe, R. P., & Calantone, R. (2003). Assessing the Impact of Environmental Management Systems on Corporate and Environmental Performance. *Journal of Operations Management, 21*(3), 329–351. doi:10.1016/S0272-6963(02)00109-2

Mishra, A. R., Rani, P., Pardasani, K. R., & Mardani, A. (2019). A Novel Hesitant Fuzzy WASPAS Method for Assessment of Green Supplier Problem Based on Exponential Information Measures. *Journal of Cleaner Production, 238*(November), 117901. doi:10.1016/j.jclepro.2019.117901

Morrow, D., & Rondinelli, D. (2002). Adopting Corporate Environmental Management Systems. *European Management Journal, 20*(2), 159–171. doi:10.1016/S0263-2373(02)00026-9

Nawrocka, D. (2008). Environmental Supply Chain Management, ISO 14001 and RoHS. How Are Small Companies in the Electronics Sector Managing? *Corporate Social Responsibility and Environmental Management, 15*(6), 349–360. doi:10.1002/csr.176

Nawrocka, D., Brorson, T., & Lindhqvist, T. (2009). ISO 14001 in Environmental Supply Chain Practices. *Journal of Cleaner Production, 17*(16), 1435–1443. doi:10.1016/j.jclepro.2009.05.004

Nishitani, K. (2010). Demand for ISO 14001 Adoption in the Global Supply Chain: An Empirical Analysis Focusing on Environmentally Conscious Markets. *Resource and Energy Economics, 32*(3), 395–407. doi:10.1016/j.reseneeco.2009.11.002

Oberthür, S., & Kelly, C. R. (2008). EU Leadership in International Climate Policy: Achievements and Challenges. *The International Spectator, 43*(3), 35–50. doi:10.1080/03932720802280594

Orcos, R., & Palomas, S. (2019). The Impact of National Culture on the Adoption of Environmental Management Standards: The Worldwide Diffusion of ISO 14001. *Cross Cultural & Strategic Management, 26*(4), 546–566. doi:10.1108/CCSM-10-2018-0168

Oreskes, N. (2004). The Scientific Consensus on Climate Change. *Science, 306*(5702), 1686–1686. doi:10.1126cience.1103618 PMID:15576594

Prajogo, D., Chowdhury, M., Yeung, A. C. L., & Cheng, T. C. E. (2012). The Relationship between Supplier Management and Firm's Operational Performance: A Multi-Dimensional Perspective. *International Journal of Production Economics, 136*(1), 123–130. doi:10.1016/j.ijpe.2011.09.022

Prajogo, D., Tang, A. K., & Lai, K. H. (2014). The Diffusion of Environmental Management System and Its Effect on Environmental Management Practices. *International Journal of Operations & Production Management, 34*(5), 565–585. doi:10.1108/IJOPM-10-2012-0448

Prasad, D. S., Pradhan, R. P., Gaurav, K., Chatterjee, P. P., Kaur, I., Dash, S., & Nayak, S. (2018). Analysing the Critical Success Factors for Implementation of Sustainable Supply Chain Management: An Indian Case Study. *Decision (Washington, D.C.), 45*(1), 3–25.

Razak, A. A., Rowling, M., White, G., & Mason-Jones, R. (2016). Public Sector Supply Chain Management: A Triple Helix Approach to Aligning Innovative Environmental Initiatives. *Foresight and STI Governance, 10*(1), 43–52. doi:10.17323/1995-459x.2016.1.43.52

Salim, H. K., & Padfield, R. (2017). Environmental Management System in the Food and Beverage Sector: A Case Study from Malaysia. *Chemical Engineering Transactions, 56*(April), 253–258.

Salim, H. K., Padfield, R., Lee, C. T., Syayuti, K., Papargyropoulou, E., & Tham, M. H. (2018). An Investigation of the Drivers, Barriers, and Incentives for Environmental Management Systems in the Malaysian Food and Beverage Industry. *Clean Technologies and Environmental Policy, 20*(3), 529–538. doi:10.100710098-017-1436-8

Santos, B. M., Godoy, L. P., & Campos, L. M. S. (2019). Performance Evaluation of Green Suppliers Using Entropy-TOPSIS-F. *Journal of Cleaner Production*, *207*(January), 498–509. doi:10.1016/j.jclepro.2018.09.235

Scur, G., & Barbosa, M. E. (2017). Green Supply Chain Management Practices: Multiple Case Studies in the Brazilian Home Appliance Industry. *Journal of Cleaner Production*, *141*(January), 1293–1302. doi:10.1016/j.jclepro.2016.09.158

Shojaei, P., & Bolvardizadeh, A. (2020). Rough MCDM Model for Green Supplier Selection in Iran: A Case of University Construction Project. *Built Environment Project and Asset Management*, *10*(3), 437–452. doi:10.1108/BEPAM-11-2019-0117

Smith, C. (1993). BS 7750 and Environmental Management. *Journal of the Society of Dyers and Colourists*, *109*(9), 278–279. doi:10.1111/j.1478-4408.1993.tb01574.x

Sommer, B., & Hain, A. (2017). Europe as a Green Leader? A Brief Evaluation of Both the European Union's Climate and Energy Policy and Common Agricultural Policy. *Culture, Practice & Europeanization*, *2*(2), 33–45. doi:10.5771/2566-7742-2017-2-33

Sturm, A., & Upasena, S. (1998). *ISO 14001 – Implementing an Environmental Management System*. Ellipson Management Consultants.

To, W. M., & Lee, P. K. C. (2014). Diffusion of ISO 14001 Environmental Management System: Global, Regional and Country-Level Analyses. *Journal of Cleaner Production*, *66*(March), 489–498. doi:10.1016/j.jclepro.2013.11.076

Urban, B., & Govender, D. P. (2012). Empirical Evidence on Environmental Management Practices. *The Engineering Economist*, *23*(2), 209–215.

Van der Veldt, D. (1997). Case Studies of ISO 14001: A New Business Guide for Global Environmental Protection. *Environmental Quality Management*, *7*(1), 1–19. doi:10.1002/tqem.3310070102

Wiengarten, F., Pagell, M., & Fynes, B. (2013). ISO 14000 Certification and Investments in Environmental Supply Chain Management Practices: Identifying Differences in Motivation and Adoption Levels between Western European and North American Companies. *Journal of Cleaner Production*, *56*(October), 18–28. doi:10.1016/j.jclepro.2012.01.021

Wu, J. Z., Santoso, C. H., & Roan, J. (2017). Key Factors for Truly Sustainable Supply Chain Management: An Investigation of the Coal Industry in Indonesia. *International Journal of Logistics Management*, *28*(4), 1196–1217. doi:10.1108/IJLM-07-2014-0103

Yiridoe, E. K., & Marett, G. E. (2004). Mitigating the High Cost of ISO 14001 EMS Standard Certification: Lessons from Agribusiness Case Research. *International Food and Agribusiness Management Review*, *7*, 37-62.

Younis, H., & Sundarakani, B. (2019). The Impact of Firm Size, Firm Age and Environmental Management Certification on the Relationship between Green Supply Chain Practices and Corporate Performance. *Benchmarking*, *27*(1), 319–346. doi:10.1108/BIJ-11-2018-0363

Zhu, Q., Cordeiro, J., & Sarkis, J. (2012). International and Domestic Pressures and Responses of Chinese Firms to Greening. *Ecological Economics*, *83*(November), 144–153. doi:10.1016/j.ecolecon.2012.04.007

Zhu, Q., Tian, Y., & Sarkis, J. (2012). Diffusion of Selected Green Supply Chain Management Practices: An Assessment of Chinese Enterprises. *Production Planning and Control*, *23*(10–11), 837–850. doi:10.1080/09537287.2011.642188

Zilahy, G. (2017). Environmental Management Systems—History and New Tendencies. In *Encyclopedia of Sustainable Technologies* (pp. 23–31). Elsevier. doi:10.1016/B978-0-12-409548-9.10529-9

Zimon, D. (2020). Iso 14001 and the Creation of SSCM in the Textile Industry. *International Journal of Qualitative Research*, *14*(3), 739–748. doi:10.24874/IJQR14.03-06

Zimon, D., & Madzik, P. (2019). Impact of Implementing ISO 14001 Standard Requirements for Sustainable Supply Chain Management in the Textile Industry. *Fibres & Textiles in Eastern Europe*, *27*(6), 8–14. doi:10.5604/01.3001.0013.4462

Zscheischler, J., & Seneviratne, S. I. (2017). Dependence of Drivers Affects Risks Associated with Compound Events. *Science Advances*, *3*(6), e1700263. doi:10.1126ciadv.1700263 PMID:28782010

APPENDIX

BS - British Standard.
CASCO - Committee on Conformity Assessment.
CEO - Chief Executive Officer.
CO_2 - Carbon Dioxide.
CSR - Corporate Social Responsibility.
EMAS - Eco-Management and Audit Scheme.
EMS - Environmental Management System.
EU - European Union.
GATT - General Agreement on Tariffs and Trade.
GHG - Green House Gases.
GSCM - Green Supply Chain Management.
ISO - International Organization for Standardization.
NGO - Non-Governmental Organisation.
PDCA - Plan-Do-Check-Act.
RoHS - Restrictions of Hazardous Substances Directive.
SDGs - Sustainable Development Goals
SME - Small and Medium Enterprise.
SSCM - Sustainable Supply Chain Management.
U.S. - United States.
UK - United Kingdom.
WTO - World Trade Organization.

Chapter 5

The Reporting of Non-Financial Information by SMEs:

Assessing the Answers to the Project of the Revised Directive 2014/95/EU

Fábio Albuquerque
https://orcid.org/0000-0001-8877-9634
Lisbon Accounting and Business School, Instituto Politécnico de Lisboa, Portugal

Maria Albertina Barreiro Rodrigues
https://orcid.org/0000-0002-7858-5358
Universidade Europeia and ISCAL - Lisbon Accounting and Business School, Instituto Politécnico de Lisboa, Portugal

Paula Gomes dos Santos
https://orcid.org/0000-0003-2192-8855
Lisbon Accounting and Business School, Instituto Politécnico de Lisboa, Portugal & COMEGI, Portugal

Ana Morais
https://orcid.org/0000-0001-7251-6418
ISEG – Lisbon School of Economics and Management, Universidade de Lisboa, Portugal

ABSTRACT

Until now, small and medium enterprises (SMEs) have largely been left out of non-financial information (NFI) disclosures. However, this will change significantly with the Non-Financial Reporting Directive. This chapter analyses the stakeholders' perspective about three core aspects related to this: the usefulness of a simplified NFI standard and/or reporting format for SMEs; if that is an effective means of limiting the burden on SMEs; and, finally, in case of implementing it, if it should be mandatory or voluntary. Data were obtained from the stakeholders' answers provided during the public consultation conducted by the European Union about possible revisions of the Directive 2014/95/EU, from February to June 2020. The results show that the most controversial topic is the mandatory or voluntary nature of the simplified standard for SMEs. Furthermore, there are different opinions about the usefulness and effectiveness of a simplified standard for SMEs, particularly between capital providers versus other entities.

DOI: 10.4018/978-1-7998-9410-0.ch005

INTRODUCTION

Corporate reporting has undergone relevant changes to meet the demands of an increasing range of stakeholders and to respond to external events. Historically, it has moved from a reporting that was essentially financial to a reporting that, in a second phase, also includes disclosures of non-financial information (NFI).

However, concerns about the harmonisation of either financial or non-financial reporting seems to firstly cover larger or listed companies, including small and medium enterprises (SMEs) only in a second phase. Regarding financial information, the effective inclusion of smaller or unlisted entities within the scope of the harmonisation process happened only in the early years of this century, following a more advanced stage of harmonisation for larger or listed ones. This can be demonstrated by the work initiated by the International Accounting Standards Board (IASB), recovering the project started by the previous standard-setter International Accounting Standards Committee (IASC) in 2001, to later develop the International Financial Reporting Standard (IFRS) for Small and Medium Entities (IFRS for SMEs) in 2004.

In this context, it is worth remembering that, previously, the full International Accounting Standards (IAS) and IFRS (full IAS/IFRS) had been already adopted by European Union (EU) through Regulation 1606/2002 of the European Parliament and of the Council of July 19, 2002, which included in its scope, particularly, the entities whose securities are admitted to trading on a regulated market of any Member State, as a mandatory requirement for their consolidated accounts (European Commission, 2002). This relevant step was achieved after almost thirty years, considering that IASC was formed in 1973.

The IFRS for SMEs emerged in a subsequent phase of the IASB's harmonisation goals, aiming to establish more simplified requirements for such entities. Its scope includes entities that do not have public accountability and publish general purpose financial statements for external users. The benefits of the IFRS for SMEs are multiple, and the following can be highlighted: international comparability between financial reports by using the same accounting language; easier preparation and presentation of consolidated financial statements; and facilitating cross border trade and international growth of SME (Sellami & Gafsi, 2018).

To reduce the reporting burden for SMEs, the IFRS for SMEs is less complex than the full set of IAS and IFRS (full IAS/IFRS), having simplifications not only in terms of recognition and measurement requirements but also in terms of disclosures. However, the IFRS for SME is separate from full IAS/IFRS and consequently is available for any jurisdiction to adopt regardless of whether it adopts full IAS/IFRS or not. For instance, in the EU case particularly, IFRS for SMEs was not adopted.

Concerning NFI, the debate on the need for harmonisation is more recent, despite increasing interest in the last twenty years. As societies are increasingly attentive to socially and environmentally responsible entities, non-financial reporting is becoming a key component of corporate communication and report, and some models have gained prominence, such as the Global Reporting Initiative (GRI) and, more recently, the Integrated Reporting (IR) (Bartolacci et al., 2020; De Villiers et al., 2014; Dumay et al., 2016).

In any case, the perspective of disclosing NFI, as well as the debate on the harmonisation of the non-financial report, has been particularly focused on larger companies, especially on listed entities (Stolowy & Paugam, 2018). This may be explained by the fact that investors' needs regarding the relevance of this complementary information for making the best decisions on allocating their investments in sustainable companies are understood in a long-term perspective. Consequently, this is a matter mostly devoted in literature to large and/or listed entities, as it includes, among other aspects, the mitigation of reputational risks, and then SMEs have largely been left out from NFI disclosures until now.

According to the European Financial Reporting Advisory Group (EFRAG, 2021a), it is not known how many SMEs disclose NFI (due to the trickle-down effect or voluntarily), what they disclose, and what are the stakeholders needs. While in countries with a rule-based governance system, sustainability report is viewed by most SMEs as an economic tool and as an opportunity, in countries with a relationship-based governance system, NFI is mainly seen by SMEs as a constraint (El Baz et al., 2016). Having in mind the EU case specifically, the disclosure of NFI for smaller or unlisted companies was almost exclusively subject to the mandatory requirements provided for in local legislations. Then, national legislations have been including, over time, several topics to be especially disclosed in the management report, covering corporate governance, environment, as well as social responsibility matters (employees-related measures, in particular).

Also, many SMEs are under increasing pressure to provide certain NFI to other businesses, namely if they are suppliers of large companies. In addition, financial institutions are increasingly likely to request certain NFI from companies to whom they provide capital, including SMEs. In this respect, SMEs that do not provide NFI may experience a negative impact on their commercial opportunities as suppliers of larger companies or their access to capital and may not be able to benefit from new sustainable investment opportunities. In this sense, researchers have found evidence of increasing transparency by SMEs concerning NFI, which may be possibly explained by the pressure from different stakeholders (Stekelorum et al., 2018; Shields & Shelleman, 2020; Fernandez-Feijoo et al., 2014).

Notwithstanding, the pressure from a particular group of relevant stakeholders, namely the financial institutions, justifies the need for reopening the debate on the inclusion of SMEs in the requirements for publishing NFI (Altman et al., 2010).

The European criteria for the SMEs definition is based on, among other requirements, "enterprises which employ fewer than 250 persons and which have an annual turnover not exceeding EUR 50 million, and/or an annual balance sheet total not exceeding EUR 43 million" (European Commission, 2003). In this context, it is worthwhile to stress that SMEs represent approximately 23 million or 99.8% of all EU non-financial businesses (European Investment Bank Group, 2020) and the majority of banks borrowers in the U.S. and Europe (Altman et al., 2010). Therefore, it is reasonable that the EU strategies for SMEs include making the funding accessible and diversifying the sources of funding to finance their investment needs for the transition to a sustainable and digital economy (European Commission, 2020a).

The Directive 2013/34/EU of the European Parliament and of the Council of 26 June 2013, transposed by member states by 2015, introduced two themes to the European agenda, namely the need of simplifying the corporate reporting for SMEs and, simultaneously, harmonising the financial and NFI in Europe.

However, the directive view of the reporting process as an excessive cost for SMEs led to an inconsistent application of the EU strategies for SMEs. In other words, the perspective of requirements with reduced benefits for SMEs conducted to an overall rejection of NFI implementation for these types of entities.

Specifically, this directive introduced the need for more simplified reporting requirements according to the entities' size, which even led to the exemption of preparing the management report by smaller entities "where a copy of all or part of any such report can be easily obtained upon request at a price not exceeding its administrative cost". According to it, "users of financial statements typically have a limited need for supplementary information from small undertakings, and it can be costly for small undertakings to collate that supplementary information. A limited disclosure regime for small undertakings is, therefore, justified". Thus, the directive seemed to have the purpose to eliminate what was seen as an excessive cost to produce information, given that its usefulness or relevance has been identified as frequently reduced.

Notwithstanding, the directive expanded the content of NFI in the management report, requiring the disclosure of information about environmental matters, human rights, anti-corruption and bribery, diversity of corporate boards in terms of gender, age, educational and professional background, and social and employees' topics. The directive also presented the view that "having regard to the potential burden placed on small and medium-sized undertakings, it is appropriate to provide that the Member States may choose to waive the obligation to provide non-financial information in the management report of such undertakings". Consequently, the twofold objectives proposed in this directive seemed to be unfeasible.

Later, the Directive 2013/34/EU was amended by the Directive 2014/95/EU of the European Parliament and of the Council of 22 October 2014 (European Parliament, 2014), called the non-financial reporting Directive (NFRD). This latter directive required a non-financial statement to be provided as part of their annual public reporting obligations, from 2018 onwards, by large Public-Interest Entities (PIEs) with more than 500 employees, which means the inclusion of large listed companies and large banks and insurance companies (whether listed or not) (Corrado et al., 2019; Hojnik, 2017).

As regards the content, the NFRD comprises four sustainability issues, namely environment, social and employee issues, human rights, and bribery and corruption. Companies are required to disclose information about their business model, policies (including implemented due diligence processes), outcomes, risks, and risk management, and key performance indicators (KPIs) relevant to the business in respect to those sustainability issues, promoting sustainability report quality (Fernandez-Feijoo et al., 2015). However, it does not introduce or require the use of a non-financial reporting standard or framework, nor does it impose detailed disclosure requirements such as lists of indicators per sector or define rules related to the assurance provider selection (Fernandez-Feijoo et al., 2015).

More recently, the public consultation process of this directive brought as novelties, not only the discussion around the need for increasing the requirements regarding NFI, but also the inclusion of SMEs in this context (European Parliament, 2021).

Then, the consultation document highlights which seemed to be an apparent contradiction (European Commission, 2020b), namely the inclusion of SME in the scope of the directive, despite recognising that SMEs often do not have the technical expertise nor resources necessary to prepare reports based on sophisticated standards (Baumann-Pauly et al., 2013). This may imply that requiring SMEs to apply the same standards as large companies may be a disproportionate burden for SMEs.

In this context, it is relevant to assess the stakeholders' perspective on the recent interest in including SMEs in the scope of NFI disclosure. For this purpose, SMEs may be defined as "undertakings which on their balance sheet dates does not exceed at least two of the three following criteria: a) balance sheet total: EUR 20 000 000; (b) net turnover: EUR 40 000 000; (c) average number of employees during the financial year: 250" (Baumann, 2013). This concept is based on a residual approach, which has the concept of a large undertaking within the Directive 2013/34/EU as a reference. Additionally, it is relevant to stress that this definition is not consistent with the one previously indicated, proposed by the European Commission (2003).

Therefore, this chapter aims to assess the answers to the project of the revised Directive 2014/95/EU in what concerns three different topics, namely: i. the usefulness of a simplified NFI standard and/or reporting format for SMEs; ii. the burden faced by SMEs in the non-financial report; iii. and, finally, the mandatory or voluntary nature of the simplified standard.

Additionally, it is relevant to assess and compare the different perspectives of the stakeholders involved in this process of public consultation. For this purpose, the views from different stakeholders were assessed, namely: i. SMEs (or entities representing SMEs) *versus* non-SMEs; ii. SMEs (or entities representing SMEs) within *versus* out of the scope of the NFRD; iii. and, finally, capital providers *versus* other entities.

With these aims, this chapter will assess, through descriptive analysis and the chi-square test, the answers given by stakeholders in the public consultation on the review of NFRD, conducted by the EU from February to June 2020.

The structure of this chapter is as follows. The next section reviews relevant literature. Section three presents the objectives and the proposed methodology, from what the analysis for this chapter was performed and the main findings discussed. Finally, the last section presents the conclusions from this research.

LITERATURE REVIEW

The NFRD requires companies to disclose NFI but does not set the use of a non-financial reporting standard or framework. Regarding SMEs, the public consultation process of that directive introduced the discussion of including them in this context (European Parliament, 2021). As it is still a recent topic for SMEs, research on the content and consequences of NFI disclosures for SMEs are scarce, mainly on the aspects related to the discussion around the advantages and disadvantages of a mandatory or voluntary NFI standard for SMEs, as well as on the specific content of an SME-oriented NFI report to increase the usefulness and to reduce the burden of that disclosure. Research in this field may generally cover different types of NFI, with particular emphasis on the corporate social responsibility (CSR) disclosures and the several reports that may include this information, such as sustainability reports, and, more recently, the IR.

EFRAG (2021b, p.17) states that the purpose of sustainability reporting is to provide "relevant, faithful, comparable and reliable information: on (i) material sustainability impacts of the reporting entity on affected stakeholders (including the environment) and (ii) material sustainability risks and opportunities for its own value creation; enabling users of information (i) to understand the reporting entity's sustainability objectives, position and performance and (ii) to inform their decisions relating to their engagement with the entity".

Although SMEs have faced less pressure to prepare NFI (Krawczyk, 2021), they are also being affected by the growing demand for this information.

In the EU, NFI disclosure by SMEs is largely due to the national legislation's requirements, namely regarding several matters to be disclosed in the management report. Furthermore, many SMEs are increasingly disclosing NFI possibly due to the pressure of different stakeholders, such as capital providers as well as other companies with which they have trade relationships (Stekelorum et al., 2018; Shields & Shelleman, 2020; Fernandez-Feijoo et al., 2014; Calace, 2014). The European Federation of Accountants and Auditors for SMEs also concluded that larger SMEs were most likely to have a trickle-down effect in financial reporting, being "fair to assume that non-financial reporting obligations towards large entities will also lead to a trickle-down effect for non-financial aspects" (EFRAG, 2021a, p. 71). Being pressured by other businesses, larger SMEs have also begun pressuring their supply chain partners, particularly if they are suppliers of large companies (Shields & Shelleman, 2020).

Besides the relevant and direct influence on sustainable development (Weber, 2014), capital providers play an important role in that process through their lending activities. Their risk management activities should be properly organized to adequately assess the climate and environmental risks, ensuring the financing of activities that contribute to greater sustainability. Also, capital providers must separate SMEs from large companies and use procedures and models adapted to the singularities and specific needs of SMEs to assess and manage their credit risk (Altman & Sabato, 2007).

Therefore, NFI is important to assess SMEs credit risk, complementing financial data, as introducing additional non-financial data better predicts specific credit risk models to SMEs (Altman et al., 2010), and is useful to predict small business failure (Andreeva et al., 2016), existing a two-directional association between NFI disclosure and sustainability performance (Rezaee & Tuo, 2017).

The NFRD anticipated the potential burden placed on SMEs on disclosing NFI. Since, in some cases, sustainability issues go beyond the reporting entity boundaries, the system to assure compliance with the standards may be extremely costly, particularly for smaller companies (Christensen et al., 2021). Considering this, it is possible to conclude that the need to limit the burden for SMEs on the production of sustainable information, justifies the relevance of a simplified standard for SMEs.

Furthermore, most sustainability standards were developed contemplating large companies' needs and for national and international issues, the difficulty and complexity of sustainability standards are frequently stated as an obstacle for SMEs (Johnson & Schaltegger, 2016) and the need for proportionality is mentioned (Dinh et al., 2021). The characteristics of SMEs require specific standards, which focus on the contents of an SME sustainability report, with sustainability, social and environmental topics related to the firm's activities (Borga et al., 2009).

Then, it is relevant that SMEs participate in the definition of NFI reporting in Europe for three main reasons: they represent an important part of the economy; they impact society and the environment; and, finally, they are facing sustainability-related opportunities and risks. Regardless of the size, preparers are concerned with the difficulty of balancing their non-financial reporting needs and the stakeholders' needs, all at reasonable cost and effort. Being non-financial reporting requirements focused on large and/or listed enterprises, they are not suitable to SMEs, which have limited resources and capabilities, so specific consideration on how to include SMEs in the non-financial reporting landscape appears to be missing (EFRAG, 2021b).

Shields and Shelleman (2020) identified the lack of resources as one of the main barriers to the implementation of sustainable management by SMEs, proposing the development of low-cost tools in this field that could facilitate both decision making and monitoring. The scarcity of resources led SMEs to only disclose the simplest possible reporting, on the limit to comply with the minimums required (Ortiz-Martínez & Marín-Hernández, 2021; Corazza, 2017).

However, Torugsa et al. (2012) suggest that the lack of resources is a relevant but not deterministic factor for implementing a proactive CSR in SMEs. Also, for Fernandez-Feijoo et al. (2014), and Rodríguez-Gutiérrez et al. (2021), communicating a transparent image, reputation, and legitimacy are the most relevant criteria for SMEs when engaging in the sustainability report, being the cost the least relevant factor.

The constraints, on one hand, and the purposes and usefulness of NFI from the SMEs' perspective, on the other hand, have been simultaneously considered in the literature as explanatory factors of voluntary adoption. The debate on the balance between costs and benefits for SMEs when providing NFI through, for instance, CSR disclosures, as well as different tools to manage this type of information, is a usual perspective of analysis in literature. Therefore, the discussion is usually focused on the reasons that led SMEs to voluntarily provide NFI, instead of on the reasons to be considered by standard-setters to issue and impose a mandatory standard for SMEs.

On the mandatory or voluntary adoption of a standard by SMEs, it may be found in the literature a more specific discussion regarding financial information, particularly related to the adoption of IFRS for SMEs. Some studies have focused on the determinants of the IFRS for SMEs adoption (Kaya & Koch, 2015; Zahid & Simga-Mugan, 2018) and the stakeholders' perceptions about the costs and benefits of such standard (Albu, 2013; Albu et al., 2013; Kiliç et al., 2014; Litjens et al., 2012; Uyar & Güngörmüş, 2013). In this context, for instance, Zahid and Simga-Mugan (2018) found that common law countries, with lower regulatory efficiency, market openness, and economic growth are more likely to adopt IFRS for SMEs. To Uyar & Güngörmüş (2013), the education level and experience (e.g., employees of BIG 4) plays an important role in the knowledge about accounting standards. Litjens et al. (2012) found that the decision to voluntarily adopt IFRS for SMEs is a non-linear process since the preparers' perceptions show a stronger association with the cost than the benefits of its adoption. The potential nonlinearity of a preparer's cost-benefit analysis may imply significant complexity for standard setters in disseminating accounting and financial reporting standards. It might also lead to less efficient contracting if benefits are not perceived as such by preparers, particularly if they focus primarily on the perceived costs.

Considering that the goal of NFI reporting may be to mitigate firm behaviour and externalities, or to influence companies' sustainability activities, sustainability information with more quality and quantity, may change firm's actions and benefit capital markets by providing better liquidity, improved capital allocation, and lower cost of capital. Therefore, it is reinforced the importance of achieving mandatory NFI reporting (Christensen et al., 2021).

Even though sustainability reporting is voluntary, the possibility to define a common guideline has several advantages and benefits both to preparers and stakeholders since they enable and facilitate comparability, understandability, also reducing costs (Borga et al., 2009). According to Barker and Eccles (2018), besides being based on a standard, sustainability reporting should also be mandatory to be most useful for investors, as long as the necessary preconditions are implemented, as the availability of metrics, and the benefits prevalence over the costs.

Sustainability reporting should also be standardised to be more useful for stakeholders since they face some difficulties related to the existence of different, often irreconcilable, assessments of a company's disclosed NFI (Barker & Eccles, 2018). The existence of a diverse NFI reporting ecosystem creates inconsistencies and gaps applied to data format specifications, access controls, and validate rules, challenges the usefulness of a compliance ecosystem, and increases its complexity (EFRAG, 2021b).

In light of the increase in the requirements for SMEs in social, environmental, and economic areas, some studies have developed and proposed frameworks and mechanisms for sustainability reports for SMEs (Borga et al., 2009; Burke & Gaughran, 2007; Massa et al., 2015; Shields & Shelleman, 2020), while others looked for the best choice between the existing frameworks to allow SMEs to improve communication with stakeholders and to enhance management strategies (Medel et al., 2011; Rodríguez-Gutiérrez et al., 2021). Thus, there is a lack of consistency among sustainable reports in SMEs, which implies a demand for harmonisation to allow comparability of sustainability performance (Steinhofel et al., 2019).

The inconsistency in NFI standards may inflate costs, creates operational and compliance risks, and also involves a higher resources consumption for both the preparers and the stakeholders. Therefore, the participation of SMEs in the definition of sustainability reporting content must have a proportionate approach in a way that may be possible to balance the need for a relevant sustainability report for their stakeholders, on one hand, and the specific SMEs resources, governance, and organisation, on the other hand (EFRAG, 2021b). Consequently, a mandatory simplified standard for SMEs on NFI could be an effective means to achieve these objectives.

In this sense, the increase in disclosures that may come from a mandatory sustainability report is now likely greater for small firms (Christensen et al., 2021). Regarding mandatory sustainability reports, the authors highlight that disclosure has both direct costs (as the preparation, certification, and communication) and indirect costs, namely in the form of proprietary costs, considering that several stakeholders have access to the information (which often explains the firms' reluctance to provide information voluntarily). Concerning this subject, it has been mentioned that higher proprietary costs may occur for smaller companies (given that it often involves information that firms regard as 'the soul of business') (Christensen et al., 2021).

A specific standard for SMEs will be also useful to capital providers considering that, from the credit risk point of view, SMEs may not be compared to the large ones. Previous studies show that two of the causes of the non-comparability of NFI are the lack of standardisation and the plurality of reporting guidelines (Wensen et al., 2011) and the use of different NFI standards (Barker & Eccles, 2018). Also, the voluntary nature of NFI seems to contribute to the lack of cross-firm comparison (Boiral & Henri, 2017). Since one of the most effective ways to achieve comparable information is with standards, it may be expected that capital providers may have a more favourable perception towards the usefulness of a simplified standard and its mandatory adoption.

Based on the literature review, it may be highlighted that issues on a simplified standard and/or reporting format for SMEs have been discussed through different approaches and perspectives, sometimes simultaneously. Those issues include different but somewhat related topics, such as the constraints and reasons (cost/benefit analysis) that entities may have to disclose NFI, the usefulness of this information from the perspective of preparers and stakeholders, the need for a simplified standard or reporting format for SMEs, as well as if its adoption should be voluntary or mandatory.

The next section presents the empirical study proposed for this chapter, including the objectives, methodology, as well as the result analysis, and the main findings.

EMPIRICAL STUDY

Objectives and Methodology

The chapter aims to assess the answers to the project of the revised Directive 2014/95/EU in what concerns the stakeholders' perspective about: i. the usefulness of a simplified NFI standard and/or reporting format for SMEs; ii. if a simplified standard for SMEs is an effective means of limiting the burden on SMEs; iii. and, if a simplified standard for SMEs should be mandatory or voluntary. Data were obtained from the answers provided within the process of public consultation conducted by the EU, from February to June 2020, related to possible revisions of the Directive 2014/95/EU.

Stakeholders from several countries, representatives of entities of different sizes, sectors, and within or out the scope of the provisions of NFRD provided their perspective on the questions launched by the EU on this theme. At the end of the process, 588 answers were gathered. From those, about 80% is comprised of respondents from the EU, which is certainly explained by the scope of the NFRD.

To assess if there are divergences of perspectives regarding the characteristics of reporting NFI by SMEs, the answers gathered from the survey proposed by the EU were analysed by groups and matters under assessment.

Regarding the groups, it was first selected the SMEs or entities representing SMEs. Following, it is provided the perspective of the SMEs that are within the scope of NFRD. Finally, it comes the group of the capital providers, given the perceived pressure from financial institutions and other capital providers on SMEs to report additional non-financial data (Altman & Sabato, 2007; Altman et al., 2010; Fernandez-Feijoo et al., 2014; Andreeva et al., 2016; Stekelorum et al., 2018; Shields & Shelleman, 2020). This classification is based on the data from the question "field of activity or sector", being gathered the answers from investment and investment management (e.g., hedge funds, private equity funds, venture capital funds, money market funds, securities), pension provision, banking, and insurance for the same topics above mentioned.

The analysis is made by comparing the complementary information of the opposite group, which means that each group was then divided into two subgroups: the first with the main characteristic selected and the second one with the respondents who do not accomplish it. As respondents were able to insert the option "I don't know/I don't have an opinion/not relevant" or even do not insert any option in certain cases, missing answers were also possible. Then, depending on the group under assessment, the total answers used for a given analysis may be lower than the total sample (588 answers).

Table 1 presents, by group, the number, and percentage of the entities included in each subgroup that will be used to perform the analysis proposed in this section.

Table 1. Demographic information

Classification	Yes		No		Total
	No.	In %	No.	In %	
SMEs or entities representing SMEs	131	28	341	72	472
SMEs or entities representing SMEs within the scope of NFRD	73	37	126	63	199
Capital providers	108	18	480	82	588

By size, respondents that are or represent entities that are SMEs comprise 28% of the sample in this group. Following, respondents were asked to answer, in the set of demographic variables, if they are in the scope of NFRD. Considering only the respondents that are or represent entities that are SMEs within the scope of NFRD, 37% of the responses indicate the inclusion in that case. Finally, answers from respondents that can play a role as capital providers represent 18% of that sample.

The matters under assessment are comprised of the three questions from the survey that are specifically related to SMEs, namely the questions 13 to 15 below indicated:

- **Question 13 (Q13):** In your opinion, would it be useful for there to be a simplified standard and/or reporting format for SMEs?
- **Question 14 (Q14)**: To what extent do you think that a simplified standard for SMEs would be an effective means of limiting the burden on SMEs arising from information demands they may receive from other companies, including financial institutions?

- **Question 15 (Q15)**: If the EU were to develop a simplified standard for SMEs, do you think that the use of such a simplified standard by SMEs should be mandatory or voluntary?

Q13 and Q15 have the characteristic of a dichotomic variable, as the answers provided assumed the main options "yes" or "no". The Q14, on the other hand, has a characteristic of a four-item Likert scale, beginning with "not at all" and finishing with "to a very great extent", passing through "to some extent but not much" and "to a reasonable extent". Then, and to make these answers easily comparable with others, the answers "not at all" and "to some extent but not much", on one hand, and "to a reasonable extent" and "to a very great extent", on the other hand, were gathered and transformed into a dichotomic variable as well. Thus, it was possible to perform the analysis using contingency tables and the non-parametric chi-square test, which requires two non-related answers as samples and dichotomic variables.

Finally, it is important to stress, again, that respondents also had the options "I don't know / I don't have an opinion / not relevant" and blanks/missing values for the Q13 to Q15, which were, despite that, residual cases if globally considered. These options, however, were exclusively considered for computing the total answers gathered in relative frequencies. Then, for the chi-square test performed, those answers were not incorporated.

Results Analysis

Globally, 73% of respondents agree that it would be useful to have a simplified standard and/or reporting format for SMEs (Q13). Additionally, 76% of respondents also understand to a reasonable or very great extent that a simplified standard for SMEs would be an effective means of limiting the burden on SMEs arising from information demands they may receive from other companies, including financial institutions (Q14). Then, to a reasonable or very great extent, the majority of the answers provided agreed with those two questions. Conversely, less than half of respondents (46%) agree with the mandatory use of a simplified standard for SMEs (Q15).

As mentioned, the matters under assessment were analysed by grouping the stakeholders into three categories: SMEs or entities representing SMEs, SMEs or entities representing SMEs within the scope of NFRD, and capital providers. The objective is to assess the different perspectives between these subgroups, also using the chi-square test to verify if these differences are statistically significant.

SMEs or Entities Representing SMEs

Q13: The usefulness of having a simplified standard and/or reporting format for SMEs is an opinion relatively shared between the group of respondents who are or represent companies that are SMEs (74%) and the opposite one (71%). Consequently, no statistically significant differences were found through the chi-square test.

Q14: It is also relatively consensual between these two subgroups that a simplified standard for SMEs would be an effective means of limiting the burden on SMEs arising from information demands they may receive from other companies, including financial institutions. More specifically, 63% of respondents who are or represent companies that are SMEs agreed with that, which can be compared with 60% for the opposite one. Again, the chi-square test did not reveal any statistically significant difference.

Q15: Conversely, assessing the answers through a breakdown analysis by SMEs *versus* non-SMEs, a significant divergence appears, it can be observed that less than one-third (27%) of the group of respondents who are or represent SMEs agree with the mandatory use of a simplified standard for SMEs. On the other hand, almost half of respondents (49%) have this opinion in the opposite group. This different perspective can be identified as relevant based on the results from the chi-square test at a significance level of 1%.

Figure 1 summarises the percentage of positive answers for Q13 to Q15 by respondents who are SMEs or are representing SMEs *versus* the opposite group.

Figure 1. Positive answers (in percentage) by SMEs or representing SMEs versus non-SMEs or not representing SMEs

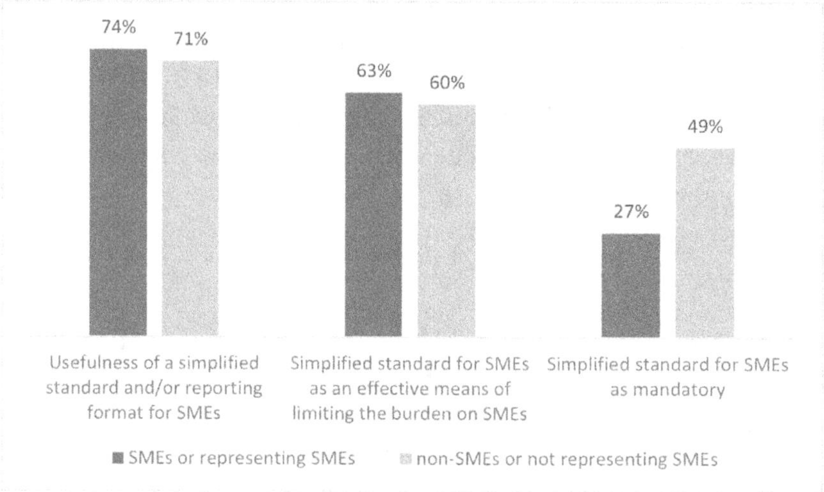

The next point provides the perspective by SMEs or entities representing SMEs *versus* non-SMEs or not representing SMEs in the scope of NFRD.

SMEs or Entities Representing SMEs within the Scope of NFRD

Q13: Considering the answers from respondents who are (or represent) SMEs and, simultaneously, are within the scope of the provisions of NFRD, it is recorded that 74% of this group agree with the usefulness of a simplified standard and/or reporting format for SMEs, which can be compared with 68% of respondents from the opposite group (non-SMEs within the scope of NFRD) that share this understanding. Based on the data gathered for these two groups, a statistically significant difference can be found through the chi-square test, but only at a 10%-level.

Q14: Following, 63% of respondents who are (or represent) SMEs and, simultaneously, are within the scope of the provisions of NFRD agree, to a reasonable or very great extent, that a simplified standard for SMEs would be an effective means of limiting the burden on SMEs. Again, this percentage is lower than the one obtained from the answers provided by the group of respondents who are not (or do not represent) SMEs within the scope of the provisions of NFRD (55%). The chi-square test did not point out statistically significant differences.

Q15: Finally, only 23% of respondents who are (or represent) SMEs and, simultaneously, are within the scope of the provisions of NFRD agree with the implementation of a standard for SMEs as mandatory. This may be compared with 37% of opinions in this sense for the non-SMEs within the scope of the provisions of NFRD. Given this relevant difference of opinions between the two groups, the chi-square test identified it as statistically significant, based on a significance level of 1%.

Figure 2 summarises the percentage of positive answers for Q13 to Q15 by entities who are within the scope of the provisions of NFRD with a breakdown by SMEs and non-SMEs.

Figure 2. Positive answers (in percentage) by entities within the scope of the provisions of NFRD

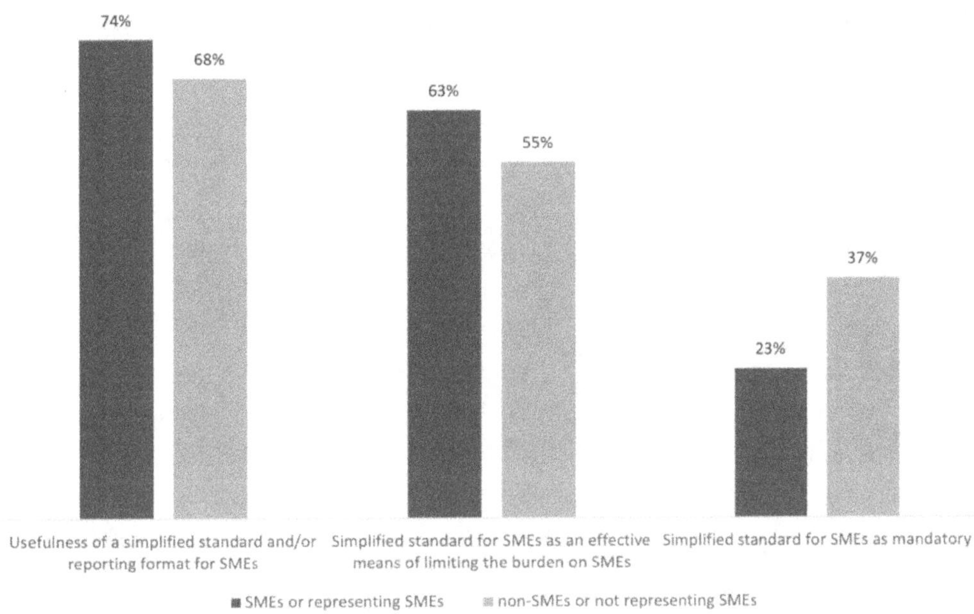

The next point provides the perspective by type of users, selected through the answers from the capital providers.

Capital Providers

Another perspective of analysis may be explored, namely the one from the capital providers.

Q13: Now, and based on the answers from respondents classified as capital providers, 86% agree that it would be useful for there to be a simplified standard and/or reporting format for SMEs, which may be compared with 70% of the agreement level computed for the group comprised of entities other than capital providers. The chi-square test confirms that there is a statistically significant difference between the two subgroups under assessment at a 5%-level.

Q14: Following, 76% of capital providers agree, to a reasonable or very great extent, that a simplified standard for SMEs would be an effective means of limiting the burden on SMEs. On the other hand, 59%

of entities other than capital providers have a different understanding. In this context, the chi-square test confirms, at a 1%-level, a statistically significant difference between the two subgroups under assessment.

Q15: Finally, only 56% of these respondents present the view that a simplified standard for SMEs should be mandatory. Conversely, less than half (44%) of entities other than capital providers agreed with this topic. However, the difference between these subgroups for this theme is the lowest (only 12 percentage points). Then, the chi-square test confirms that there is a statistically significant difference between the two subgroups under assessment, but only at a 10%-level.

Figure 3 summarises the percentage of positive answers for Q13 to Q15 by respondents who are capital providers *versus* the opposite group.

Figure 3. Positive answers (in percentage) by capital providers versus entities other than capital providers

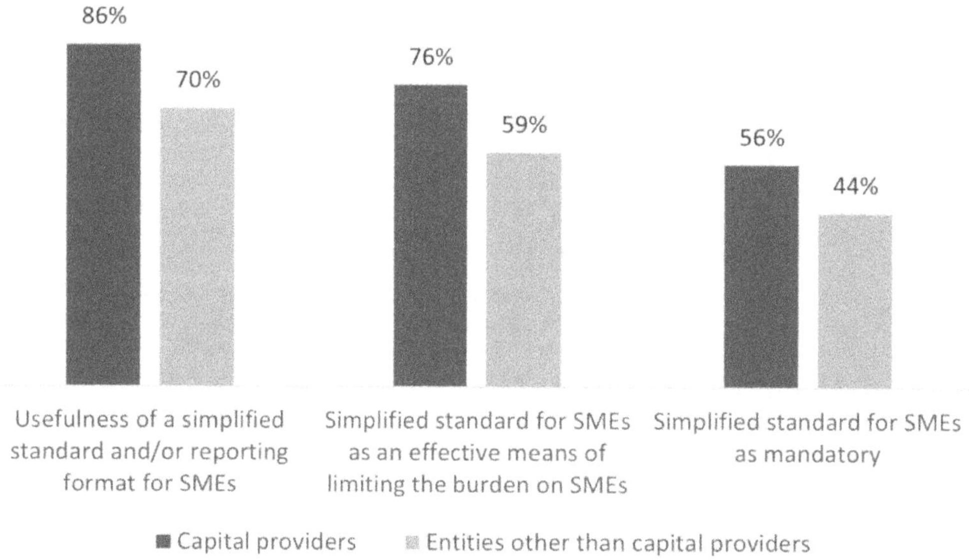

Next, it is provided an overview regarding the main findings from the different comparisons previously proposed.

Main Findings

Figure 4 summarises the main results obtained among the different subgroups analysed for the positive answers provided to Q13 to Q15.

The most relevant difference between the groups (higher than 20 percentage points) is related to the positive answers provided to Q15, on the mandatory use of a possible simplified standard for SMEs, with the SMEs or entities representing them presenting a lower level of agreement with that. Also concerning this matter, the group of capital providers has the highest level of agreement with it (with 56% of positive answers), being followed by the non-SMEs or entities not representing SMEs overall (49%). On the other hand, SMEs within the scope of NFRD and SMEs overall represent the groups with the lowest percentage of positive answers (23% and 27%, respectively).

Figure 4. Answers regarding Q13 to Q15 by different groups

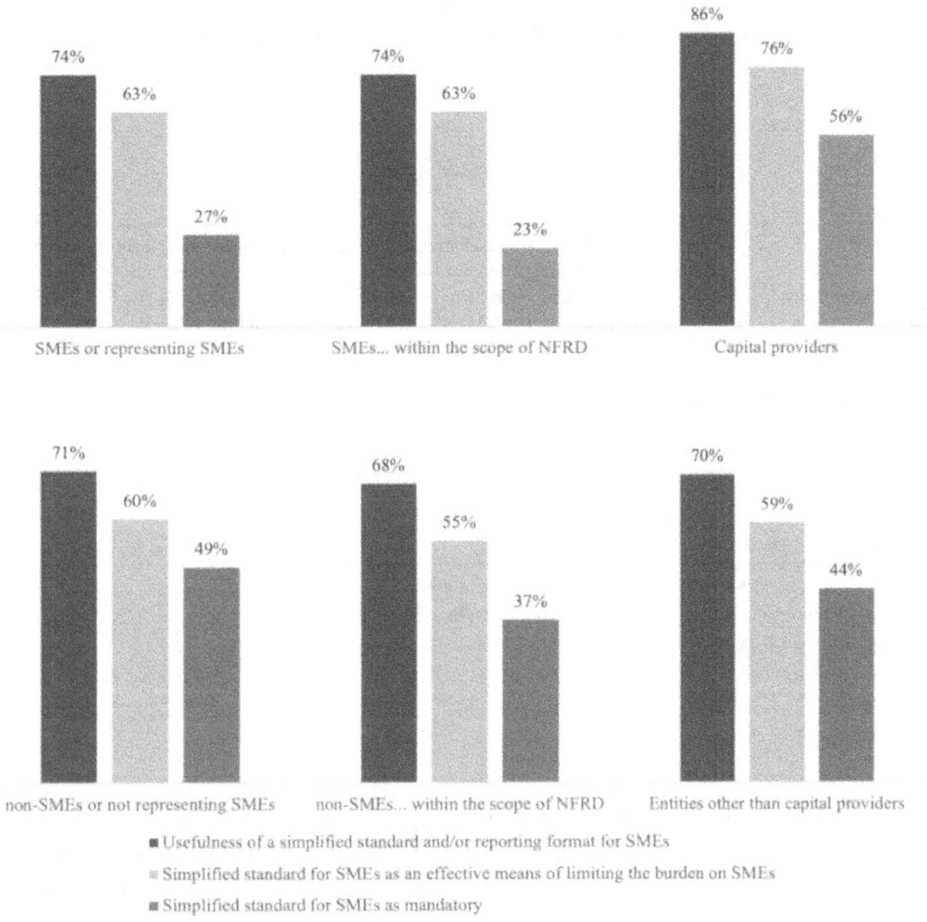

Generally, it is possible to rank, in descending order, the percentage of positive answers from Q13 to Q15, regardless of the group under assessment. Then, the usefulness of a simplified standard for SMEs (Q13) has a minimum of 68% and a maximum of 86%, which is the lowest range (18 percentage points). Following, with a minimum of 55% and a maximum of 76%, the topic on a simplified standard for SMEs as an effective means of limiting the burden on SMEs (Q14) presents the second-lowest range (21 percentage points). Finally, the mandatory use of a possible simplified standard for SMEs (Q15) presents, as already mentioned, the highest range (33 percentage points), with a minimum of 23% and a maximum of 56%.

Following, Table 2 identifies the differences, in percentage points, between the subgroups shown in Figure 5, as well as those cases in which the chi-square test revealed a statistically significant difference.

Based on the data provided in Table 2, the higher differences for the positive answers provided by groups can be found for Q15, on a simplified standard for SMEs as mandatory. The breakdown by SMEs *versus* non-SMEs presents the highest difference (22 percentage points), which is followed by SMEs within the scope of NFRD *versus* non-SMEs within the scope of NFRD (14 percentage points).

The distinction between capital providers *versus* entities other than capital providers produces a less significant difference for the range (12 percentage points).

The next section presents a summary of the main conclusions obtained during the development of this analysis.

Table 2. Differences, in percentage points, between subgroups

Subgroups	Q13: The usefulness of a simplified standard and/or reporting format for SMEs	Q14: A simplified standard for SMEs as an effective means of limiting the burden on SMEs	Q15: A simplified standard for SMEs as mandatory
SMEs or representing SMEs *versus* non-SMEs or not representing SMEs	3	3	**-22*** **
SMEs or representing SMEs within the scope of NFRD *versus* non-SMEs or not representing SMEs within the scope of NFRD	**6***	8	**-14*** **
Capital providers *versus* entities other than capital providers	**16** **	**17*** **	**12***

* Statistically significant difference based on the chi-square test at a 10% of significance level
** Statistically significant difference based on the chi-square test at a 5% of significance level
*** Statistically significant difference based on the chi-square test at a 1% of significance level

CONCLUSION

SMEs have largely been left out from NFI disclosures. However, this may change with possible revisions of the Directive 2014/95/EU (NFRD). The answers provided within the process of public consultation conducted by the EU about those revisions raised some divergences of perspectives regarding the characteristics of reporting NFI by SMEs, depending on the matters and groups under assessment.

Most respondents agree about the usefulness of a simplified standard and/or reporting format for SMEs, being the capital providers the stakeholders' group who have the highest level of agreement, having a statistically significant difference towards the opposite one (not capital providers). This is in line with several authors (Christensen et al., 2021; Johnson & Schaltegger, 2016; Borga et al., 2009) for whom the characteristics of SMEs requires a specific and simplified standard.

Furthermore, most respondents also understand to a reasonable or very great extent that a simplified standard for SMEs would be an effective means of limiting the burden on SMEs, being the capital providers the stakeholders with the highest level of agreement. Again, this is the group in which there are statistically significant differences between the perspectives. Also, these findings corroborate the literature, as the lack of resources has been identified as limiting SMEs in implementing sustainable management (Shields and Shelleman, 2020; Ortiz-Martínez & Marín-Hernández, 2021; Corazza, 2017).

Finally, less than half of the respondents agree with the mandatory use of a simplified standard for SMEs. Once more, it is the capital providers who most agree with this mandatory use. Given that higher proprietary costs may occur for smaller companies (Christensen et al., 2021) it is not surprising that SMEs are not willing for a mandatory standard, conversely to the capital providers for whom NFI is of most importance (Stekelorum et al., 2018; Shields & Shelleman, 2020; Fernandez-Feijoo et al., 2014; Calace, 2014; Andreeva et al., 2016; Altman et al., 2010).

As mentioned, a relevant difference is identified between the groups of capital providers and entities other than capital providers, namely concerning the positive answers on a simplified standard for SMEs as an effective means of limiting the burden on SMEs, and on the usefulness of a simplified standard for SMEs. These findings also confirm the literature review in what concerns the relevance of SMEs' NFI for the capital providers (for instance, Fernandez-Feijoo et al., 2014).

Globally, as in what concerned the adoption of the IFRS for SMEs (pointed out by the literature as a non-linear process), the matter on the mandatory *versus* voluntary NFI reporting based on a simplified standard for SMEs was revealed as the mostly non-consensual question between the stakeholders' groups.

Despite that, these conclusions should be considered considering data that were obtained exclusively from the answers provided to the EU public consultation, which represents the main limitation of this study that cannot be left out. Then, they may not be representative of each group of stakeholders under assessment.

At the end of this research, some perspectives can be suggested in terms of their exploration in future research projects. For instance, further analysis could explore other explanatory factors, including different social, economic, and cultural variables, that may be useful to explain the differences of opinion on the use of a simplified standard by SMEs as mandatory, either using the same data source or exploring other alternatives sources.

REFERENCES

Albu, C. N. (2013). How Relevant is Size for Setting the Scope of The IFRS for SMEs. *Journal of Accounting and Management Information Systems*, *12*(3), 424–439.

Albu, C. N., Albu, N., Pali-Pista, S. F., Gîrbină, M. M., Selimoglu, S. K., Kovács, D. M., & Strouhal, J. (2013). Implementation of IFRS for SMEs in Emerging Economies: Stakeholder Perceptions in the Czech Republic, Hungary, Romania and Turkey. *Journal of International Financial Management & Accounting*, *24*(2), 140–175. doi:10.1111/jifm.12008

Altman, E. I., & Sabato, G. (2007). Modelling Credit Risk for SMEs: Evidence from the U.S. Market. *Abacus*, *43*(3), 332–357. doi:10.1111/j.1467-6281.2007.00234.x

Altman, E. I., Sabato, G., & Wilson, N. (2010). The value of non-financial information in SME risk management. *The Journal of Credit Risk*, *6*(2), 95–127. doi:10.21314/JCR.2010.110

Andreeva, G., Calabrese, R., & Osmetti, S. A. (2016). A comparative analysis of the UK and Italian small businesses using Generalised Extreme Value models. *European Journal of Operational Research*, *249*(2), 506–516. doi:10.1016/j.ejor.2015.07.062

Barker, R., & Eccles, R. G. (2018). Should FASB and IASB Be Responsible for Setting Standards for Nonfinancial Information? SSRN *Electronic Journal*, (October). doi:10.2139/ssrn.3272250

Bartolacci, F., Caputo, A., & Soverchia, M. (2020). Sustainability and financial performance of small and medium sized enterprises: A bibliometric and systematic literature review. *Business Strategy and the Environment*, *29*(3), 1297–1309. doi:10.1002/bse.2434

Baumann-Pauly, D., Wickert, C., Spence, L. J., & Scherer, A. G. (2013). Organizing corporate social responsibility in small and large firms: Size matters. *Journal of Business Ethics*, *115*(4), 693–705. doi:10.100710551-013-1827-7

Boiral, O., & Henri, J. F. (2017). Is Sustainability Performance Comparable? A Study of GRI Reports of Mining Organizations. *Business & Society*, *56*(2), 283–317. doi:10.1177/0007650315576134

Borga, F., Citterio, A., Noci, G., & Pizzurno, E. (2009). Sustainability report in small enterprises: Case studies in Italian furniture companies. *Business Strategy and the Environment*, *18*(3), 162–176. doi:10.1002/bse.561

Burke, S., & Gaughran, W. F. (2007). Developing a framework for sustainability management in engineering SMEs. *Robotics and Computer-integrated Manufacturing*, *23*(6), 696–703. doi:10.1016/j.rcim.2007.02.001

Calace, D. (2014). Non-Financial Reporting in Italian SMEs: An Exploratory Study on Strategic and Cultural Motivations. *International Journal of Business Administration*, *5*(3), 34–48. doi:10.5430/ijba.v5n3p34

Christensen, H. B., Hail, L., & Leuz, C. (2021). Mandatory CSR and sustainability reporting: Economic analysis and literature review. *Review of Accounting Studies*, *26*(3), 1176–1248. Advance online publication. doi:10.100711142-021-09609-5

Corazza, L. (2017). The standardization of down-Streamed Small Business Social Responsibility (SBSR): SMEs and their sustainability reporting practices. *Information Resources Management Journal*, *30*(4), 39–52. doi:10.4018/IRMJ.2017100103

Corrado, M., Demartini, P., & Dumay, J. (2019). *Assurance on Integrated Reporting: A Critical Perspective*. doi:10.1007/978-3-030-01719-4_11

De Villiers, C., Rinaldi, L., & Unerman, J. (2014). Integrated Reporting: Insights, gaps and an agenda for future research. *Accounting, Auditing & Accountability Journal*, *27*(7), 1–47. doi:10.1108/AAAJ-06-2014-1736

Dinh, T., Husmann, A., & Melloni, G. (2021). The role of non-financial performance indicators and integrated reporting in achieving sustainable value creation. Publication for the committee on Economic and Monetary Affairs, Policy Department for Economic, Scientific and Quality of Life Policies, European Parliament.

Dumay, J., Bernardi, C., Guthrie, J., & Demartini, P. (2016). Integrated reporting: A structured literature review. *Accounting Forum*, *40*(3), 166–185. doi:10.1016/j.accfor.2016.06.001

El Baz, J., Laguir, I., Marais, M., & Staglianò, R. (2016). Influence of National Institutions on the Corporate Social Responsibility Practices of Small- and Medium-sized Enterprises in the Food-processing Industry: Differences Between France and Morocco. *Journal of Business Ethics*, *134*(1), 117–133. doi:10.100710551-014-2417-z

European Commission. (2002). Regulation (EC) No. 1606/2002 of the European Parliament and of the Council of 19 July 2002 on the application of International Accounting Standards. *Official Journal of the European Communities*. L243/1 (11.9.2002). Retrieved from https://eur-lex.europa.eu/legal-content/PT/TXT/?uri=celex%3A32002R1606

European Commission. (2003). Commission Recommendation of 6 May 2003 concerning the definition of micro, small and medium-sized enterprises. *Official Journal of the European Communities*. L124 (20.5.2003). Retrieved from https://eur-lex.europa.eu/legal-content/EN/TXT/?uri=CELEX:32003H0361

European Commission. (2020a). *Communication from the commission to the European parliament, the council, the European economic and social committee and the committee of the regions: An SME Strategy for a sustainable and digital Europe*. European Commission. Retrieved from https://ec.europa.eu/info/sites/default/files/communication-sme-strategy-march-2020_en.pdf

European Commission. (2020b). *Summary report on the public consultation on the Evaluation and Review of the ePrivacy Directive*. Retrieved from https://ec.europa.eu/info/law/better-regulation/have-your-say/initiatives/12129-Revision-of-Non-Financial-Reporting-Directive/public-consultation_en

European Financial Reporting Advisory Group. (2021a). *Current non-financial reporting formats and practices. Efrag, (February), Stream A6 Assessment Report*. Retrieved from https://www.efrag.org/Assets/Download?assetUrl=%2Fsites%2Fwebpublishing%2FSiteAssets%2FEFRAG%2520PTF-NFRS_A6_FINAL.pdf

European Financial Reporting Advisory Group. (2021b). *Proposals for a Relevant and Dynamic Eu Sustainability Reporting Standard-Setting*. Retrieved from https://www.efrag.org/Assets/Download?assetUrl=%2Fsites%2Fwebpublishing%2FSiteAssets%2FEFRAG%2520PTF-NFRS_MAIN_REPORT.pdf

European Investment Bank Group. (2020). *Small and medium enterprises overview*. Retrieved from https://www.eib.org/attachments/thematic/small_and_medium_enterprises_overview_2020_en.pdf

European Parliament. (2013). Directive 2013/34/EU amending Directive 2006/43/EC and repealing Council Directives 78/660/EEC and 83/349/EEC on the annual financial statements, consolidated financial statements and related reports of certain types of undertakings. *Official Journal of the European Union, 2013*(June), L 182/19-75. Retrieved from https://eur-lex.europa.eu/legal-content/EN/TXT/?uri=celex:32013L0034

European Parliament. (2014). Directive 2014/95/EU of the European Parliament and of the Council of 22 October 2014 amending Directive 2013/34/EU as regards disclosure of non-financial and diversity information by certain large undertaking and groups. *Official Journal of the European Union, 2014*(April), 1–9. Retrieved from https://eur-lex.europa.eu/legal-content/EN/TXT/?uri=CELEX:32014L0095

European Parliament. (2021). Proposal for a Directive of the European Parliament and of the Council amending Directive 2013/34/EU, Directive 2004/109/EC, Directive 2006/43/EC and Regulation (EU) No 537/2014, as regards corporate sustainability reporting. *Official Journal of the European Union*, (April), 1–65. Retrieved from https://eur-lex.europa.eu/legal-content/EN/TXT/?uri=CELEX:52016PC0198

Fernandez-Feijoo, B., Romero, S., & Ruiz, S. (2014). Effect of Stakeholders' Pressure on Transparency of Sustainability Reports within the GRI Framework. *Journal of Business Ethics*, *122*(1), 53–63. doi:10.100710551-013-1748-5

Fernandez-Feijoo, B., Romero, S., & Ruiz, S. (2015). Multilevel Approach to Sustainability Report Assurance Decisions. *Australian Accounting Review*, *25*(75), 346–358. Advance online publication. doi:10.1111/auar.12104

Hojnik, J. (2017). Environmental Corporate Reporting under eu Law: Historic Achievement or Just a Moderate Step Forward? *Journal for European Environmental & Planning Law*, *14*(1), 41–75. doi:10.1163/18760104-01401004

Johnson, M. P., & Schaltegger, S. (2016). Two Decades of Sustainability Management Tools for SMEs: How Far Have We Come? *Journal of Small Business Management*, *54*(2), 481–505. doi:10.1111/jsbm.12154

Kaya, D., & Koch, M. (2015). Countries adoption of the International Financial Reporting Standard for Small and Medium-sized Entities (IFRS for SMEs) - Early empirical evidence. *Accounting and Business Research*, *45*(1), 93–120. doi:10.1080/00014788.2014.969188

Kiliç, M., Uyar, A., & Ataman, B. (2014). Preparedness for and perception of IFRS for SMEs: Evidence from Turkey. *Accounting and Management Information Systems*, *13*(3), 492–519.

Krawczyk, P. (2021). Non-Financial Reporting—Standardization Options for SME Sector. *Journal Risk Financial Management*, *14*(9), 417. doi:10.3390/jrfm14090417

Litjens, R., Bissessur, S., Langendijk, H., & Vergoossen, R. (2012). How Do Preparers Perceive Costs and Benefits of IFRS for SMEs? Empirical Evidence from the Netherlands. *Accounting in Europe*, *9*(2), 227–250. doi:10.1080/17449480.2012.720875

Massa, L., Farneti, F., & Scappini, B. (2015). Developing a sustainability report in a small to medium enterprise: Process and consequences. *Meditari Accountancy Research*, *23*(1), 62–91. Advance online publication. doi:10.1108/MEDAR-02-2014-0030

Medel, F., García, L., Enriquez, S., & Anido, M. (2011). Reporting models for corporate sustainability in SMEs. *Environmental Science and Engineering*, 407–418. doi:10.1007/978-3-642-19536-5_32

Ortiz-Martínez, E., & Marín-Hernández, S. (2021). European SMEs and non-financial information on sustainability. *International Journal of Sustainable Development and World Ecology*, *00*(00), 1–13. doi:10.1080/13504509.2021.1929548

Rezaee, Z., & Tuo, L. (2017). Voluntary disclosure of non-financial information and its association with sustainability performance. *Advances in Accounting*, *39*, 47–59. doi:10.1016/j.adiac.2017.08.001

Rodríguez-Gutiérrez, P., Guerrero-Baena, M. D., Luque-Vílchez, M., & Castilla-Polo, F. (2021). An approach to using the best-worst method for supporting sustainability reporting decision-making in SMEs. *Journal of Environmental Planning and Management*, *0*(0), 1–24. doi:10.1080/09640568.2021.1876003

Sellami, Y. M., & Gafsi, Y. (2018). What drives developing and transitional countries to adopt the IFRS for SMEs? An institutional perspective. *Journal of Corporate Accounting & Finance*, *29*(2), 34–56. doi:10.1002/jcaf.22331

Shields, J. F., & Shelleman, J. M. (2020). SME sustainability dashboards: An aid to manage and report performance. *Journal of Small Business Strategy*, *30*(2), 106–114.

Steinhofel, E., Galeitzke, M., Kohl, H., & Orth, R. (2019). Sustainability reporting in German manufacturing SMEs. *Procedia Manufacturing*, *33*, 610–617. doi:10.1016/j.promfg.2019.04.076

Stekelorum, R., Laguir, I., & Elbaz, J. (2018). CSR disclosure and sustainable supplier management: A small to medium-sized enterprises perspective. *Applied Economics*, *50*(46), 5017–5030. doi:10.1080/00036846.2018.1466990

Stolowy, H., & Paugam, L. (2018). The expansion of non-financial reporting : an exploratory study The expansion of non- fi nancial reporting : an exploratory study. *Accounting and Business Research ISSN*, *48*(5), 525–548. doi:10.1080/00014788.2018.1470141

Torugsa, N. A., O'Donohue, W., & Hecker, R. (2012). Capabilities, Proactive CSR and Financial Performance in SMEs: Empirical Evidence from an Australian Manufacturing Industry Sector. *Journal of Business Ethics*, *109*(4), 483–500. doi:10.100710551-011-1141-1

Uyar, A., & Güngörmüş, A. H. (2013). Perceptions and knowledge of accounting professionals on IFRS for SMEs: Evidence from Turkey. *Research in Accounting Regulation*, *25*(1), 77–87. doi:10.1016/j.racreg.2012.11.001

Weber, O. (2014). The financial sector's impact on sustainable development. *Journal of Sustainable Finance & Investment*, *4*(1), 1–8. doi:10.1080/20430795.2014.887345

Wensen, K., Van Broer, W., Klein, J., & Knopf, J. (2011). The state of play in sustainability reporting in the EU. *European Union.*

Zahid, R. M. A., & Simga-Mugan, C. (2018). An Analysis of IFRS and SME-IFRS Adoption Determinants: A Worldwide Study. *Emerging Markets Finance & Trade*, *55*(2), 391–408. doi:10.1080/1540496X.2018.1500890

Chapter 6
Multidimensional Concept of Integrated Thinking and Proposal of Indicators for Its Identification

Isabelle Caroline Bevilaqua
University of Brasília, Brazil

Marguit Neumann
https://orcid.org/0000-0003-1246-3769
State University of Maringá, Brazil

Valter da Silva Faia
State University of Maringá, Brazil

ABSTRACT

The objective of the chapter is to understand the multidimensional concept of integrated thinking and propose a set of indicators that enable organizations to identify integrated thinking. The chapter is characterized as qualitative, descriptive, and exploratory. Seventeen semi-structured interviews were conducted with researchers who have developed dissertations and/or theses on integrated reporting. The results were analyzed from content analysis and allowed comparing the dimensions proposed for integrated thinking. The results show that the concept of integrated thinking is multidimensional and has four dimensions: 1) connectivity and systemic view, 2) integrated planning, 3) communication and cooperation, 4) understanding stakeholders. Based on that, it was possible to develop a set of 27 indicators. The findings contribute to the academic literature on integrated reporting, filling research gaps, and the development of integrated thinking indicators contributes to organizational practice, as they enable the identification of integrated thinking.

DOI: 10.4018/978-1-7998-9410-0.ch006

INTRODUCTION

Integrated Reporting (IR) was proposed by the International Integrated Reporting Council (IIRC) with the purpose of providing a simple, reliable and qualified narrative, enhancing accountability and improving dialogue with stakeholders and internal communication (Cheng, Green, Conradie, Konishi & Romi, 2014; Lodhia, 2014; Lai, Melloni, & Stacchezzini, 2018; Adhariani & Villiers, 2018). For IR to be prepared with quality, Integrated Thinking must be rooted in organizational activities (Lodhia, 2014). According to the IIRC (2013; 2021), Integrated Thinking is the effective consideration of the relationships between organizational units and the capitals that the organization uses or affects.

Integrated Thinking can be defined in different ways: an organizational mindset that encompasses the habit of dealing with uncertainties and disturbances (Al-Htaybat & von Alberti-Alhtaybat, 2018); a senior-management attribute of constructively handling tensions between various capitals in strategy, resource allocation, performance measurement and control (Oliver, Vesty & Brooks, 2016); the systematic management of capitals with the aim of providing sustainable profitability (Knauer & Serafeim, 2014); a process that incorporates strategies from different perspectives (financial, environmental, sustainability, governance) and extends the habits of managers and departments, ensuring awareness of the strategy in practice (Guthrie, Manes-Rossi & Orelli, 2017); an internal process that organizations follow to improve the reporting disclosure level (Adhikariparajuli, Hassan, Fletcher & Elamer, 2020).

The presence of Integrated Thinking in organizations reflects their ability to show and communicate the integration of the economic, socio-environmental and governance dimensions in daily decision-making processes (Venter, Stiglingh & Smit 2017; Esch, Schulze & Wald, 2019). In addition to helping with the provision of information about value creation processes and internal and external business relationships (Mio, Marco & Pauluzzo, 2016). Bevilaqua, Neumann and Faia (2020) present a review of the main benefits of Integrated Thinking in organizations, among which the following stand out: a) integration of socio-environmental information; b) reduction of silos; c) assistance in decision making; d) fluidity of information; e) improvements related to strategic aspects.

It is increasingly frequent that organizations seek to adopt integrated thinking in their practices. The motivation to incorporate Integrated Thinking is top-down, through strategies carried out by senior management that consider social, environmental and governance aspects for reporting and decision-making. The execution of Integrated Thinking in practice occurs from the bottom up (bottom-up) as the commitment and engagement between departments and teams makes it possible to spread it across all organizational layers (Bevilaqua et al., 2020).

For it to be effectively adopted and disseminated, Integrated Thinking must be understood at different levels and can be measured reliably, using metrics and instruments capable of measuring its construct. However, even in organizations that adopt IR, it is observed that Integrated Thinking can be confusing for stakeholders (Gunarathne & Senaratne, 2017), as its concept is obscure and challenging, as it is not discussed in depth (Feng et al, 2017; Al-Htaybat & von Alberti-Alhtaybat, 2018).

Therefore, it becomes difficult for organizations to understand, classify, identify and measure Integrated Thinking in practice (Del Baldo, 2017; Churet & Eccles, 2014). Consequently, the results of this difficulty are the faulty or incomplete incorporation of thinking that does not reflect the organizational reality and does not reach the capacity proposed by Integrated Thinking to connect capitals (IIRC, 2013; 2021). Therefore, the issue that drives the discussion of the chapter is "Which dimensions characterize the concept of Integrated Thinking?".

The objective consisted of understanding the multidimensional concept of Integrated Thinking and proposing a set of indicators that enable organizations to identify Integrated Thinking. To this end, seventeen semi-structured interviews were conducted with researchers who have developed dissertations and/or theses on Integrated Reporting. The results were analyzed from content analysis and allowed comparing the dimensions proposed for Integrated Thinking. By understanding the multidimensional concept of Integrated Thinking, it was possible to develop the set of indicators.

In this way, the chapter contributes to the academic literature on IR by filling gaps that suggest the involvement of students in the development of forms of communication that consider Integrated Thinking (Adams, 2015), in addition to advancing research from a focus on Integrated Thinking (Perego, Kennedy & Whiteman, 2016). Finally, the chapter contributes to organizational practice by proposing a set of indicators to identify and map Integrated Thinking, which can be applied by organizations (Bevilaqua, Neumann & Faia, 2020).

BACKGROUND

In 2013, the International Integrated Reporting Council (IIRC) introduced Framework 1.0 for IR, reviewed in 2021 the document highlights seven basic principles and eight content elements aimed at ensuring the preparation and presentation of IR and the content to be disclosed. The principles are: i) consistency and comparability; ii) trust and completeness; iii) strategic focus and future orientation; iv) information connectivity; v) relationship with stakeholders; vi) materiality; and vii) conciseness. The content elements consist of: a) organizational vision and external environment; b) governance; c) future perspectives; d) business model; e) strategy and resource allocation; f) performance; g) risks and opportunities; h) bases for presentation.

The IIRC (2013; 2021) also suggests the presence of six capitals, that is, the resources and inputs used and/or the affected relationships, which can increase, decrease or change in the course of activities (Villiers & Sharma, 2018; IIRC, 2013; 2021; Thomson, 2015). The capitals are: i) financial, ii) manufactured, iii) intellectual, iv) human, v) social and relationship and vi) natural. The use of the capitals in the organizational business model represents a joint effort to determine and present relevant financial and non-financial information to stakeholders (Rupley, Brown & Marshall, 2017).

However, even though six classifications and concepts of the capitals are presented in Framework 1.0 for IR, organizations can use different terminology and define them in different ways, since not all six capitals may be equally relevant or applicable to all organizations (IIRC, 2013a; SAICA, 2015). In addition, IR writers can categorize and define the capitals for their own circumstances, as well as the appropriate narrative and metrics to be reported (IIRC, 2013a).

In this sense, IR provides evidence on the base of capitals and on changes in the organization's strategy for the long term (Stubbs & Higgins, 2014). Supported by a cohesive and efficient approach (IIRC, 2013; 2021), IR represents a tool capable of improving the transparency of corporate reporting, presenting the available information with quality, providing a more efficient and productive capital allocation (Kılıç & Kuzey, 2018; Thomson, 2015) and evidencing value creation.

From the principle of connectivity, available information allows an efficient and productive capital allocation, promoting Integrated Thinking as a way to break internal silos (IIRC, 2013; 2021; Chartered Global Management Accountant – CGMA, 2014; WICI, 2016). WICI (2013) presents connectivity as the essence of Integrated Thinking, since it enables the organization to understand the relationship between its different units, helping break down internal barriers, monitoring and managing information.

Integrated Thinking is defined as the administration's attribute or ability to manage the relationship between the organization's operational units and the capitals it uses or affects (IIRC, 2013; 2021; Oliver et al., 2016). In accordance with its definition, the IIRC (2013; 2021) states that Integrated Thinking aims to break the silos between different functions and organizational units, in order to improve internal communication through fluid and efficient information flows, thus resulting in changes in organizations, in addition to directing the focus during the preparation of the report.

In this sense, Integrated Reporting (IR) is understood as a process that, when considering Integrated Thinking and disseminating it in organizational practices, leads to a sustainable change in the organization and to a periodic reporting on long-term value creation (IIRC, 2013; 2021; Adams, 2015; Thomson, 2015; Villiers & Sharma, 2018). However, the concept of Integrated Thinking is obscure and challenging for stakeholders (Feng et al., 2017). This is because it is not discussed in depth (Feng et al., 2017; Al-Htaybat & von Alberti-Alhtaybat, 2018), even in the organizations that prepare IR (Gunarathne & Senaratne, 2017).

Some ways to classify Integrated Thinking found in the literature refer to Integrated Thinking as a connection between different types of capital, planning and decision making for value creation, which demands a broad level of connection between people, functions and information for its strategic objectives to be achieved (Macias & Farfan-Lievano, 2017). In particular, Stacchezzini, Florio, Sproviero and Corbella (2019) state that the connection between intellectual capital and corporate strategy in the business model represents the components of Integrated Thinking.

Integrated Thinking is a process that incorporates strategies from different perspectives (financial, environmental, sustainability, governance) and extends the habits of managers and departments (Guthrie, et al., 2017). Higgins, Stubbs, Tweedie and McCallum (2019) evidence that the organizational strategic process influences Integrated Thinking, which is seen as the result of a strategic communication. Additionally, integrated thinking was deemed to be similar to the emerging mentality in the organization, responsible for encompassing the habit of dealing with uncertainties and interruptions in the business model adopted by the organization (Al-Htaybat & von Alberti-Alhtaybat, 2018).

Moreover, Integrated Thinking represents a way used to tell the value creation story of the organization through the use of its capitals and can be guided by the structure of corporate reporting, such as IR (Adams, 2017). Maniora (2017) states that Integrated Thinking is defined as the integration of values, beliefs, norms, attitudes, cultural standards that emphasize the importance of Environmental, Social and Corporate Governance (ESG) issues. Furthermore, it is observed that Integrated Thinking is classified as multidimensional. Dumay and Dai (2017) measured Integrated Thinking in nine sub-dimensions: departmental communication; incorporation of sustainability; elimination of silos; need for silos (contrast); risk integration; existence of silos; materiality; where the pieces fit; and interdepartmental cooperation.

Tirado-Valencia, Cordobés-Madueño, Ruiz-Lozano and Vicente-Lama (2019) followed the classification proposed by the International Federation of Accountants (IFAC) (2015) and classified Integrated Thinking from five dimensions: information connectivity; external focus on value creation; integrated planning and connection to strategy; effective governance and oversight; and integrated communication. Bevilaqua, Neumann and Faia (2020) emphasize that Integrated Thinking is conceptualized from five dimensions: connection; strategic communication; organizational mindset; strategic process; internalization. Created from that, Table 1 is responsible for grouping the components of Integrated Thinking.

Table 1. Components of integrated thinking

Components	Description
Connectivity	Guarantees the connectivity of all aspects involving people, processes and systems and enables the development of integrated planning.
Incorporation of sustainability	Connects sustainability initiatives, socioenvironmental information.
Integrated Planning Risk Integration	Allows the board, senior management and management to be aware of the risks and opportunities that require proactive management for decision making.
Strategic Process	Process that incorporates strategies from different perspectives and extends organizational habits.
Materiality	Identification and understanding of relevant issues, that is, materials for decision making.
Effective Governance AND Oversight	Appropriate governance and oversight structures ensure effectiveness for organizational processes and responsibilities and lead to better performance.
Strategic Communication and Cooperation	Result of a strategic communication, used to tell the value creation story. Integrated communication uses integrated thinking for effective reports to be prepared with connected, useful information. Internally, it provides fluidity, dialogue and cooperation between members and teams.
Silos	Presence, absence, reduction, removal and creation of silos.
Internalization	Process that internalizes integrated reporting, through integration of norms, beliefs, values and culture.

Source: Adapted from Dumay and Dai (2017); IFAC (2015); Tirado-Valencia et al. (2019); Bevilaqua et al. (2020).

The difficulty in understanding the concept of Integrated Thinking (Del Baldo, 2017) limits its classification and measurement (Churet & Eccles, 2014). Integrated Thinking is mainly measured by means of proxies, such as item: Corporate Governance: Vision and Strategy (CGVS) adopted by Venter et al. (2016). It is opportune to understand the multidimensional concept of Integrated Thinking in order to propose indicators that enable organizations to identify Integrated Thinking.

MAIN FOCUS OF THE CHAPTER

The chapter is characterized as qualitative, descriptive and exploratory (Gibbs, 2009; Cooper & Schindler, 2003). The exploratory aspect stems from the understanding of the concept of Integrated Thinking by means of the dimensions that characterize it, initially presented in Table 1. The descriptive aspect was delimited by the description of the phenomenon (Integrated Thinking) obtained through its characteristics, that is, the dimensions and indicators developed to measure Integrated Thinking.

The data collection procedures adopted in the chapter were semi-structured interviews (Cooper & Schindler, 2016) with specialists on the topic under study: Integrated Reporting and Integrated Thinking. The interview was conducted in accordance with the steps described by Creswell (2014), consisting of 5 phases: Phase 1: Choosing participants; Phase 2: Gaining access; Phase 3: Collecting data; Phase 4: Recording information; Phase 5: Storing data.

The interviewees were selected (phase 1) in March 2020 from a search through the Theses and Dissertations Catalog of the Coordination for the Improvement of Higher Education Personnel (Capes) and totaled 41 files. Capes's catalog was delimited, as it covers all postgraduate works defended in Brazil, facilitating access to researchers who work with the IR theme. The following criteria were used: (i) Search term: "Integrated Reporting", with the initials of the word in capital letters and between quotation marks, as it was found to cover a larger number of files; (ii) Year: without time delimitation, aiming at a broader scope of research; (iii) Major Area of Knowledge: Applied and Multidisciplinary Social Sciences, as it is the field in which IR studies delimited by the IIRC are found.

From the selection, contact was sought (phase 2) with the 41 researchers, which started with the search for the electronic addresses of these researchers by means of a verification of their Lattes Curriculum available on the Lattes Platform. At that time, 28 e-mails were obtained; the other ones were contacted through invitation sent using the tool called "Contact", available on the Lattes Platform. There were 22 responses, 4 of which declining participation, totaling 18 participants from the catalog, who were interviewed between April 15 and 27, 2020 (Table 1).

Besides these, 1 additional interview (pre-test) was conducted with the participant named "SPE01", in order to validate the interview script and verify the existence of difficulties in understanding the questions. Also, during the interview, it was found that specialist "SPE19" did not develop their dissertation on the "Integrated Reporting" topic. Therefore, both (SPE01 and SPE19) were removed from the sample in order for the data to be validated, which resulted in 17 valid interviews, totaling nine hours, twenty-nine minutes and fifty-nine seconds (09:29:59) in length.

The interviews were conducted via Skype and WhatsApp (phase 3) and followed a script designed for this purpose, with the guiding question being: "How do you define Integrated Thinking?". In addition, the script included questions about the characteristics, actions and behaviors of an organization that presents Integrated Thinking; the relationship of IR and the six capitals proposed by Framework 1.0 for IR with Integrated Thinking; and the relationship between the incorporation process and the benefits of Integrated Thinking.

It is noteworthy that the interviewees were duly informed and signed the Free and Informed Consent Term; meeting an ethical requirement, their names were not revealed. In addition, their right to anonymity and confidentiality of responses was ensured, and after being asked, they authorized the interview to be recorded (phase 4). The interviews were recorded in two ways, using the audio and video recording tool available on Skype, and using the audio recording feature of the researcher's smartphone. Additionally, the recordings were uploaded to Google Drive with a view to data security (phase 5) for further processing and analysis.

Table 2 shows that the interviewees' profile is mostly composed of masters who have earned their degree in the last three years (2017-2019). The others who defended their dissertations/theses between the years 2014 and 2016 currently have their PhD in progress or are professors, thus representing an engagement with research.

The data obtained were submitted to content analysis in three stages: pre-analysis, material exploration, and treatment and inference of results (Bardin, 2016). The pre-analysis consisted of preparing the material, which began with the transcription of the interviews using Google Docs' voice typing tool, with the aid of the CABLE Input/Output tool (VB-Audio Virtual Cable). The transcripts were performed simultaneously with the interviews and totaled thirty hours and fifteen minutes (30:15:00). After this process was completed, all the content was grouped into a single Word file (Version 2016), called *corpus*.

Additionally, the transcripts were sent to the interviewees for them to validate the meaning of their speeches, as suggested by Szymansk (2011). Feedbacks were given by 14 of the 17 interviewees, and the transcripts were confirmed, with adjustments being made only to punctuation and speech intonation.

Subsequently, the material was explored, which is the second stage of content analysis (Bardin, 2016). The technique used was categorization, responsible for gathering the elements under a generic title, and the criterion adopted was semantic, thematic. The analyzed categories were those that emerged as dimensions of Integrated Thinking in the literature review (Table 2). Table 3 was prepared with the purpose of presenting the categories.

Table 2. Details of the semi-structured interviews

Interviewee	Interviewee's profile	Interviewee's education focused on IR	Interview length	Transcript time
(SPE01)	Master's in progress	Master's in progress in 2020	00:25:54	01:42:00
(SPE02)	Teacher	Master's dissertation defended in 2018	00:25:24	01:37:00
(SPE03)	Managing Partner	Master's dissertation defended in 2019	00:25:11	01:57:00
(SPE04)	Accounting Analyst	Master's dissertation defended in 2017	00:28:59	02:03:00
(SPE05)	Accountant	Master's dissertation defended in 2018	00:14:57	00:44:00
(SPE06)	Teacher	Master's dissertation defended in 2017	00:34:31	01:55:00
(SPE07)	Teacher	Master's dissertation defended in 2016	00:22:50	01:36:00
(SPE08)	Distance-Education Professor	Master's dissertation defended in 2019	00:32:55	01:53:00
(SPE09)	PhD in progress	Master's dissertation defended in 2019	00:15:18	01:15:00
(SPE10)	PhD in progress	Master's dissertation defended in 2015	00:29:48	01:33:00
(SPE11)	PhD in progress	Master's dissertation defended in 2016	00:23:53	01:17:00
(ESP12)	Full Professor	PhD thesis defended in 2019	00:20:49	00:54:00
(SPE13)	Budget Analyst	Master's dissertation defended in 2019	00:25:52	01:32:00
(SPE14)	PhD in progress/Report Writer	Master's dissertation defended in 2018	01:09:02	02:57:00
(SPE15)	Teacher	Master's dissertation defended in 2015	00:38:14	02:37:00
(SPE16)	Teacher	PhD thesis defended in 2019	00:37:49	01:42:00
(SPE17)	Teacher	Master's dissertation defended in 2017	00:36:07	01:56:00
(SPE18)	Teacher	Master's dissertation defended in 2017	00:23:30	01:05:00
(SPE19)	PhD in progress	Master's dissertation defended in 2014	00:38:56	-
Length of the interviews and transcripts (in hours, minutes and seconds)			09:29:59	30:15:00

Source: Survey data (2020)

Table 3. Analyzed categories

CATEGORIES	
I	Connectivity
II	Incorporation of sustainability
III	Integrated Planning/Risk Integration
IV	Strategic Process
V	Materiality
VI	Effective Governance AND Oversight
VII	Strategic Communication and Cooperation
VIII	Silos
IX	Internalization

Source: Prepared by the authors (2021)

Finally, inferences were made regarding the results – the third stage of content analysis (Bardin, 2016). The analysis of the interviews (4.1) resulted in the dimensions of Integrated Thinking (4.1 and 4.2); then, the inferences helped propose the indicators (4.2).

RESULTS AND DISCUSSION

Mapping the Concept of Integrated Thinking

To understand the concept of Integrated Thinking, nine categories were initially proposed (Table 2), which were compared with the interviews carried out with the specialists and grouped by similarity. This section will present the mapping of the concept and the dimensions of Integrated Thinking identified from the data obtained.

a) Connectivity (I) and Incorporation of Sustainability (II) Dimensions

The interviewees argued that Integrated Thinking is intertwined with connectivity, which includes information, capitals and people. Integrated Thinking consists of joining the organization's financial and economic information with subjective, qualitative information, strategies that used to be disclosed separately (SPE18), maintaining sustainability from the understanding of the employees who consider their actions, as well as the causes and effects of their actions from the environmental, social and economic perspective (SPE06).

Integrated Thinking is the integrated perception of the company's internal and external impacts (SPE12). Based on that, Integrated Thinking consists of a systemic view capable of seeing the whole, that is, it is responsible for expanding the organizational understanding beyond the niches, as exposed by this interviewee (SPE11).

It is the "[...] connection or interconnection of organizational and managerial processes and practices from different areas, so it was about thinking of the organization from a holistic viewpoint, from a systemic viewpoint" (SPE17). Integrated Thinking is the "organization's holistic view and mainly how the organization interacts with stakeholders, how stakeholders affect the organization and also identify new ways of creating value" (SPE02).

It's thinking about different factors, different capitals together (SPE04). And it encompasses understanding how the company works with the interdependencies of all factors (capitals, people, reporting process and management) (SPE16).

When you can understand the transformations, increases and decreases that occur within these capitals from value generation, this is also an Integrated Thinking, because you're managing to understand how the company actually generates value, how one capital interferes with the other, how you investing in one will boost the other, or whatever other transformations that may occur (SPE05).

Integrated Thinking is the consideration of all people, individuals, employees, when defining the business model, goals, objectives and short-, medium- and long-term strategies (SPE07).

[...] so what is Integrated Thinking? It is about bringing together strategic, I mean, people within the company to discuss this issue, you know, and not just these people, you know, these leaders. But it also often involves the teams themselves, right (...) so it's a collective thinking always aiming at improving decision making (SPE13).

Based on the interviews, it can be observed that Integrated Thinking is seen as a concept that encompasses the connectivity of people, information and capitals. This connectivity expands and presents a new dimension for Integrated Thinking, which is the systemic view, as presented by the interviewees; integrated thinking is linked to the ability to see the whole and fully understand the activities and relationships that take place in the organization.

b) Integrated Planning/Risk Integration (III), Strategic Process (IV), Materiality (V) and Internalization (IX) Dimensions

It is noteworthy that the specialists saw Integrated Thinking from a perspective capable of integrating and internalizing impacts, risks and strategies. Integrated Thinking is the integrated perception of the company's internal and external impacts (SPE12).

[...] for me, Integrated Thinking is about you better understanding these impacts of both risks and opportunities within your area (...) understanding that you're part of a system network and that everything is interrelated (SPE11).

Integrated Thinking is a proactive strategy that considers all angles of the company, that is, the way to follow (SPE03).

when you have an Integrated Thinking, you already have to see ahead of you how that decision can impact you as to all capitals, so for me to have this vision of all capitals I need to know what the vision of each one, of that strategy or of that future decision of mine is (SPE03).

Based on the interviews, it is observed that Integrated Thinking is seen as the organization's strategy that, together with a systemic and holistic view, enables the identification of risks and opportunities, besides providing a background for the integration of these elements.

c) Strategic Communication/Cooperation (VII) and Silos (VIII) Dimensions

Integrated Thinking is the integrated perception of the company's internal and external impacts, and this perception is obtained through conversation between all different areas (SPE12). For instance, they can be related to "strategic information, I mean, information about goals, or whatever, so that somehow everyone could have knowledge of these strategies, you know" (SPE18).

I understand Integrated Thinking... as one department talking to another department, one department understanding what the other does, and unifying all of this within the company's objective, which is the business model. (pause) That is, within the main practices that the company has (...). It is about understanding what the business model is by defining all the capitals it encompasses and developing the practice of integrated thinking in all departments based on these definitions (SPE09).

I think that much stronger than companies that don't have this type of thinking, I think that this matter of belonging and effective and affective relationships at work becomes much more evident in organizations that work with Integrated Thinking than in others, it's the idea of me knowing why that decision was made, ...where it's going, where I'm going, so consequently it also brings greater engagement (SPE17).

It is observed that Integrated Thinking was also described by the interviewees as an effort that requires the formation of teams and helps propose solutions to internal problems, break silos, and strengthen the feeling of belonging. Furthermore, Integrated Thinking provides fluid communication in day-to-day activities when it comes to strategic information and the goals of the organization that adopts it.

d) Effective Governance and Oversight (VI) Dimensions

Integrated Thinking is presented as information connectivity, a collective action between people and areas that guides actions and decisions (SPE13).

a company that has an approach that manages to look outside the company's walls and recognizes its main dependencies and its main impacts on its value creation process, which is the business model (...). Integrated Thinking would be about mapping everything I depend on, everything I have an impact on, but not just mapping, knowing alone it's no use, but it's also about, from this mapping, seeing which risks are most important, then defining perspectives and scenarios, then starting to design the strategy as to where I am headed to, where I want to go, and so on (SPE14).

holistic view of the organization and mainly how the organization interacts with stakeholders, how stakeholders affect the organization, and also identifies new ways of creating value (SPE02).

Integrated thinking is similar to a process that takes place within the company and helps understand the needs of both internal and external users, as well as their impact when they interact and transform the capitals (SPE08). Given what was exposed by the interviewees, it is observed that the understanding of Integrated Thinking shows a new dimension, which is about understanding stakeholders.

The results obtained by the interview allowed mapping the dimensions of Integrated Thinking. From nine initial categories that represented the components of Integrated Thinking (Tables 2 and 3), four final dimensions were obtained (Table 4).

It is noteworthy that the "connectivity" category presented in the literature was evidenced by the interviewees, showing that it can be considered a fundamental element in the definition of Integrated Thinking. This fact corroborates with Macias and Farfan-Lievano (2017), who classify integrated thinking as a connection between different types of capital, which relate to each other and progress over time (Quarchioni, Ruggiero & Damiano, 2020). In addition, it corroborates with the definition for Integrated Thinking proposed by Framework 1.0 for IR (IIRC, 2013) and with Oliver et al. (2016), who define integrated thinking as the active consideration of the relationships between operational units and the capitals that they use or affect.

Table 4. Construct and dimensions of integrated thinking

Categories	Interviews	Final dimension	Delimitation
Connectivity	Connectivity	Connectivity and Systemic View	Integrated thinking is responsible for the connectivity between capitals, people, functions and information, including about sustainability. From this connection, it is possible to expand the understanding of organizational characteristics as a whole.
Incorporation of Sustainability	Incorporation of Sustainability		
-	Systemic View		
Integrated Planning/ Risk Integration	Risk Integration	Integrated Planning	Integrated thinking is responsible for dealing with the uncertainties, interruptions, risks and opportunities of the organizational business model. It also ensures an integration of norms, beliefs, values, culture and material information for decision making.
Strategic Process	-		
Materiality	-		
Internalization	-		
Strategic Communication and Cooperation	Communication	Communication and Cooperation	Integrated thinking is a process that incorporates strategies from different perspectives and extends organizational habits. It can be used as a form of communication for the organization to tell its value creation story and establish dialogue and cooperation between members of internal teams.
Silos	-		
Effective Governance and Oversight	-	Understanding of *Stakeholders*	Integrated thinking is responsible for connecting the organization to stakeholders (internal and external), helping the former to understand the needs of the latter.
-	Understanding Stakeholders		

Source: Prepared by the authors (2021)

It is observed that connectivity is responsible for covering aspects related to information, including about sustainability. Therefore, the "incorporation of sustainability" category can be incorporated into the connectivity category. According to Busco, Malafronte, Pereira and Starita (2019), commitment to financial and non-financial aspects is responsible for the levels of integration of companies, that is, the adoption of Integrated Thinking and IR (article 24).

From the interviews, a new component called "systemic view" was found, which encompasses the expansion of the understanding of organizational characteristics to all stakeholders. It was observed that systemic view results from the connectivity of the elements presented by the connectivity dimension – information, capitals and people –, and it is possible to group them in a single dimension called "Connectivity and Systemic View".

The "strategic process, integrated planning/risk incorporation and materiality" categories observed in the literature review were compared with the information made available from the interviews. It was found that Integrated Thinking is responsible for dealing with uncertainties, interruptions, risks and opportunities in the organizational business model. The organizational mindset that encompasses the habit of dealing with uncertainties and interruptions in the business model adopted by the organization is presented by Al-Htaybat and von Alberti-Alhtaybat (2018).

It also ensures an integration of norms, beliefs, values, culture and material information for decision making. This fact shows Integrated Thinking as a form of management used for better organizational decisions and results (Dimes & Villiers, 2020). The categories presented are complementary; for this reason, they were grouped into a single dimension – "Integrated Planning". Additionally, the "internalization" dimension, observed both in the literature and in the interviews, refers to the integration of strategic and cultural elements; thus, it is complementary and was grouped into said dimension.

It is inferred that the "strategic communication and cooperation" and "silos" categories were grouped into a single dimension of Integrated Thinking called "Communication and Cooperation". The final dimension comprises integrated thinking as a process that considers communication as a key element for telling the value creation story and improving the internal dialogue between employees and teams, corroborating with Lai et al. (2018), who state that Integrated Thinking improves the dialogue between departments and employees as they share thoughts and opinions.

Additionally, it was observed that the relationship between Integrated Thinking and the "governance and oversight" category was not evidenced. However, from the interviews, a new dimension was found, called "Understanding Stakeholders", which comprises the organization's relationship with its internal and external stakeholders, as well as the understanding of their needs. It is noteworthy that the dimension emerged in the interviews with the specialists, showing a previous lack of interest from the literature on IR and management of organizational stakeholders.

The results of the interviews shown in section 4.1 of the chapter enabled the creation of 27 indicators for identifying Integrated Thinking. The indicators are presented in Table 4 together, in accordance with the dimensions proposed for Integrated Thinking.

As shown in Table 5, the "connectivity and systemic view" dimension presents a set of seven indicators (01 to 07) related to the disclosure of information about the capitals (financial, manufactured, natural, human, social and relationship or intellectual), as well as the nature of the information (financial, non-financial, environmental and governance). In addition, this set encompasses aspects that enable the organization to claim that it has a systemic view of itself.

The second dimension, "integrated planning", is responsible for housing eight indicators (08 to 15) that inform whether the organization is able to identify and manage its uncertainties, risks and opportunities. Furthermore, the indicators make it possible to identify whether the organization presents its mission, vision and values aligned with each other and informs the set of actions to achieve long-term goals. The "communication and cooperation" dimension presents six indicators (16 to 21), which cover value generation, the participation of employees in decision making, and the communication and relationship between them.

Finally, the "understanding stakeholders" dimension comprises six indicators (22 to 27) related to the identification of the organization's relationship with its internal and external stakeholders. It is noteworthy that the set of indicators was designed with the purpose of providing the organization with the ability to identify evidence of Integrated Thinking; to this end, each indicator represents an action that can be taken by the organization that practices integrated thinking.

FUTURE RESEARCH DIRECTIONS

Integrated thinking, in the context of Integrated Reporting, is the target of criticism because its concept is broad and not specifically delimited. With the data obtained for the elaboration of the chapter, it is observed that there are theoretical advances that help policy makers and report writers to understand aspects of integrated thinking. However, there is still a way to go in this theme.

Table 5. Indicators for identifying Integrated Thinking

DIMENSIONS OF INTEGRATED THINKING	
CONNECTIVITY AND SYSTEMIC VIEW	
01	The organization evidences financial information along with non-financial information (social and/or environmental and/or governance)
02	The organization evidences at least one of the following capitals (financial, manufactured, natural, human, social and relationship or intellectual)
03	The organization evidences the business model
04	The organization informs the way in which capitals (financial, manufactured, natural, human, social and relationship or intellectual) are related
05	The organization informs the involvement of teams for discussions and activities
06	The organization informs its concern with employee integration
07	The organization evidences the impacts caused by its activities
INTEGRATED PLANNING	
08	The organization presents how it handles internal and/or external uncertainties
09	The organization informs how uncertainties impact value creation
10	The organization shows how it adapts in order to respond to uncertainties
11	The organization evidences mapped financial and/or non-financial risks
12	The organization informs that it manages financial and/or non-financial risks
13	The organization informs that it is concerned with future opportunities
14	The organization presents its mission, vision and values aligned with each other
15	The organization informs the set of actions necessary to achieve long-term goals
COMMUNICATION AND COOPERATION	
16	The organization shows how it generates value
17	The organization informs that decision-making considers information from employees
18	The organization evidences how communication between employees is
19	The organization shows how the relationship between employees occurs
20	The organization projects its future perspectives and scenarios
21	The organization presents its strategies from different perspectives (financial, environmental, sustainability or governance)
UNDERSTANDING STAKEHOLDERS	
22	The organization evidences that it maps the demands of internal employees
23	The organization evidences to its external audience what it does to generate value
24	The organization evidences that it cares about its employees
25	The organization informs how it considers the demands of internal employees
26	The organization informs how it considers the demands of its external audience
27	The organization identifies its internal and external audiences

Source: Prepared by the authors (2021)

First, the concept of integrated thinking must be clearly explained by the organizations responsible for the IR. For this, future research should serve as a basis, establish partnerships, and provide quality and timely information.

In the academic field, it is necessary to develop new academic field research to identify how integrated thinking is understood within organizations, which can be done using techniques such as surveys and interviews. In particular, future studies could extend the sample of interviewees to: researchers who are in the process of preparing studies on IR and their advisors; professionals from organizations that already prepare IR; professionals from regulatory bodies. Another suggestion for further studies is the practical application of the indicators to validate them and map the presence of Integrated Thinking, thus contributing to the academy and the organizational practice.

CONCLUSION

The chapter was conducted with two purposes. The first is to understand the multidimensional concept of Integrated Thinking. From this, the second purpose was to propose a set of indicators that enable organizations to identify Integrated Thinking. The first objective was met through the semi-structured interviews held with the specialists on the subject and covered the mapping of the concept of Integrated Thinking. Corroborating with what was presented by the IFAC (2015), Bevilaqua et al. (2020) and Dumay and Dai (2017), the concept of Integrated Thinking is multidimensional. It is concluded that the concept has four dimensions: a) connectivity and systemic view; b) integrated planning; c) communication and cooperation; d) understanding stakeholders.

It is noteworthy that dimension (a) presents Integrated Thinking as responsible for the connectivity between capitals, people, functions and information, including about sustainability. From this connection, it is possible to expand the understanding of organizational characteristics as a whole. Dimension (b) defines that Integrated Thinking is responsible for dealing with uncertainties, interruptions, risks and opportunities in the organizational business model. It also ensures an integration of norms, beliefs, values, culture and material information for decision making.

Dimension (c) classifies Integrated Thinking as a process that incorporates strategies from different perspectives and extends organizational habits. It can be used as a form of communication for the organization to tell its value creation story and establish dialogue and cooperation between members of internal teams. According to dimension (d), Integrated Thinking is responsible for interconnecting the organization with stakeholders (internal and external), helping their needs to be understood.

Additionally, 27 indicators were proposed, which were grouped around the four identified dimensions. Dimension (a) presents a set of seven indicators (01 to 07), while dimension (b) groups eight indicators (08 to 15). Dimensions (c) and (d) evidence six indicators each (16 to 21) and (22 to 27), respectively. It is claimed that the indicators were developed with the purpose of providing the organization with the ability to identify evidence of Integrated Thinking.

The results of the chapter contribute to the IR literature by broadening discussions on the topic and filling the gap as to the scarcity of research that seeks to understand the concept of Integrated Thinking, which is still the target of multiple interpretations (Adams, 2015; Perego et al., 2016). In this sense, the results are relevant to the regulatory body (IIRC) as they advance concepts understood by Integrated Reporting.

The chapter contributes to managerial practice, starting with an understanding and presentation of the general concept for Integrated Thinking. Based on the managers' understanding and familiarity with the topic, Integrated Thinking can be prioritized by management, and organizational mentality can change to incorporate and disseminate integrated thinking at all hierarchical levels and in different sectors.

In addition, the indicators can be applied by organizations to verify the existence or not of Integrated Thinking internally, or identify evidence of Integrated Thinking. Subsequently, it is possible to carry out constant monitoring of integrated thinking, this fact can contribute to the creation of goals and strategic plans, team engagement, communication improvement, among others.

The qualitative delimitation of the chapter does not allow data generalization, so the interviews conducted represent the individuals' personal inferences. Another limitation is related to the selection of interviewees; initially, all researchers with dissertations and theses on Integrated Reporting in Brazil were invited to join. Finally, the indicators developed to identify Integrated Thinking have not been empirically validated, which thus limits their explanatory capacity. In addition, the intersection between the categories of integrated thinking with questionnaires anchored in the indicators proposed in this chapter are a scenario for the development of future research in companies that already adopt Integrated Reporting.

REFERENCES

Adams, C. A. (2015). The International Integrated Reporting Council: A call to action. *Critical Perspectives on Accounting*, *27*, 23–28. doi:10.1016/j.cpa.2014.07.001

Adams, C. A. (2017). Conceptualizing the contemporary corporate value creation process. *Accounting, Auditing & Accountability Journal*, *30*(4), 906–931. doi:10.1108/AAAJ-04-2016-2529

Adhariani, D., & Villiers, C. (2018). *Integrated reporting: perspectives of corporate report preparers and other stakeholders. Sustainability Accounting, Management and Policy Journal.* doi:10.1108/SAMPJ-02-2018-0043

Adhikariparajuli, M., Hassan, A., Fletcher, M., & Elamer, A. A. (2020). Integrated reporting in higher education: Insights from Scotland, Northern Ireland and Wales. *Social Responsibility Journal*. Advance online publication. doi:10.1108/SRJ-01-2019-0031

Al-Htaybat, K., & von Alberti-Alhtaybat, L. (2018). Integrated thinking leading to integrated reporting: Case study insights from a global player. *Accounting, Auditing & Accountability Journal*, *31*(5), 1435–1460. doi:10.1108/AAAJ-08-2016-2680

Bardin, L. (2016). *Análise de Conteúdo*. Edições 70.

Bevilaqua, I. C., Neumann, M., & Faia, V. S. (2020). Fatores que Favorecem e/ou Limitam o Pensamento Integrado na Prática: Uma análise do Conceito, Incorporação e Benefícios apresentados pela Literatura do Relato Integrado. *XX USP International Conference in Accounting*.

Busco, C., Malafronte, I., Pereira, J., & Starita, M. G. (2019). The determinants of companies' levels of integration: Does one size fit all? *The British Accounting Review*, *51*(3), 277–298. doi:10.1016/j.bar.2019.01.002

CGMA. (2014). *Integrated thinking: The next step in integrated reporting*. CIMA & American Institute of CPAs. Available in: https://competency.aicpa.org/media_resources/206473-integrated-thinking-the-next-step-in-integrated-re/detail

Cheng, M., Green, W., Conradie, P., Konishi, N., & Romi, A. (2014). The International Integrated Reporting Framework: Key Issues and Future Research Opportunities. *Journal of International Financial Management & Accounting*, *25*(1), 90–119. doi:10.1111/jifm.12015

Churet, C., & Eccles, R. G. (2014). Integrated Reporting, Quality of Management, and Financial Performance. *Journal of Applied Corporate Finance*, *26*(1), 56–65. doi:10.1111/jacf.12054

Cooper, D. R., & Schindler, P. S. (2003). *Métodos de Pesquisa em Administração* (7th ed.). Bookman.

Creswell, J. W. (2014). *Investigação Qualitativa e Projeto de Pesquisa: escolhendo entre cinco abordagens*. Penso Editora.

Del Baldo, M. (2017). The implementation of integrating reporting <IR> in SMEs: Insights from a pioneering experience in Italy. *Meditari Accountancy Research*, *25*(4), 505–532. doi:10.1108/MEDAR-11-2016-0094

Dimes, R., & De Villiers, C. (2020). How management control systems enable and constrain integrated thinking. *Meditari Accountancy Research*. Available in: https://papers.ssrn.com/sol3/papers.cfm?abstract_id=3750429

Dumay, J., & Dai, T. (2017). Integrated thinking as a cultural control? *Meditari Accountancy Research*, *25*(4), 574–604. doi:10.1108/MEDAR-07-2016-0067

Esch, M., Schulze, M., & Wald, A. (2019). The dynamics of financial information and non-financial environmental, social and governance information in the strategic decision-making process. *Journal of Strategy and Management*. doi:10.1108/JSMA05-2018-0043

Feng, T., Cummings, L., & Tweedie, D. (2017). Exploring integrated thinking in integrated reporting – an exploratory study in Australia. *Journal of Intellectual Capital*, *18*(2), 330–353. doi:10.1108/JIC-06-2016-0068

Gibbs, G. (2009). *Análise de dados qualitativos: coleção pesquisa qualitativa*. Bookman.

Gunarathne, N., & Senaratne, S. (2017). Diffusion of integrated reporting in an emerging. *Managerial Auditing Journal*, *32*(4–5), 524–548. doi:10.1108/MAJ-01-2016-1309

Guthrie, J., Manes-Rossi, F., & Orelli, R. L. (2017). Integrated reporting and integrated thinking in Italian public sector organizations. *Meditari Accountancy Research*, *25*(4), 553–573. doi:10.1108/MEDAR-06-2017-0155

Higgins, C., Stubbs, W., Tweedie, D., & McCallum, G. (2019). Journey or toolbox? Integrated reporting and processes of organisational change. *Accounting, Auditing & Accountability Journal*, *32*(6), 1662–1689. doi:10.1108/AAAJ-10-2018-3696

International Federation of Accountants (IFAC). (2015). *Creating Value with Integrated Thinking. The Role of Professional Accountants*. Recovered in: https://www.ifac.org/system/files/publications/files/Creating-value-with-iIntegrated-thinking-role-of-accountants.pdf

International Integrated Reporting Council (IIRC). (2013). *The International <IR> Framework*. Available in: https://integratedreporting.org/wp-content/uploads/2015/03/13-12-08-THE-INTERNATIONAL-IR-FRAMEWORK-2-1.pdf

International Integrated Reporting Council (IIRC). (2013a). *Capitals Background paper for <IR>*. Recuperado em: https://integratedreporting.org/wp-content/uploads/2013/03/IR-Background-Paper-Capitals.pdf

International Integrated Reporting Council (IIRC). (2021). *International <IR> Framework*. IIRC.

Kılıç, M., & Kuzey, C. (2018). Assessing current company reports according to the IIRC integrated reporting framework. *Meditari Accountancy Research*, *26*(2), 305–333. doi:10.1108/MEDAR-04-2017-0138

Knauer, A., & Serafeim, G. (2014). Attracting Long-Term Investors Through Integrated Thinking and Reporting: A Clinical Study of a Biopharmaceutical Company. *Journal of Applied Corporate Finance*, *26*(2), 57–64. doi:10.1111/jacf.12067

Lai, A., Melloni, G., & Stacchezzini, R. (2018). Integrated reporting and narrative accountability: The role of preparers. *Accounting, Auditing & Accountability Journal*, *31*(5), 1381–1405. doi:10.1108/AAAJ-08-2016-2674

Lodhia, S. (2014). Exploring the Transition to Integrated Reporting Through a Practice Lens: An Australian Customer Owned Bank Perspective. *Journal of Business Ethics*, *129*(3), 585–598. doi:10.100710551-014-2194-8

Macias, H. A., & Farfan-Lievano, A. (2017). Integrated reporting as a strategy for firm growth: Multiple case study in Colombia. *Meditari Accountancy Research*, *25*(4), 605–628. Advance online publication. doi:10.1108/MEDAR-11-2016-0099

Maniora, J. (2017). Is Integrated Reporting Really the Superior Mechanism for the Integration of Ethics into the Core Business Model? An Empirical Analysis. *Journal of Business Ethics*, *140*(4), 755–786. doi:10.100710551-015-2874-z

Mio, C., Marco, F., & Pauluzzo, R. (2016). Internal application of IR principles: Generali's Internal Integrated Reporting. *Journal of Cleaner Production*, *139*, 204–218. doi:10.1016/j.jclepro.2016.07.149

Oliver, J., Vesty, G., & Brooks, A. (2016). Conceptualising integrated thinking in practice. *Managerial Auditing Journal*, *31*(2), 228–248. doi:10.1108/MAJ-10-2015-1253

Perego, P., Kennedy, S., & Whiteman, G. (2016). A lot of icing but little cake? Taking integrated reporting forward. *Journal of Cleaner Production, 136*(July), 53–64. doi:10.1016/j.jclepro.2016.01.106

Quarchioni, S., Ruggiero, P., & Damiano, R. (2020). *Flows of information and meaning: a vocabulary approach to integrated thinking and reporting*. Meditari Accountancy Research. doi:10.1108/MEDAR-01-2020-0677

Rupley, K. H., Brown, D., & Marshall, S. (2017). Evolution of corporate reporting: From stand-alone corporate social responsibility reporting to integrated reporting. *Research in Accounting Regulation*, *29*(2), 172–176. doi:10.1016/j.racreg.2017.09.010

SAICA. (2015). *Integrated thinking: An exploratory survey*. Disponível em: https://www.saica.co.za/Portals/0/Technical/Sustainability/SAICAIntegratedThinkingLandscape.pdf

Stacchezzini, R., Florio, C., Sproviero, A. F., & Corbella, S. (2019). An intellectual capital ontology in an integrated reporting context. *Journal of Intellectual Capital*, *20*(1), 83–99. doi:10.1108/JIC-05-2018-0090

Stubbs, W., & Higgins, C. (2014). Integrated Reporting and internal mechanisms of change. *Accounting, Auditing & Accountability Journal*, *27*(7), 1068–1089. doi:10.1108/AAAJ-03-2013-1279

Szymansk, H. (2011). *A entrevista na educação: a prática reflexive*. B. L. Livro, Ed.

Thomson, I. (2015). "But does sustainability need capitalism or an integrated report" a commentary on "The International Integrated Reporting Council: A story of failure" by Flower, J. *Critical Perspectives on Accounting*, *27*, 18–22. doi:10.1016/j.cpa.2014.07.003

Tirado-Valencia, P., Cordobés-Madueño, M., Ruiz-Lozano, M., & De Vicente-Lama, M. (2019). Integrated thinking in the reporting of public sector enterprises: A proposal of contents. *Meditari Accountancy Research*, *28*(3), 435–453. doi:10.1108/MEDAR-03-2019-0458

Venter, E. R., Stiglingh, M., & Smit, A. R. (2017). Integrated Thinking and the Transparency of Tax Disclosures in the Corporate Reports of Firms. *Journal of International Financial Management & Accounting*, *28*(3), 394–427. doi:10.1111/jifm.12064

Villiers, C., & Sharma, U. (2018). A critical reflection on the future of financial, intellectual capital, sustainability and integrated reporting. *Critical Perspectives on Accounting*, *70*, 101999. Advance online publication. doi:10.1016/j.cpa.2017.05.003

World Intellectual Capital Initiative (WICI). (2013). *Connectivity: Background Paper for <IR>*. Disponível em: https://integratedreporting.org/wp-content/uploads/2013/07/IR-Background-Paper-Connectivity.pdf

Chapter 7
Institutionalization of the Process for External Assurance on Integrated Reporting:
Perceptions of Both the IIRC and Social Actors

Juliane Campoe Correa
https://orcid.org/0000-0002-2074-255X
State University of Maringá, Brazil

Anne Carolina dos Santos
https://orcid.org/0000-0002-5214-1273
State University of Maringá, Brazil

Marguit Neumann
State University of Maringá, Brazil

ABSTRACT

This study aimed to analyze the perception of the International Integrated Reporting Council (IIRC) and social actors about the external assurance process involving integrated reporting, based on Zilber's institutionalization model. Two documents made available for public consultation by the IIRC, as well as the 62 documents with responses from social actors, were analyzed by means of the content analysis technique. The results indicate that the IIRC seeks to institutionalize its guidelines, as it has a favorable stance for believing that the theme is substantial, even though the actors with an interest in the subject demand guidance due to the subjectivity involving external assurance and some contents of Framework 1.0. Thus, Framework 1.0 for IR and its external assurance must evolve in view of the needs exposed by the social actors. The opinions of the social actors indicate that they seek to meet their own specific interests, rather than critically contributing to adjustments in the meanings of external IR assurance.

DOI: 10.4018/978-1-7998-9410-0.ch007

INTRODUCTION

Integrated Reporting (IR) was created in response to the complexity of corporate and sustainability reports (Cheng, Green, Conradie, Konishi & Romi, 2014; Villiers, Rinaldi & Unerman, 2014) and to divergences between information contained in them and financial statements (Simnett & Huggins, 2015). Since its launch in 2013, it is not known how many companies prepare Integrated Reporting, but its website (Value Reporting Foundation, 2021) lists 496 companies that claim to follow Framework 1.0 for IR (herein simply referred to as framework).

With the trend towards the regulation of voluntary reporting (Maroun, 2019a, Simnett & Huggins, 2015; Steyn, 2014) and the increase in applications of the first version of the framework for IR (Framework 1.0), the International Integrated Reporting Council (IIRC, 2013a) believes it will become the standard for corporate reporting (before June 2021, the coalition was named IIRC, so the coalition is so called). Thus, independent assurance is essential for its long-term credibility (IIRC, 2014a). For this reason, the IIRC, together with the Assurance Technical Collaboration Group (ATCG), started a discussion about the importance of independent external assurance to attest to the credibility and reliability of IR (IIRC, 2014b).

This discussion resulted in the release of two documents entitled "Assurance on<IR>: An introduction to the discussion" (IIRC, 2014a) and "Assurance on<IR>: An exploration of issues" (IIRC, 2014b). Both documents were released for public consultation and provided an overview of IR assurance as they discussed the benefits and challenges that arise in this field. Thus, those who were interested submitted their responses and opinions, which were made available on their electronic address (IIRC, 2014d).

Since then, different studies have emerged involving IR assurance (Briem & Wald, 2018; Hoang & Phang, 20210; Kiliç, Kuzei & Uyar, 2021; Maroun, 2017, 2018, 2019a, 2019b; Oprisor, 2015; Prinsloo & Maroun, 2020; Reimsbach, Hahn & Gürtürk, 2017; Richard & Odendaal, 20210; Rivera-arrubla, Zorio-grima & García-benau, 2017), but these studies did not analyze the responses of those interested in the subject. Only the studies by Simnett, Zhou and Hoang (2016) and Corrado, Demartini and Dumay (2019) synthesized the responses received from this public consultation, but without a theoretical lens to interpret it.

Considering that IR and its assurance represent a new social practice, Institutional Theory becomes useful in clarifying how new organizational practices are institutionalized from the negotiations of meanings attributed by social actors (Zilber, 2002) and allows us to understand how institutions are formed. from an ontological perspective (Coraiola, Suddaby & Foster, 2018) and not just as a result of institutional isomorphism, as observed in the current literature (Briem & Wald, 2018; Kiliç, Kuzei & Uyar, 2021). In this study, institutionalization is approached as a process in which social actors, by assigning meanings, create and recreate institutions, that is, social or organizational practices (Zilber, 2002). Thus, the objective of this study is to analyze the perception of both the IIRC and social actors about the process for external IR assurance, based on Zilber's institutionalization model (2002, 2006), whose assumptions indicate that institutionalization involves a political process, as the interpretation and meanings attributed by actors are different and influenced by individual interest and power relations (Zilber, 2002).

To achieve the proposed objective, the content analysis technique was used in documents made available for public consultation, with the employment of a qualitative and interpretive approach. The documents examined were "Assurance on<IR>: An introduction to the discussion" and "Assurance on<IR>: An exploration of issues", authored by the IIRC, along with the 62 responses received from different interest groups from different countries (IIRC, 2014b; IIRC, 2014c; IIRC, 2014d).

The results reveal that the IIRC has a favorable stance on the IR external assurance process; however, the actors who responded to the public consultation negotiate different meanings on the matter. They even demand guidance on subjective issues as to external assurance and on some contents of the framework, such as basic principles, capitals and content elements. From these results, some problems that emerge from the application of the framework proposed by the IIRC were identified (a second version was published in January 2021, applicable as of January 2022, see https://integratedreporting.org/resource/international-ir-framework/), which should provide more precise guidance on its content in subsequent versions. This indicates that there must be negotiation of meanings for a reflection in the actions of these actors and reveals that the institutionalization of IR assurance depends primarily on the institutionalization of IR, whose process is still in the negotiation phase.

From a theoretical point of view, this study contributes to the understanding of the opinions of those involved regarding the IR assurance process, through the meanings assigned both by the IIRC and several actors who responded to its public-consultation document (IIRC, 2014d). Furthermore, it contributes to the literature given the absence of studies aimed at the analysis of documents within the scope of this research. From a practical point of view, it makes it possible for organizations to be aware of the criticisms inherent to IR (Dumay, *et al.*, 2016). In addition, the present study contributes with possible solutions to problems related to the institutionalization of IR and to future versions of its framework.

The study has this structure: section 2 provides an overview of the literature on integrated reporting, its assurance and presents the theoretical framework, considering institutionalization from the perspective of the relationship between actors, actions and meanings. Section 3 describes the methodological procedures employed. The results found and their interpretation are exposed in section 4. The last section concludes the study by providing recommendations and suggestions for future research based on the limitations found.

THEORETICAL FRAMEWORK

Integrated Reporting and its Assurance

Assurance is the process carried out by a competent and independent external professional, who expresses their written opinion on a given report. This process encompasses both financial and non-financial information and, therefore, tends to raise the users' level of trust in the disclosed information (Simnett, Vanstraelen & Chua, 2009; Briem and Wald, 2018; Simnett and Huggins, 2015). Unlike auditing on financial information, external assurance of sustainability reporting and integrated reporting involves less extensive processes and testing. This results in different assurance practices by assurance bodies (Bepari & Mollik, 2016).

Among the best known global standards, audit firms use ISAE 3000, by the International Auditing and Assurance Standards Board (2020), while consulting firms use AA1000AS, by Accountability (2018). Both standards have been an object of study in the literature and presented varied results as to their applications. However, in terms of integrated reporting, Maroun (2019a) provides evidence that the assurance process conducted by large audit firms (Big 4) is associated with better quality in information integration.

The first version of the framework for IR (there is a version applicable as of January 2022, see https://integratedreporting.org/resource/international-ir-framework/) identifies the information to be included in IR, but does not set standards (Adams, 2015; Brown & Dillard, 2014; Chaidali & Jones, 2017). That said, assurance represents a way for the organization to legitimize itself before its stakeholders (Dimaggio & Powell, 1983), as it brings credibility to disclosed information (Junior *et al.*, 2014; KPMG, 2017). Briem and Wald (2018) found evidence that German companies face coercive pressure from their stakeholders to obtain external assurance for their reports. Kiliç, Kuzei and Uyar (2021) found that the institutional environment is significantly associated with the adoption of independent assurance on IR, in addition to arguing that the level of sustainable development in a country has a positive association with assurance quality.

When investigating how the choice of the reporting format interacts with voluntary assurance on sustainability information, Reimsbach *et al.* (2017) reported that the investors in the sample highlighted the relevance of external assurance in the context of voluntary disclosure, which resulted in the weighting of information. However, this effect was reduced in the case of IR compared to reporting of a financial nature. Stubbs and Higgins (2014) argue that the content of Integrated Reporting focuses mainly on the supply side, that is, it includes the perception of report writers, leaving out the perspectives of external users. However, the IIRC does not require, but encourages, through the framework, external IR assurance. Simnett *et al.* (2016) add that the current challenge is that the traditional assurance model is changed for IR, as its content is broad, complex and requires specific skills for this engagement.

Considering that the framework asserts, by means of the principle of reliability, that disclosed information depends on balance and exemption from significant errors, the IIRC (2014a) advised the existence of internal mechanisms that increase the credibility of the information present in the RI, namely: internal-control system, relationship with stakeholders, internal audit, and corporate governance (Figure 1).

Figure 1. IR reliability
Source: Based on the Committee of Sponsoring Organizations of the Treadway Commission (COSO, 2013); Brazilian Institute of Corporate Governance (IBGC, 2009); IIRC (2014a); Meintjes and Grobler (2014); The Institute of Internal Auditors (IIA, 2009)

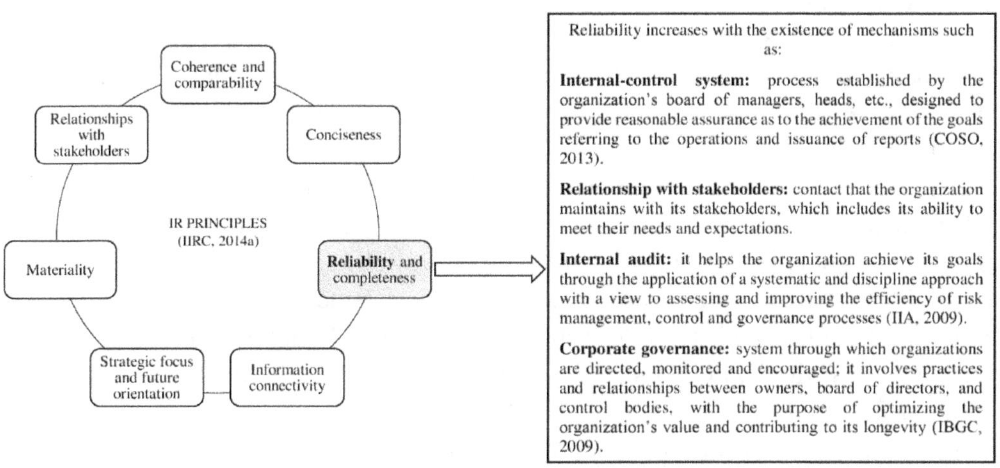

As shown in Figure 1, in addition to external assurance, IR reliability depends on internal mechanisms, which also support the opinion of auditors and assurance practitioners (IIRC, 2014b). These internal mechanisms were the object of analysis in the study by Richard and Odendaal (2021). Through two case studies, the results support the concept that these mechanisms increase the credibility of the information disclosed and are alternative (genuine and economic) means of assurance in IR. Hoang and Phang (2021) argue that a combined assurance – audit committee, internal audit and external audit – improves the reliability perceived by investors when there are significant reliability risks. Eccles, Krzus and Watson (2012) argue that only an integrated guarantee, that is, covering all IR information, will add value to the process and the report. Maroun (2018) went further and developed elements for an interpretive assurance model based on interviews with auditors and integrated report preparers, emphasizing that there must be rigor in the interpretation and analysis of qualitative and prospective information.

For Chaidali and Jones (2017), the IR principles result from the IIRC's effort to expand its territory, by promoting a reporting without specific rules; thus, it is based on fragile guidelines and allows for discretion. According to this perspective, writers must trust the concepts and credibility of IR, as these actors report problems about lack of adequate and clear guidance, high preparation costs, and the format and length of the reporting (Chaidali & Jones, 2017), in addition to the limited number of methodologies or the assurance structure (Prinsloo & Maroun, 2020).

In view of this, to promote a discussion on assurance issues pending after the approval of the first version of the framework, the IIRC released two discussion documents, namely "Assurance on<IR>: An introduction to the discussion" and "Assurance on<IR>: An exploration of issues" (IIRC, 2014b, IIRC, 2014c). Said documents mention other mechanisms that reinforce the credibility and trust of IR (as shown in Figure 1), making room for concepts to be re-signified. The next subsection deals with this subject, that is, the theoretical foundation of this research.

Institutionalization from the Perspective of the Relationship Between Actors, Actions and Meanings

Organizations and individuals are part of a network of values, norms, rules and beliefs, which, in their turn, are partially the result of their own creation (Meyer & Rowan, 1977). In this context, institutions are shared rules and typifications that identify categories of social actors, their activities and relationships. Therefore, institutions represent restrictions on the actions of individuals, even though they can be changed over time (re-signification); in this way, they arise as a result of the social interaction between actors (Zucker, 1987).

From this perspective, social actors create institutions through negotiations that lead to shared typifications. This means that institutionalization depends on the incorporation of knowledge constituted through the social interaction of actors (Dimaggio & Powell, 1983; Machado-da-Silva, Fonseca & Crubellate, 2005). With that, shared typifications are historical additions of memories, reproduced over time until they acquire the status of ontological reality (Coraiola, Suddaby & Foster, 2018). Interactions allow observing how institutional fields are understood, as actors manipulate their meanings in order to understand them (Leibel, Hallett & Bechky, 2018).

Power eliminates the need for social control and facilitates transmission, that is, institutionalized actions are taken naturally by individuals (Zucker, 1987). In accordance with this reasoning, Zilber's institutionalization model (2002, 2006) seeks to understand the symbolic dynamics of institutionalization, as institutions are incorporated by means of a dialectical interaction between: actions (practices and structures), meanings and actors. Social structures are built by individuals, and practices are created by them. The meanings, spread across society, comprise the combination of institutionalized and legitimized practices and structures. However, it is through interpretation that structures and practices are given meaning. From this perspective, meanings attract actors to actions (Zilber, 2002).

It turns out that actors can be considered passive when accepting structures and practices. On the contrary, actors can become active by choosing and infusing actions through the interpretation of meanings; therefore, social actions create, reproduce and change institutions (Zilber, 2002). For instance, Edelman, Abraham and Erlanger (1992) evidence that an ambiguous law was interpreted by communities of professionals, and this determined its institutionalization. Thus, the importance of meanings for a given subject to become an institutionalized cultural convention is highlighted (Gray, Purdy & Ansari, 2015).

Individuals' interpretations must be broadened in order to include the point of views of all members of society with an interest in a particular matter, with the aim of sustaining or rejecting institutional meanings. Even if the interpretation is automatic, individual power and interests must be considered (Zilber, 2002, 2006). Social actors who have resources and, consequently, power are unlikely to participate in changes that alter the current order (Machado-da-Silva *et al.*, 2005). For this reason, on the one hand, influential actors fight for the perpetuation of institutions that already have a reputation, while on the other hand, there are peripheral actors intentionally fighting for change, aiming to defend their own interests (Machado-da-Silva *et al.*, 2005).

From this perspective, reputation is seen as a record of past actions. The trust or distrust of individuals and organizations depends on their past experience (Chaidali & Jones, 2017). In short, subgroups of society can negotiate which structures and practices will be institutionalized, which makes it necessary to consider joint actions between members of society (Zilber, 2006). Therefore, this process is the result of interactions between the individual parts, and finally, the practice or structure is transmitted (institutionalized). Nonetheless, this process is constant in the sense that meanings are frequently adjusted (re-signification) (Zilber, 2006).

It is noted that IR and its assurance is a new communication paradigm (Simnett & Huggins, 2015), the adoption of which depends on the perspectives of report writers, assurance practitioners and other interested actors who can influence this process when creating, changing and rejecting meanings. In this regard, there is evidence that weak institutional environments increase the need for IR assurance (Kılıç *et al.*, 2021), that the main concerns of stakeholders refer to the difficulty in assuring qualitative information, and that there is discretion in IR concerning the reporting process (Corrado, Demartini & Dumay, 2019).

METHODOLOGICAL PROCEDURES

This documentary and qualitative research, in line with its objective, used the content analysis technique (Bardin, 1977) with the aim of evidencing nuclei of meaning of communication, thus using an interpretive approach to analyze the results. The following documents, authored by the IIRC, were analyzed: "Assurance on<IR>: An introduction to the discussion" and "Assurance on<IR>: An exploration of issues". The details of the questions proposed in the first document are shown in Table 1.

Table 1. Details of the questions proposed in the IIRC document

Document: Assurance on<IR>: An introduction to the discussion
It introduces the debate on IR assurance and proposes the following questions for discussion: Q1: What priority should be given to assurance in the context of increasing credibility and trust in IR? Q2: What are the key assurance resources that will best meet the needs of integrated-reporting users in the coming years? Q3: Is the availability of duly qualified and experienced assurance professionals an issue in your jurisdiction and, if so, what needs to be done, and by whom, to remedy the situation? Q4. What needs to be done, and by whom, to ensure that IR quality assurance is maintained at a high level, including staff compliance with adequate educational, ethical (including independence), quality control, and performance standards? Q5. Is the robustness of internal systems an issue, and if so, what needs to be done, and by whom, to remedy the situation? Q6. Is assurance likely to be a cost-effective mechanism to ensure credibility and trust (a) in the short/medium term; (b) in the long term? Q7. If so, what needs to be done, and by whom, to maximize the benefits of assurance? Q8. Assurance standard setters must develop one or both: (a) a new assurance standard; (b) guidance, in order to ensure consistency for such issues to be addressed? Q9. Should any standard/guidance be specific to IR, or should it cover topics that are also relevant to other forms of reporting and assurance? For example, should there be a standard/guidance on how to ensure narrative information in integrated reporting? Q10. What are the (a) key challenges and (b) proposed approaches that assurance standard setters should consider with respect to: Materiality? Reporting limit? Connectivity? Narrative reports and future-oriented information? Q11. What other technical issues, if any, specific to IR should assurance standard setters address? Q12. What are the (a) key challenges and (b) proposed approaches that assurance standard setters should consider with respect to: Reasonable assurance? Limited assurance? Hybrid engagements? Engagements as to agreed procedures? Other approaches? Q13. What are the (a) key challenges and (b) proposed approaches that should be considered, and by whom, to ensure that IR pays due attention to other assurance processes?

Source: IIRC (2014b)

The responses of the actors to the public consultation, obtained from the IIRC website (IIRC, 2014b, 2014c, 2014d), totaled 63 documents, which, in their turn, make up 426 pages (Table 2). A document was excluded from the content analysis carried out; it refers to the "Writers" group and does not provide feedback to the IIRC, it only mentions a case study about assurance at the Philips company.

It is noteworthy that, as of June 2021, the IIRC coalition is called Value Reporting Foundation, but this study uses documents dating back to 2014 and, for this reason, the IIRC nomenclature was maintained when referring to the coalition in charge of IR. Another point is that this study was part of a research maturation process, so the references and analyses were updated over time.

The content analysis of this study followed the pre-analysis, coding, categorization and inference steps (Figure 2).

The indicators for coding and categorization were defined from the concepts related to the principle of reliability of the framework (Figure 1), which indicates that IR assurance depends on internal mechanisms. Therefore, steps 2 and 3 are based on these indicators defined *a priori* (Assurance, Corporate Governance, Internal Control, Internal Audit, and Stakeholder Management) used to unveil units of analysis (coding) and categories (categorization) in view of the data. The categories and units of analysis emerged from the data analysis (they were not defined a priori), considering the agglutination of related themes. Bardin (1977) describes that this fact indicates that the researcher conducted an analysis to see what is in the established context (Bardin, 1977). Such steps were taken using the ATLAS.ti software, version 8.

Table 2. Characteristics of the public-consultation documents

ID* of the groups of social actors	Actors	Number of responses	Country/region (quantity)
(A) Academy	Academy	2 (3.2%)	Argentina (1) and Australia (1)
(B) Standard setters	Stock Exchange	1 (1.6%)	South Africa (1)
	Issuers of accounting, audit or assurance standards	12 (19.4%)	South Africa (2); Australia (1); China (1); USA (3); England (2); New Zealand; Singapore and several (1 each)
(C) Writers	Companies	2 (3.2%)	Germany (1); Denmark (1)
(D) Accounting professionals	Audit firms	9 (14.5%)	Brazil (1); USA (2); France (1) and England (5)
	Consulting firms	10 (16.1%)	South Africa; Australia; Canada; China (1 each); England (3); Ireland (1) and several (2)
	Organizations of accounting or finance professionals	18 (29.0%)	South Africa (2); Australia (4); Germany; Belgium; Brazil; Canada; Scotland; England; Italy; Malaysia; Kenya; Sweden; Tokyo and European Union (1 each)
	Independent professionals	3 (4.8%)	Australia (1) and Brazil (1)
	Roundtable	3 (4.8%)	Brazil; Denmark and several (1 each)
(E) Information users	Investor organizations	2 (3.2%)	Australia (1) and Japan (1)
TOTAL	-	**62 (100%)**	**Africa (7); North America (7); South America (5); Asia (6); Europe (23); Oceania (10) and several (4)**

Source: Research data (2021)

ID*: Identification of the groups of social actors

Figure 2. Content analysis steps and details
Source: Based on Bardin (1977)

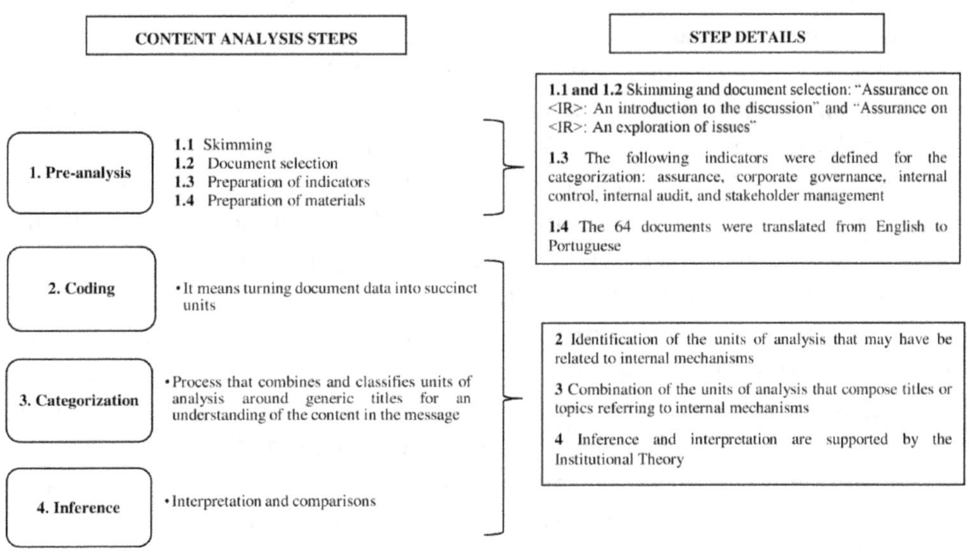

RESEARCH RESULTS

Presentation of Results

The results of the documentary analysis are segregated into tables, in accordance with the indicators used to conduct the content analysis – Assurance, Corporate Governance, Internal Control, Internal Audit, and Stakeholder Management (Tables 3 to 7, respectively). All IIRC comments (F) are attributed to the reference of this organization (IIRC, 2014b, 2014c), whereas Academy (A); Standard setters (B); Report writers (C); Accounting professionals (D) and Information users (E) refer to the content present in IIRC (2014d).

The Assurance theme resulted in 4 categories and 16 units of analysis, as shown in Table 3.

It is observed that all actors, except for the investors, who did not comment on the matter, agree on the degree of subjectivity of IR and demand clarification from the IIRC on non-detailed matters, such as the reporting limit, the capitals and the IR principles (Challenges category). This is consistent with Chaidali and Jones (2017) and Corrado *et al.* (2019), that is, that IR writers report problems about lack of guidance. The clarifications that the IIRC must provide, in addition to allowing interested parties to interpret IR and external assurance, allow assigning meanings to these practices (Zilber, 2002, 2006). Consequently, the adoption of external assurance enables the institutionalization of IR, as it attracts the development of the guidelines set out in the framework. According to Rivera-Arrubla *et al.* (2017), performing external assurance on IRs has a positive and significant impact on the level of disclosure, increasing reliability.

Corrado *et al.* (2019) point out the difficulty in assuring qualitative information, but the IIRC states that subjectivity is inherent to the process of preparing IR and engaging in the practice of external assurance (U.A. # 11), in addition to reinforcing the need for integration and participation of internal mechanisms to support the preparation of external IR assurance, since internal users would be engaged in the search for integrated thinking. However, the IIRC admits that the assurance challenges will be overcome by the existence of robust internal controls (U.A. # 13), qualified assurance practitioners (U.A. # 15) and the development of an integrated assurance methodology for IR (U.A. # 16).

Prinsloo and Maroun (2020) state that there is no adequate assurance methodology. Because external assurance procedures for IR must evolve, the IIRC admits that it is necessary to develop assurance techniques of an integrated nature (U.A. # 16). Eccles *et al.* (2012) says that for the creation of an integrated assurance methodology, the level of assurance of non-financial information must be approximated to the level of assurance of financial information (U.A. # 9), because only in this way will the assurance process add value to the report. However the establishment of such techniques and procedures must include the perspectives of all members with an interest in the subject (Zilber, 2002). Since interest groups from different countries have divergent point of views (U.A. # 6 and 16), for this depends on the reality of each one (Chatham *et al.*, 2010; Yen *et al.*, 2007). The development of the assurance methodology demands qualified professionals, according to Gomes *et al.* (2015), usually the auditors who occupy this place.

Because IR is mainly aimed at financial capital providers, the market perceives the cost-benefit ratio of IR assurance (U.A. # 6). For instance, it was identified that the investors (E) show distrust as to the benefits of external IR assurance (U.A. # 1, 2 and 3), because its content falls short of the information required by the market (Stubbs & Higgins, 2014). This distrust may be linked to nationality, since the IIRC did not receive responses from investors (E) in Africa, where IR is a legal requirement for companies listed on the Johannesburg Stock Exchange.

Table 3. Assurance

Categories Units of analysis (UA)	(A) Academy; (B) Standard setters; (C) Report writers; (D) Accounting professionals; (E) Information Users	(F) IIRC
Benefits **1.** Internal to the organization; **2.** External to the organization;	**1.** External assurance on IR is beneficial to the organization's internal processes and provides assistance for value creation, thus enabling the rooting of the IR practice **(A, B, C, D)**; **2.** External assurance provides trust and credibility to IR **(A, B, C, D)**;	**1.** It enables corporate governance to believe in the integrity of IR; **2.** Credibility and trust on the part of users regarding the information, so that they can assess the risk of investing in a particular company;
Demand **3.** External users' needs; **4.** IR preparation organizations;	**3.** Respondents from groups **B** and **E** state that the need for external assurance with investors must be identified, as there are doubts about its benefits for the credibility of IR. However, an independent practitioner **(D)** claims that external users are not in a position to question the work of assurance practitioners; **4.** An audit firm and a consulting firm **(D)** say that it is useful to hear the opinion of writers to understand the limitations of this process;	**3.** IR is rules by the market, that is, assurance requires validation by users or, otherwise, it becomes dispensable;
Challenges **5.** Capitals; **6.** Cost of the assurance process; **7.** Non-financial information; **8.** Reporting limit; **9.** Assurance level; **10.** Obtaining demand; **11.** IR basic principles; **12.** Trained professionals;	**5.** The absence of more detailed information about capitals in the framework for IR decreases the assurance integrity **(C, D)**; **6.** The cost of assurance will be lower in the long run, after organizations' internal controls become more robust and assurance professionals obtain the necessary qualifications **(B, D)**. A point raised by an audit firm **(D)** is that the cost of external assurance should be seen as an investment; **7.** The internal controls on this information tend to be less robust **(A, B, C, D)**. For this reason, in the investors' perception **(E)**, assurance practitioners must emphasize how the organization come to the statements present in the IR; **8.** it can affect assurance and information connectivity, so guidance on the subject is required **(A, B, C, D)**; **9.** The hybrid or integrated assurance level is a challenge **(B, D)**. However, two audit firms **(D)** agree on the application of the hybrid assurance, due to the scope of IR; **10.** Users must value and demand IR assurance due to its cost **(D)**; **11.** The principles of information connectivity, materiality, reliability and completeness are subjective, which makes it difficult to standardize external assurance, so guidance on such principles is required **(A, B, C, D)**; **12.** The qualification of assurance practitioners will be surpassed along with the evolution of IR **(B, D)**. Organizations of professionals **(D)** suggest that universities introduce disciplines and give lectures on IR and its assurance;	**6.** For this process to be viable, it is conditioned to its generated benefits, as the cost-benefit ratio of preparing IR is perceived by the market; **8.** The body of the IR text may have hyperlinks, for example, containing information that affects the value creation capacity and make the assurance process difficult; **11.** With regard to narrative and prospective information, the challenge is to ensure connectivity, completeness and conciseness. Therefore, it will require judgment by both report writers and assurance practitioners;
Assurance methodology **13.** Relationship with other mechanisms; **14.** Legal requirement; **15.** Professional's profile; **16.** Assurance procedures.	**13.** Mechanisms such as governance, robust internal controls, internal audit, and stakeholder involvement must be developed at this early stage of IR preparation **(B, D)**. Likewise, for investors **(E)**, there should not only be dependence on external assurance, but other mechanisms; **14.** IR should be assured only if it was a legal requirement set by regulators **(E)**. According to organizations of professionals and auditors **(D)**, IR assurance should be mandatory; **15.** Extensive preparation and professional judgment, in addition to multidisciplinary teams, are required **(A, B, C, D, E)**; **16.** Actors in groups **B** and **D** argue that standards such as ISAE 3000, ISA 350, and ISA 330 provide a sufficient basis for limited or reasonable IR assurance engagements to be carried out. For others **(A, B, D)**, only the principles existing in the current IAASB standards or the development of new standards by this body can contribute to IR assurance. Other respondents **(B, D)**, in their turn, suggest only the development of guidelines by the IIRC, including the evolution of the framework for IR. Finally, actors from groups C and E suggest that the market (stakeholders) should decide on the assurance procedures.	**13.** The external assurance engagement must be related to other mechanisms. It depends on robust internal controls to assure information, and specific guidance or standards for IR assurance should be required. Internal auditors and financial statement auditors can provide assistance in the sense of reducing the cost of the assurance engagement; **15.** They must have new skills, such as understanding the business model, the company's environment, and being familiar with the guidelines of the framework; **16.** Assurance methods such as ISAE 3000 and ISA 720 support some IR contents. However, for specific matters established in the framework for IR, new assurance methodologies are needed. However, the ideal would be an assurance of an integrated nature, by encompassing all aspects of an IR. However, the assurance engagement being related to certain parts of IR contradicts the objectives of IR and would confuse users.

Source: Research data (2021)

The Corporate Governance theme (Table 4) comprises 3 categories and 12 units of analysis.

Table 4. Corporate governance

Categories **Units of analysis**	**(A) Academy; (B) Standard setters; (C) Reporter writers; (D) Accounting professionals; (E) Information Users**	**(F) IIRC**
Responsibilities **17.** Alignment of stakeholder interests; **18.** Definition of the IR preparation basis; **19.** Development of a comprehensive assurance plan; **20.** Establishment of the internal-control environment; **21.** IR integrity assurance **22.** Assurance on the information generation process; **23.** IR preparation; **24.** Establishment of the assurance level;	**17.** Governance and stakeholders must align their objectives **(A)**; **18.** Since governance must provide a basis for IR to be prepared and presented **(B, D, E)**; **19.** Actors in groups **B** and **D** assign to governance the responsibility for developing a comprehensive assurance plan with the help of internal and external audit; **20.** In the opinion of a standards issuer **(B)**, the establishment and oversight of the internal-control environment is the responsibility of governance; **21.** It is up to governance to assure IR integrity through a written statement **(B, D, E)**; **22.** As a consequence, governance becomes responsible for assuring the information generation process **(B, D, E)**; **23.** An actor from group **B** mentions that IR must be prepared by independent, non-executive advisers from the organization's audit committee, in order to reduce the bias of IR information;	**20.** Corporate governance is responsible for establishing the internal-control environment, in order to ensure IR integrity. Therefore, the organization must keep an audit trail so that the disclosed information can be verified and thus ensure trust in the report; **21.** The framework for IR requires a statement from those in charge of governance that includes an opinion on the compliance of the reporting with the framework itself; **24.** Corporate governance monitors this process and reviews the reporting, which is why it is also responsible for defining the level of assurance to be executed by external professionals.
Relationship with other mechanisms **25.** Relationship with internal audit; **26.** Relationship with internal controls; **27.** Relationship with other mechanisms;	**25.** For actors in group **B**, governance is related to internal audit; **26.** And internal control for monitoring risks and adequacy of the robustness of internal controls **(B)**; **27.** In addition, a consulting firm **(D)** suggests that corporate governance representatives should collaborate with an independent body that oversees and trains assurance professionals;	-
Complementary **28.** Skills/knowledge; **29.** Governance regulation.	**28.** Actors in group **C** judge that people linked to corporate governance must develop new skills; **29.** There should also be an increase in the regulation of the board of directors in order to improve corporate governance **(C)**.	-

Source: Research data (2021)

Just as the IIRC, actors in groups B, D and E agree that corporate governance plays a strong role in assuring IR integrity and the information generation process (U.A. # 21 and 22). The actors in groups B and D, due to their experience, and because they are more involved in the assurance process, perceive that governance depends on other internal mechanisms (U.A. # 25 and 26), since they use them to perform their assignments. This interdependence between governance and internal mechanisms for IR assurance denotes that joint actions must be negotiated in view of this new social context. However, the institutionalization of new structures and practices depends on how each group of actors will interpret this new reality and legitimize practices (Gray *et al.*, 2015; Zilber, 2006).

That is, as organizations adopt internal mechanisms and external assurance, there will be negotiations on how such practices will be conducted and the reason for them to take place. Social interaction (Dimaggio & Powell, 1983; Machado-da-Silva, Fonseca & Crubellate, 2005) between professionals in corporate governance, internal audit, stakeholder management and external assurers constitute sources of meaning (Leibel *et al.*, 2018). Therefore, the IIRC must constantly capture such meanings, monitoring whether practices interfere with the institutionalization process of external IR assurance, and whether, over time, social actors are instituting new meanings, changing them or rejecting them. Just as what Kiliç *et al.* (2021) found, that is, the institutional environment is significantly associated with the adoption of independent assurance on IR.

The Internal control theme covers 4 categories and 8 units of analysis (Table 5).

Table 5. Internal control

Categories Units of analysis	(A) Academy; (B) Standard setters; (C) Report writers; (D) Accounting professionals; (E) Information Users	(F) IIRC
State **30.** Maturity;	**30.** Non-financial systems do not support the implementation of IR and its assurance, so they need to evolve **(B, C, D)**;	**30.** The maturity of the internal-control system is linked to its maintenance, in order to increase the trust of corporate governance regarding the information contained in the IR;
Model **31.** Company's own model; **32.** Existing models;	**31.** With respect to its own internal-control system model, a company **(C)** informs that its processes for disclosing corporate, social and environmental information are aligned and ensure a high level of quality; **32.** However, respondents from groups **A**, **C** and **D** mention existing frameworks to improve the quality and robustness of internal controls, such as COSO, CoCo, King Report and Vienot Report;	-
Relationship with other mechanisms **33.** External audit; **34.** Internal audit; **35.** Audit committee; **36.** Management;	**33.** Regardless of the model used, some actors in groups **B**, **D** and **E** consider that external audit; **34.** and internal audit contribute to improving internal control **(B, D, E)**; **35.** The academy **(A)** and organizations of professionals **(D)** further add that the audit committee; and **36.** the organization's management must also contribute to the robustness of the internal control **(A, D)**;	-
Complementary **37.** New control systems; **38.** Information technology.	**37.** In a complementary way, there is a need for new systems and controls **(A, D, E)**; **38.** For this reason, organizations will have to invest in information technology to help internal control verify the information required by the framework for IR **(A, D, E)**.	-

Source: Research data (2021)

Regarding this mechanism, the IIRC only mentions that internal-control systems need to mature in order to ensure the reliability of non-financial information (U.A. # 30) and does not present details about this issue. Groups B, C and D, because they are close to the process of preparing and assuring IR, also have the same understanding and, together with groups A and E, make more comments because they are closer to the practice. These groups highlight the need to implement frameworks for internal controls and that the quality and robustness of such controls depend on interaction with other internal mechanisms. These stances make the re-signification of external IR assurance relevant, in the sense that the practice interferes with this process.

Nonetheless, the Danish company Novo Nordisk (C) informs that its information disclosure processes are aligned and ensure the quality of IR (U.A. # 31). In this case, it is stated that it is not necessary to implement improvements in internal processes, which leads to the idea that they fight for specific internal-control frameworks not to be imposed on IR assurance. This corroborates with Zilber's assumptions (2002, 2006) that actors fight for their own interests, based on their individual needs, and that groups from different countries have different points of view (Chatham *et al.*, 2010; Yen *et al.*, 2007). These results are attuned to the findings of Kiliç *et al.* (2021), according to which the institutional environment of each nation interferes with the form of independent IR assurance.

The Internal audit theme covered 2 categories and 7 units of analysis, as shown by the results in Table 6.

Table 6. Internal audit

Categories **Units of analysis**	**(A) Academy; (B) Standard setters; (C) Report writers; (D) Accounting professionals; (E) Information Users**	**(F) IIRC**
Attributions **39.** Involvement of internal audit; **40.** Risk assessment; **41.** Assessment and recommendation of systems and controls; **42.** Process assurance; **43.** IR assurance; **44.** Interaction with external-assurance professionals;	**40.** Internal audit can provide insight into organizations' potential risks **(B, D)**; **41.** Internal audit has the role of assessing and recommending systems and internal controls in accordance with their position in the organizations **(B, D)**; **42.** However, internal audit is seen by some **(B, D)** as the most adequate mechanism to assess the process of building integrated thinking, that is, internal auditors must be involved in the project for implementing and preparing IR; **43.** Therefore, it is the role of internal audit to ensure that IR becomes reliable **(B, D)**; **44.** Also, according to **B** and **D**, the interaction of internal audit with external-assurance professionals avoids engagement duplication and reduces costs;	**39.** Internal audit must be involved in the IR preparation process to validate information generation and attribute confidence.
Professional's profile **45.** Knowledge of the organization and environment; **46.** Skills.	**45.** They must have knowledge of the organization and its environment **(B, D)**; **46.** However, in **B**'s perception, the internal auditor will have to acquire skills with a focus on governance, risk management, and on IR guidelines. The organizations of professionals **(D)** themselves admit that they are responsible for evidencing the relevance of internal audit to companies.	-

Source: Research data (2021)

Groups B and D praise the role of internal-audit professionals in the process of attributing trust to IR. However, there is a lack of opinion on the part of groups A, C and E. The strong participation in the discussion about internal audit by groups B and D presupposes acts of agency interpretation, as indicated by Zilber (2002, 2006), since both assign similar meanings to internal audit in the IR external assurance process. The effort of these groups of actors reveals their willingness to assign a new meaning to this internal mechanism (Machado-da-Silva *et al.*, 2005).

The experiences and work of these two groups are close and related to the work carried out by internal audit. They defend the significant role that this internal mechanism plays, even indirectly, in external IR assurance. Thus, they seek to re-signify the role of internal audit within the organization, previously seen as a guarantor of processes, now as a participatory agent in IR preparation and in the application of integrated thinking. From this interpretation, the joint actions of these actors will be able to adjust which structures and practices will be institutionalized (Zilber, 2002, 2006).

Thus, it becomes important to identify the problems that arise from the application of the IIRC guidelines and, consequently, to set guidelines on how to solve them. Considering the argument of Adams

(2015), Brown and Dillard (2014) and Chaidali and Jones (2017) that the IIRC intends to expand its territory by promoting a reporting without specific rules, this scope should be changed so that practical experiences reported in future public consultations allow the framework to evolve in order to prevent subjectivity. This becomes a challenge, as the opinions of respondents to the public consultation show that each group seeks its own interests (Zilber, 2002, 2006), which makes the capture of meanings biased (Gray, *et al.*, 2015). Also, the undisputed acceptance of the framework can be a factor that interferes with the actors' opinions (Dumay, *et al.*, 2016).

The Stakeholder management theme comprises 1 category and 4 units of analysis, according to the results presented in Table 7.

Table 7. Stakeholder management

Categories Units of analysis	(A) Academy; (B) Standard setters; (C) Report writers; (D) Accounting professionals; (E) Information Users	(F) IIRC
Relationship management **47.** Identification of stakeholders; **48.** Transparency with stakeholders; **49.** Meeting the needs of stakeholders; **50.** Participation of stakeholders in the materiality process.	**47.** Stakeholder management, according to some respondents **(B, D, E)**, involves identification; and **48.** transparency with stakeholders, considering that it encompasses their involvement in the IR preparation process, and even allows stakeholders to assess the adequacy of independent assurance **(B, D, E)**.	**48.** There is a need for transparency on the part of the organization when engaging positively with stakeholders; **49.** Importance of meeting the needs of stakeholders. In this sense, the IR preparation process must be based on the interests of the main stakeholders and evidence the nature and quality of this relationship; **50.** Relationship management allows the participation of stakeholders in the materiality process, that is, whether the reporting covers material information, both positive and negative.

Source: Research data (2021)

The importance of stakeholders for the adoption of IR and its external assurance is highlighted, as observed in U.A. # 3 (Table 3), that is, there must be a demand from external users. However, the literature (Stubbs & Higgins, 2014) points out that companies include information in accordance with their perception of relevance, forgetting to consider the perspectives of external users. Consequently, this makes the signification process difficult (Zilber, 2006), as meaning can be assigned by stakeholders differently, compared to other social actors.

The small number of responses from Group E to the public consultation implies that financial capital providers still do not use IR for investment decisions. The framework still needs to evolve as to the application and explanation of some current concepts so that IR becomes comparable (Feng, Cummings & Tweedie, 2017; Lodhia, 2015; Maroun, 2017; Simnett & Huggins, 2015) and robust (Cheng, Green, Conradie, Konishi & Romi, 2014; Feng *et al.*, 2017; Lodhia, 2015; Zhou, Simnett & Green, 2017).

For groups B, D and E, organizations must identify who their stakeholders are and be transparent when preparing an IR, which is imperative for the users of information to trust the IR. In this way, one avoids considering only the supply side, that is, the perception of IR writers (Stubbs & Higgins, 2014).

The IIRC's perception goes further, when considering that organizations must meet the expectations and information needs of their stakeholders (Meintjes & Grobler, 2014; Stubbs & Higgins, 2014). In this context, it is assumed that the institutionalization of IR assurance depends primarily on the importance given by stakeholders to this practice, so that it becomes an institutionalized cultural standard (Gray *et al.*, 2015). The next subsection (4.2) consolidates the discussion of the results.

Interpretation of Results

Analyzing the perceptions on the subject, it is possible to observe that each group of social actors fight for their own interests when trying to negotiate the meanings proposed by the IIRC, with the aim of reconstructing them with a view to their needs. This behavior is welcome, considering that both the practice of IR and its assurance is under negotiation (Zilber, 2002, 2006).

The IIRC's perception is entirely favorable to external IR assurance so that this reporting acquires reliability before external users. From this interpretation, IR can be institutionalized and become a cultural convention. The same perception can be attributed to groups A, B, C and D, as most of them, just as the IIRC, are audit, consulting and accounting professionals. These groups of social actors see the institutionalization of IR assurance as a means of staying ahead in the corporate reporting market and thus acquiring status and power. This can be seen as a political process, since the interpretations given by these actors are a way for them to keep their interests.

The lack of response to the public consultation from information users (group E) implies that these social actors may not see, in the short term, that external assurance on IR adds value or increases the credibility of IR. In response to the public consultation, group E did not comment on the benefits that external assurance can provide, but actively participated by answering questions involving internal mechanisms. Therefore, the relevance of understanding the stance and perception of information users, especially investors, is highlighted, as the framework states that IR is targeted at this group.

The importance given to internal mechanisms as a means of supporting the reliability of external IR assurance is also a consensus among groups of social actors, including information users (group E). However, there are differences of opinion in specific units of analysis. Thus, it is inferred that the auditors (D) argue that the cost of external assurance is a necessary investment (U.A. # 6). Groups A, C and D mention existing frameworks that improve the robustness of internal controls (U.A. # 32), which would be a way to explain the current permanence of academic curricula, preventing the insertion of new frameworks (group A); consequently, writers (C) and accounting professionals (D) would not spend resources on qualification and training.

However, it is understood that social actors must critically contribute to adjusting the meanings of IR and its external assurance, so that the IIRC is aware of the actions that hinder the acceptance of its initiative. In its turn, this does not mean that the actors that responded to the public consultation are active, because, for this, they must practice, using interpretation, what is determined in the IR framework, in addition to reporting their experiences in opportunities such as public-consultation documents and thus contributing to the evolution of the IIRC guidelines.

Thus, it becomes important to identify the problems that arise from the application of the framework proposed by the IIRC and, consequently, to set guidelines on how to solve them. Considering the argument of Adams (2015), Brown and Dillard (2014) and Chaidali and Jones (2017) that the IIRC intends to expand its territory by promoting a reporting without specific rules, this scope should be changed so that practical experiences reported in future public consultations allow the framework to evolve in order to prevent subjectivity. This is a challenge, as each nation's institutional environment interferes with the form of independent IR assurance (Kiliç *et al.*, 2021).

This challenging circumstance is expanded with the predominance of opinions in the public consultation composed of professionals in the accounting field: standard setters, auditors, consultants and accountants (72.6% - Table 1). By the way, the members of the IIRC are also predominantly the same professionals. Both the IIRC and groups B and D (91.9%) have a favorable stance on external IR assurance, undeniably accepting the framework (Dumay, *et al.*, 2016) and seeking to drive its institutionalization based on their interpretations.

This dominance can render the capture of meanings biased (Gray *et al.*, 2015) and make IR assurance a legal requirement of regulatory bodies. However, its institutionalization still depends on the acceptance of the main stakeholders (mentioned by the framework), that is, financial capital providers (investors – 3.2% stake). As an inference from the results, this group of social actors should participate more actively in the public consultations promoted by the IIRC so that their interpretations are reflected in the guidelines of the framework. Moreover, auditors usually perform the work of external IR assurance (Gomes *et al.*, 2015), thus, the hypothesis is raised to investigate whether this fact is a limitation for the advancement and development of methodologies focused on the topic, because the focus of auditors is on financial reporting. Eccles *et al.* (2012) argue that only an integrated guarantee will add value to IR assurance, such a perspective has not yet been developed. Thus, the challenges related to IR assurance remain current, even though the public-consultation documents refer to 2014.

FURTHER CONSIDERATIONS AND RECOMMENDATIONS

This study analyzed the perception of both the IIRC and social actors about the IR external assurance process, based on Zilber's institutionalization model (2002, 2006). The documents made available for public consultation – "Assurance on<IR>: An introduction to the discussion" and "Assurance on<IR>: An exploration of issues", authored by the IIRC – and the 62 responses from social actors to these documents (IIRC, 2014b; IIRC, 2014c; IIRC, 2014d) were analyzed by means of content analysis.

It is concluded that each actor defends the group in which they are inserted; this occurred through different stances, and, in the same way, the IIRC defended its cause. Therefore, four recommendations are inferred with a view to institutionalizing IR assurance. As a first recommendation, the IIRC (Value Reporting Foundation, currently) should issue specific guidelines on IR assurance containing more detailed information on how to conduct this process. Despite different perceptions, the actors that responded to the public consultation demand guidance on how to carry out the IR external assurance process. Thus, from Zilber's perspective (2002, 2006), the results indicate that there must be negotiation of meanings for a reflection in the actors' actions, thus reducing the subjectivity of the reporting itself, since external IR assurance has not yet been institutionalized. In this regard, the IIRC must take a stand when indicating the meanings of external IR assurance, especially how to carry it out, as well as provide guidance on how to put into practice the Basic Principles, Capitals, and Content Elements of IR.

As a second recommendation, the IIRC and companies that publish IR should encourage investors to contribute to changes in IR and its assurance, since this stakeholder doubts its benefits. This incentive would be possible if the IIRC reconsidered the way of collecting responses, in the sense of having the participation of other groups, given the predominance of professionals in the accounting field (72.6%).

As a third recommendation, respondents to future IIRC consultations should strive to contribute with opinions and criticisms, without the bias of meeting their own needs. Also, segregating the discussions by the companies' branch of activity can provide a greater understanding of external IR assurance. In this way, future versions of the framework for IR may evidence guidelines that can be applied to all types of organizations, reducing subjectivity in the preparation of the reporting and in its external assurance.

It is surmised that the IIRC, in order to negotiate the meanings of IR assurance, receives support from social actors through feedback on the documents made available for public consultation. However, the institutionalization process requires that the theme in question be constantly re-signified, given the perception of social actors of different nationalities and interests, in order to reduce subjectivity when applying IR assumptions. In this regard, a fourth recommendation is that the process of re-signifying external IR assurance must be conducted periodically in all geographic regions. Considering that the internal mechanisms of organizations are not fully prepared to contribute to the execution of external IR assurance, problems will arise as to how to carry out this procedure; therefore, as internal mechanisms adapt, the IIRC must monitor, by means of future public consultations, the adjustments in the meanings before those who assign them (actors).

Given the constant resignifications of the practice of external IR assurance, the IIRC could make it a legal requirement to allow the institutionalization of IR itself, since it induces writers to follow the reporting preparation guidelines. To this end, it is assumed that the IIRC needs to build the reputation of its guidelines. Because the IR guidelines are subjective and based on principles, it is assumed that the institutionalization of these guidelines, understood as IR and its external assurance, is relevant. Judging by the fact that power eliminates the need for social control and facilitates the transmission of meanings, which are naturally mobilized by individuals.

The present chapter contributes to the process of institutionalizing external IR assurance when considering that those interested in this matter, that is, actors of different nationalities, are aware of the criticisms made to IR internationally. Such contributions refer to the debate on the institutionalization of external IR assurance by evidencing, through the Institutional Theory, the stance of actors with an interest in these practices.

Although the study provides important information and contributions on IR assurance from an institutional perspective, this study has limitations. Among them, the new version of the framework for IR stands out, not analyzed in this study and whose structure may present new approaches and elements that can be related to the IR assurance process.

We suggest that future investigations analyze the trajectory of the IIRC towards the establishment of the framework and its current version (effective from January 2022), as well as on the understanding the reasons that lead companies to adopt IR, including how companies are adapting to prepare the external assurance on disclosed information in IR. In this way, it would contribute to the discussion about IR institutionalization.

REFERENCES

Accountability. (2008). *AA1000 Assurance Standard.* https://www.accountability.org/standards/aa1000-assurance-standard

Adams, C. (2015). The International Integrated Reporting Council: A call to action. *Critical Perspectives on Accounting*, *27*(1), 23–28. doi:10.1016/j.cpa.2014.07.001

Bardin, L. (1977). *Análise de Conteúdo*. Lisboa: Edições 70.

Bepari, M. K., & Mollik, A. T. (2016). Stakeholders' interest in sustainability assurance process: An examination of assurance statements reported by Australian companies. *Managerial Auditing Journal, 31*(6/7), 655–687. doi:10.1108/MAJ-06-2015-1208

Briem, C. R., & Wald, A. (2018). Implementing Third-Party Assurance in Integrated Reporting: Motivação das empresas e papel dos auditores. *Accounting, Auditing & Accountability Journal, 31*(5), 1461–1485. doi:10.1108/AAAJ-03-2016-2447

Brown, J., & Dillard, J. (2014). Integrated reporting: On the need for broadening out and opening up. *Accounting, Auditing & Accountability Journal, 27*(7), 1120–1156. doi:10.1108/AAAJ-04-2013-1313

Chaidali, P., & Jones, M. J. (2017). It's a matter of trust: Exploring the perceptions of Integrated Reporting preparers. *Critical Perspectives on Accounting, 48*(1), 1–20. doi:10.1016/j.cpa.2017.08.001

Chatham, M. D., Larson, R. K., & Vietze, A. (2010). Issues affecting the development of an international accounting standard on financial instruments. *Advances in Accounting, 26*(1), 97–107. doi:10.1016/j.adiac.2010.03.002

Cheng, M., Green, W., Conradie, P., Konishi, N., & Romi, A. (2014). The international integrated reporting framework: Key issues and future research opportunities. *Journal of International Financial Management & Accounting, 25*(1), 90–119. doi:10.1111/jifm.12015

Committee of Sponsoring Organizations of the Treadway Commission. (2013). *Internal Control – Integrated Framework: Executive Summary.* https://na.theiia.org/standards-guidance/topics/Documents/Executive_Summary.pdf

Coraiola, D., Suddaby, R., & Foster, W. M. (2018) Organizational Fields as Mnemonic Communities. Knowledge and Institutions, 13(1), 45-68. doi:10.1007/978-3-319-75328-7_3

Corrado, M., Demartini, P., & Dumay, J. (2019). Assurance on Integrated Reporting: A Critical Perspective. In S. Idowu & M. Del Baldo (Eds.), *Integrated Reporting. CSR, Sustainability, Ethics & Governance.* Springer. doi:10.1007/978-3-030-01719-4_11

Dimaggio, P., & Powell, W. W. (1983). The iron cage revisited: Institutional isomorphism and collective rationality in organizational fields. *American Sociological Review, 48*(2), 147–160. doi:10.2307/2095101

Dumay, J., Bernardi, C., Guthrie, J., & Demartini, P. (2016). Integrated reporting: A structured literature review. *Accounting Forum, 40*(3), 166–185. doi:10.1016/j.accfor.2016.06.001

Dumay, J., Bernardi, C., Guthrie, J., & La Torre, M. (2017). Barriers to implementing the International Integrated Reporting Framework: A contemporary academic perspective. *Meditari Accountancy Research, 25*(4), 461–480. doi:10.1108/MEDAR-05-2017-0150

Eccles, R. G., Krzus, M. P., & Watson, L. A. (2012). Integrated reporting requires integrated assurance. *Effective auditing for corporates: Key developments in practice and procedures*, 161–178.

Edelman, L. B., Abraham, S. E., & Erlanger, H. S. (1992). Professional construction of law: The inflated threat of wrongful discharge. *Law & Society Review, 26*(1), 47–83. doi:10.2307/3053836

Feng, T., Cummings, L., & Tweedie, D. (2017). Exploring integrated thinking in integrated reporting: An exploratory study in Australia. *Journal of Intellectual Capital, 18*(2), 330–353. doi:10.1108/JIC-06-2016-0068

Gomes, S. F., Eugénio, T. C. P., & Branco, M. C. (2015). Sustainability reporting and assurance in Portugal. *Corporate Governance: The International Journal of Business in Society, 15*(3), 281–292. doi:10.1108/CG-07-2013-0097

Gray, B., Purdy, J. M., & Ansari, S. (2015). From interactions to institutions: Microprocesses of framing and mechanisms for the structuring of institutional fields. *Academy of Management Review, 40*(1), 115–143. doi:10.5465/amr.2013.0299

Hoang, H., & Phang, S. Y. (2021). How does combined assurance affect the reliability of integrated reports and investors' judgments? *European Accounting Review, 30*(1), 175–195. doi:10.1080/09638180.2020.1745659

Instituto Brasileiro de Governança Corporativa. (2009). *Código das melhores práticas de governança corporativa.* https://www.ibgc.org.br/userfiles/files/Codigo_Final_4a_Edicao.pdf

International Auditing and Assurance Standards Board. (2020). *Revised International Standard on Assurance Engagements (ISAE) 3000, Assurance Engagements Other Than Audits or Reviews of Historical Financial Information.* https://www.iaasb.org/publications/international-standard-assurance-engagements-isae-3000-revised-assurance-engagements-other-audits-or-0

International Integrated Reporting Council. (2014a). *A Estrutura Internacional para Relato Integrado.* http://integratedreporting.org/resource/international-ir-framework

International Integrated Reporting Council. (2014b). *Assurance on <IR>: An introduction to the discussion.* http://www.theiirc.org/resources-2/assurance/

International Integrated Reporting Council. (2014c). *Assurance on <IR>: An exploration of issues.* http://www.theiirc.org/resources-2/assurance/

International Integrated Reporting Council. (2014d). *Responses to assurance on <IR>.* http://integratedreporting.org/responses-to-assuranceon-ir/

Kiliç, M., Uyar, A., & Kuzey, C. (2019). The impact of institutional ethics and accountability on voluntary assurance for integrated reporting. *Journal of Applied Accounting Research, 21*(1), 1–18. doi:10.1108/JAAR-04-2019-0064

KPMG. (2017). *The road ahead. The KPMG survey of corporate responsibility reporting 2017.* https://home.kpmg/content/dam/kpmg/campaigns/csr/pdf/CSR_Reporting_2017.pdf

Leibel, E., Hallett, T., & Bechky, B. A. (2018). Meaning at the Source: The Dynamics of Field Formation in Institutional Research. *The Academy of Management Annals, 12*(1), 154–177. doi:10.5465/annals.2016.0035

Lodhia, S. (2015). Exploring the transition to integrated reporting through a practice lens: An Australian customer owned bank perspective. *Journal of Business Ethics, 129*(3), 585–598. doi:10.100710551-014-2194-8

Machado-da-Silva, C. L., Fonseca, V. S., & Crubellate, J. M. (2005). Estrutura, agência e interpretação: Elementos para uma abordagem recursiva do processo de institucionalização. *Revista de Administração Contemporânea*, *9*(1), 9–39. doi:10.1590/S1415-65552005000500002

Maroun, W. (2017). Assuring the integrated report: Insights and recommendations from auditors and preparers. *The British Accounting Review*, *49*(3), 329–346. doi:10.1016/j.bar.2017.03.003

Maroun, W. (2018). Modifying assurance practices to meet the needs of integrated reporting: The case for 'interpretive assurance'. *Accounting, Auditing & Accountability Journal*, *31*(2), 400–427. doi:10.1108/AAAJ-10-2016-2732

Maroun, W. (2019a). Does external assurance contribute to higher quality integrated reports? *Journal of Accounting and Public Policy*, *38*(4), 106670. doi:10.1016/j.jaccpubpol.2019.06.002

Maroun, W. (2019b). Exploring the rationale for integrated report assurance. *Accounting, Auditing & Accountability Journal*, *32*(6), 1826–1854. doi:10.1108/AAAJ-04-2018-3463

Meintjes, C., & Grobler, A. F. (2014). Do public relations professionals understand corporate governance issues well enough to advise companies on stakeholder relationship management? *Public Relations Review*, *40*(1), 161–170. doi:10.1016/j.pubrev.2013.10.003

Meyer, J. W., & Rowan, B. (1977). Institutionalized organizations: Formal structure as myth and ceremony. *American Journal of Sociology*, *83*(1), 340–363. doi:10.1086/226550

Oprisor, T. (2015). Auditing integrated reports: Are there solutions to this puzzle? *Procedia Economics and Finance*, *25*, 87–95. doi:10.1016/S2212-5671(15)00716-9

Prinsloo, A., & Maroun, W. (2020). An exploratory study on the components and quality of combined assurance in an integrated or a sustainability reporting setting. *Sustainability Accounting. Management and Policy Journal*, *12*(1), 1–29.

Reimsbach, D., Hahn, R., & Gürtürk, A. (2017). Integrated Reporting and Assurance of Sustainability Information: An Experimental Study on Professional Investors' Information Processing. *European Accounting Review*, 1–23.

Richard, G., & Odendaal, E. (2021). Credibility-enhancing mechanisms, other than external assurance, in integrated reporting. *The Journal of Management and Governance*, *25*(1), 61–93. doi:10.100710997-020-09509-x

Rivera-Arrubla, Y. A., Zorio-Grima, A., & García-Benau, M. A. (2017). Integrated Reports: Disclosure level and explanatory factors. *Social Responsibility Journal*, *13*(1), 155–176. doi:10.1108/SRJ-02-2016-0033

Simnett, R., & Huggins, A. L. (2015). Integrated reporting and assurance: Where can research add value? *Sustainability Accounting. Management and Policy Journal*, *6*(1), 29–53.

Simnett, R., Vanstraelen, A., & Chua, W. F. (2009). Assurance on sustainability reports: An international comparison. *The Accounting Review*, *84*(3), 937–967. doi:10.2308/accr.2009.84.3.937

Simnett, R., Zhou, S., & Hoang, H. (2016). Assurance and other credibility enhancing mechanisms for integrated reporting. In *Integrated Reporting: a new accounting disclosure* (pp. 269–286). Palgrave Macmillan UK. doi:10.1057/978-1-137-55149-8_14

Stubbs, W., & Higgins, C. (2014). Integrated Reporting and internal mechanisms of change. *Accounting, Auditing & Accountability Journal*, *27*(7), 1068–1089. doi:10.1108/AAAJ-03-2013-1279

The Institute of Internal Auditors. (2009). *Código de ética*. http://www.iiabrasil.org.br/new/images/down/02_IPPF_Intro_Cod_de_etica.pdf

Value Reporting Foundation. (2021). *<IR> Reporters*. http://examples.integratedreporting.org/reporters?start=A&page=1

Yen, A., Hirst, E., & Hopkins, P. (2007). A Content analysis of the comprehensive income exposure draft comment letters. *Research in Accounting Regulation*, *19*(1), 53–79. doi:10.1016/S1052-0457(06)19003-7

Zhou, S., Simnett, R., & Green, W. (2017). Does integrated reporting matter to the capital market? *Abacus*, *53*(1), 94–132. doi:10.1111/abac.12104

Zilber, T. B. (2002). Institutionalization as an interplay between actions, meanings, and actors: The case of a rape crisis center in Israel. *Academy of Management Journal*, *45*(1), 234–254.

Zilber, T. B. (2006). The work of the symbolic in institutional processes: Translations of rational myths in Israeli high tech. *Academy of Management Journal*, *49*(2), 281–303. doi:10.5465/amj.2006.20786073

Zucker, L. G. (1987). Institutional theories of organization. *Annual Review of Sociology*, *13*(1), 443–464. doi:10.1146/annurev.so.13.080187.002303

Chapter 8
Future Perspectives and Evolution of Citizen Participation in Government

Maria Jesus Garcia Garcia
University of Valencia, Spain

ABSTRACT

A regulatory policy based on the idea of smart regulation takes the view that legislation and regulations serve the economy, and therefore, laws must be intended to create the conditions to improve national economies and economic growth, as laid down by the OECD and the European Union itself. Such a regulatory policy is informed by a set of principles which apply to both regulatory institutions and regulations, such as transparency, cooperation, accountability, efficiency, effectiveness, participation, and openness. This regulatory approach takes the view that regulations are of better quality when they have been influenced by citizens. As a consequence, smart regulation results in participatory democracy. Citizens collaborate with the government in the lawmaking process at an early stage and throughout the lifecycle of a legislative act by means of consultations. This chapter focuses on consultation as the principal mechanism for participatory democracy in the context of smart regulation law-making in the European Union.

INTRODUCTION

The increase in the democratic participation of citizens has been a constant throughout the evolution of the European Union which has run in parallel with its progressive increase in competences as the states have ceded sovereignty to it in areas increasingly distant from those that justified the creation of the European Communities in the middle of the last century. The gradual expansion of the Community cosmos has made more evident the distance between the citizens, around whom the European Union orbits, and the institutions with the capacity to decide, which has accentuated the perception of the deficit of democratic legitimacy that has been attributed since its inception to this supranational organization.

DOI: 10.4018/978-1-7998-9410-0.ch008

To remedy this, the European Union has resorted to democratic participation, first through the mechanisms of representative democracy and the direct election of members of Parliament by the citizens and then by articulating integrated collaboration mechanisms in the field of so-called participatory democracy (Bouza Garcia, L.,2015). The legitimacy provided by the mechanisms of representative democracy is limited, and for this reason it is necessary to complement this system with other options that involve the use of participatory democracy. Such participation confers a greater degree of democratic legitimacy to the rules on which these mechanisms are projected, since, being exposed to the participation of citizens, who theoretically have been able to contribute to the contents of the rule, it is assumed that they have the support of the recipients (Prieditis, M., 2018). On the other hand, the principles postulated in the White Paper on European Governance have also influenced the projection of the premises of participatory democracy on the normative procedure.

From a more general point of view, it is also necessary to refer to a trend that is gradually consolidating and that tends towards a growing recognition of the capacity of individuals and non-governmental organizations to act on their own at the international level. This implies a growing tendency for citizens to participate in the creation of international law, or in our case of European law, although this is far from recognizing the capacity to create this right themselves.

All these perspectives converge in the need to articulate mechanisms that allow citizens to participate in the creation of the European legal order, whether or not the resulting rules affect their rights and interests.

Well, the appeal to civic participation for legitimate democratically European standards has to be placed in the context of a regulatory policy that seeks to streamline the legal system by establishing procedures to conclude the character inexcusable of the standard, the needs and objectives that you are going to attend and that articulates at the same time formal channels to integrate the public interest and individual interests converging, usually financial, in its precepts. The result is the so-called "smart legislation", a term that has been adapted to Spanish legislation and that comes from the literal translation of the expression "Smart regulation" (Baldwin, R.,2005)., which arose in the scope of certain international organizations with economic projection, such as the OECD and the European Union. From the point of view of this regulatory trend, the exercise of regulatory power is a kind of last resort, so that before approving a new rule the need for it must be verified and be unappealable. The aim is to produce "quality legislation", i.e. efficient, effective and proportionate legislation that is not unduly burdensome for its recipients, especially those with economic interests. The idea is that the legislation adopted should be able to promote growth and promote development without interfering with economic agents.

The so-called" smart legislation", a term that we will adopt hereafter to simplify the references to this regulatory trend, makes use of the mechanisms of participatory democracy and reshapes and adapts them to their needs and objectives, integrating them into the regulatory procedures. The institutional proposals on the improvement of legislation call for streamlining the drafting of legislative acts, considering that it is not a matter of legislating more, but better. Precisely, the participation of citizens in the processes of production of law is one of the elements that characterize the call legislation intelligent, composed by rules whose approval has been planned in advance, that are intended to be effective and efficient by focusing on issues that really need a European Union regulation, that they try to anticipate the effects of such regulation on the sector that is projected, and which are subject to a continuous process of evaluation to detect those contents that can be improved or that are even unnecessary.

Logically, in order to achieve some ambitious goals, it is necessary that the institutions involved in the process of production rules take into account and assess different sources of information, which should not be the importance of the contributions of the general public and those affected by the legislation. It can therefore be said that the use of the mechanisms of participatory democracy is one of the presuppositions of intelligent legislation and one of the elements that define such legislation.

Consequently, the general interest appreciated by the public authorities no longer legitimizes the rules in itself. Norms must also be legitimized through the contributions of the subjects affected by them. The public authorities have lost the monopoly to determine the legal interests served by the rules, which have to be defined with the participation of external actors through the channels offered by the mechanisms of participatory democracy.

"Smart legislation", which seeks to rationalize the legal system and norms through normative processes that make use of the mechanisms of participatory democracy, is reshaping the legal system, not only at the European level, but also, by extension, at the internal level of the member states, including Spain.

This legislative formulation gradually leads to a normative context in which interventionist and restrictive measures are increasingly exceptional (Cuesta Lopez, V.,2010). This implies the adoption of rules whose contents are increasingly less prescriptive, where they are regulated by principles and where recourse to coercive measures is increasingly limited. Alternatives to classic regulation are sought and rules are disregarded when possible in favor of other regulatory models involving self-regulation and so-called "soft legislation". Where the requirement of the rule is unappealable, it seeks to limit its interventionist and coercive impact, as we have said.

This trend is very clear when it comes to the regulation of activities and sectors that have an economic background. However, paradoxically, when it comes to the regulation of matters relating to public order, this lax regulation often contrasts with rules that restrict the rights of individual or collective exercise and grant the public authorities important powers of sanction.

Among the mechanisms of participatory democracy in the service of intelligent legislation, the European documents basically refer to consultations, integrated into impact assessments, prior to the adoption of rules or after such approval, that is, once the rules have entered into force and have been in force for a certain period of time. To these must be added the European citizens ' initiative, whose well-understood virtuality can be likened to the consultations carried out in the early stages of the regulatory process, when the Commission is considering the preparation of a regulatory proposal.

The objectives and purposes of the "smart regulation" reshape these participatory tools, so that the query is integrated into the impact assessment procedures, and performed in a very early stage of the standard-setting process, when the standard is still in gestation, in phase pre-legislative, with the aim that the subjects really have the possibility to influence the content of the standard and to integrate their interests. In addition, however, consultations are envisaged not only at the time of adoption of standards, but once they have been adopted and are in force. The consultation of citizens extends to the evaluation of the rules in force to assess the desirability of their maintenance or the need for their repeal or modification if it is found that they have become ineffective or obsolete. The consultation procedures are structured so that the so-called interest groups or pressure groups gain weight, to have the possibility of influencing the norm, to the detriment of the citizens individually considered, becoming a sort of regulators in the shade, as the records of transparency are optional and not mandatory. All these circumstances pervert the very concept of participatory democracy, conceived and designed to enhance the direct collaboration of individuals and the associations that represent them with the public authorities.

Ultimately, the legal justification for the application of participatory democracy mechanisms in standard-setting procedures has several reasons and is the product of a synergy of factors that are both cause and effect of the normative outcome. In this paper, the aforementioned mechanisms are analyzed and their practical effectiveness is evaluated, highlighting the problems and shortcomings that prevent and hinder their complete effectiveness or distract them from the main objective for which they were conceived.

BACKGROUND

The process of European creation and integration has from the beginning raised doubts of democratic legitimacy arising from the very circumstances of its creation, because the European Union owes its beginnings to constituted powers, which are the governments of the Member States. Indeed, the process of European development and integration has been made possible by the agreement of the central governments of the states, agreements that have been ratified in each country according to its own internal constitutional system. Therefore, in the process of European integration there has not been a genuine constituent power, understood as a prior power or original power and source of legitimacy of the European Union and its institutions. The absence of a sovereign European people is absent in the process of European construction. Instead, it is the Member States that constitute the mandatory reference point in the European integration process. We can therefore say that the European Union originally lacked direct democratic legitimacy, since its existence derives from a constituted power and therefore its democratic foundation rests on the legitimacy provided by the governments of the Member States. This indirect legitimation has been perceived as lacking as the European Union has assumed greater powers.

The situation has sought to be corrected in parallel with progress in the process of European integration. Thus, as states have ceded sovereignty and the exercise of powers, the need to give greater legitimacy to the institutions that make European decisions and policies has also become more evident. These institutions need to connect with citizens to get the support their decisions need.

It is necessary to point out that this democratic legitimacy implies a connection between rulers and governed, and it manifests itself basically in two ways: the first, through legal mechanisms that allow citizens to choose their representatives and control their political and government action. And a second one that focuses on the effectiveness of the results obtained, when assessing to what extent the results produced by a given political structure reflect the demands of citizens and meet their needs. This second legitimizing mechanism of public power focuses on the decision-making process, and in this sense praises the adoption of legal acts that take into account the opinion, will and opinion of citizens, in a way that enhances democratic participation in this process.

The mechanisms of participatory democracy have been progressively incorporated into the European Union, although the use of representative democracy was the first attempt to address the deficit of democratic legitimacy that has been attributed to the European Union since its inception, and for this, the attention was focused on the European Parliament. Thus, the strengthening of democratic legitimacy has long been associated exclusively with this legislative institution. European leaders understood that the direct election of MEPs would give the EU the direct legitimacy it had previously lacked. Therefore, the elections to Parliament were the first attempt to directly involve the citizens in the European process and justified the progressive increase in the powers of this institution.

However, this strengthening of the role of Parliament has not been sufficient to strengthen the democratic legitimacy of the European Union, as we will see in the next sections. That is why other mechanisms have been established to encourage citizens' participation in the legislative and decision-making process, such as those deriving from the exercise of participatory democracy.

The creation of European citizenship, with common rights of democratic participation that can be exercised in the country of residence, undoubtedly represents a further step in the process of European integration by allowing citizens exercising their right of free movement and residence to participate in the public affairs of a country of which they are not nationals.

The legal status derived from citizenship implies that the exercise of certain political rights is dissociated from the nationality of the country in which they are exercised, so that a national of a state of the Union may exercise those rights in another Member State of which he is not a national but in which he or she lives. Thus, the right to participate in public affairs through active and passive suffrage in local elections in the country of residence implies the extension of the rights of its nationals to Community citizens who are residents in the country in question and entails the exercise of the sovereignty of the country of residence. Progress in the process of European integration entails changes at an internal political level that imply an attempt to integrate European citizens, equating them in the exercise of certain rights of political participation to nationals. The exercise of national sovereignty is no longer linked exclusively to nationality, but also allows for weaker connections such as domicile or residence. This is a consequence of the application of the principle of non-discrimination and is the corollary of the right of free movement and residence provided for in Article 8a of the EC Treaty (Council Directive 93/109/EC of 6 December).

Recognition of these rights is intended to enable all citizens of the Union, whether or not they are nationals of the Member State of residence, to exercise their right to vote in local elections in that state on equal terms with nationals. These rights allow citizens to become involved in democratic life and influence the decision-making process, allowing their political and social integration in the host country. But in addition, the institutionalization of these mechanisms also has a second aspect, since they help to create a common public sphere that aims to be one more element in the process of cohesion and belonging to a political community capable of catalyzing the process of European integration.

Title II of the Treaty on European Union lays down the provisions on the democratic principles governing the European Union. Its functioning is based on representative democracy, so that European citizens, in addition to being represented in their respective national parliaments, will be directly represented by the members of the European Parliament who vote every five years.

As we have said, at the time of the constitution of the European Communities there was no sovereign European people acting as a constituent power and the subsequent concept of European citizenship does not have this sovereign connotation either. European citizenship, derived from being a national of a member state, does not create a European people, but a link between the different European peoples. It is the citizens of the Member states, as constituent power and holders of national sovereignty, who have legitimized and continue to legitimize their states to join the European Union and move forward in the integration process. The fact that now, according to the Treaties, citizens are directly represented in the European institutions through the Parliament does not imply the creation of a European people as the sovereign, but merely that the mechanisms of representative democracy have been "transplanted" to the European level, with the logical constraints that derive from the very nature of things, creating a simile that from a formal point of view is tailored to the budgets of representative democracy, but that lacks some of their most basic budgets. Representative democracy within the European Union is imperfect, since it is articulated as a sum of democratic representations of the citizens of each of the Member States, despite the literal wording of Article 14.2 of the Treaty on European Union.

But, in addition to representative democracy, the Treaties open up to other forms of democracy that complement it. Thus, Article 10 of the Treaty on European Union introduces participatory democracy by stating that every citizen has the right to participate in the democratic life of the Union. Decisions will be taken as openly as possible to citizens and the institutions will maintain an open, transparent and regular dialogue with representative associations and civil society. And according to Article 11, the institutions shall give citizens and representative associations, through appropriate channels, the opportunity to express and exchange their views publicly in all fields of action of the European Union. As a corollary of all of the above, Article 11 of the Treaty on European Union introduces the European citizens ' initiative.

The above articles demonstrate that European democracy is not only based on the rules and principles of representative democracy, but also on the interaction between government and citizens that foster participatory democracy (Karatzia, A., 2017). The idea is to strengthen European democracy and give citizens the opportunity to participate in the development and functioning of the European Union. In the section that follows we develop the way in which European standards acquire democratic legitimacy through these two forms of citizen participation.

European standards are the result of a supranational legislative process involving European Union institutions, where citizens are represented, as is the case with Parliament, and other institutions that lack this representative connotation, such as the Commission and the Council.

And the first question that arises is whether Parliament's intervention can be understood to be sufficient to democratically legitimize European rules before the citizens. Needless to recall that the European Parliament acts as co-legislator alongside the Council in the ordinary procedure to which most legislative acts of the European Union are subject? The fact that the citizens elect the MEPs could be understood enough to understand that the normative manifestations of the European Union enjoy the legitimacy they need to impose themselves on the citizens. However, if we look more closely at the composition and functioning of this legislative institution, as well as the procedures followed for the adoption of rules, we will see that this supposed legitimization that is presumed to be European laws suffers from obvious limitations that have to do with the design and institutional architecture of the European Union (Espaliú Berdud, C.,2016).

THE EUROPEAN PARLIAMENT AS A SOURCE OF DEMOCRATIC LEGITIMACY FOR EUROPEAN STANDARDS

First of all, it must be said that the elections to the European Parliament do not really have a supranational or European dimension. On the contrary, they are considered an internal electoral issue, a type of domestic election in which the vote is determined by the action of the party that is in the national government at the time. The impact of election results at European level is irrelevant to voters who are unaware of the implications of representative democracy in the European Union. This circumstance is evident in the low turnout rate in the elections to Parliament, which shows how these elections are measured as second-order elections. The feeling of disaffection towards the European Union and its institutions is demonstrated by the scarce participation rates, fostered in part by the behavior of the parties and parliamentarians themselves, that must be more to the political formations that appoint them as candidates to the voters themselves. The national affiliation of political parties is also reflected in their weakness within the European Parliament. To have this Europeanist dimension, these elections should be genuinely transnational, with European ballots and a unified electoral law. However, there is no common electoral system, but parliamentary elections have an important national component to the extent that they are based on differentiated electoral systems. The democratic representation of citizens in the EU is not carried out in an abstract way, but by assigning it to each of the Member States.

But, in addition, the defects in the architecture of the institutional system also involve other European institutions. This is evident when we consider the process of electing the members of the Commission. It is true that as a result of the latest amendments to the Treaties, its president has been elected by Parliament on a proposal from the European Council, and this means that before his appointment, the candidate must explain his government agenda and how he intends to develop it. Thus, the president of the Commission appears endowed with an indirect democratic legitimacy, conferred through his parliamentary appointment. So it is a way of giving democratic legitimacy to regulatory proposals, and therefore to European standards from the outset.

The question to be asked is whether such indirect legitimation is sufficient, especially when we consider that the Commission has almost exclusively the power of legislative initiative (Radaelli, C. M., 2005). If the legislative acts resulting from integration can only be proposed by that institution, is the indirect legitimacy provided by Parliament through the election of its president sufficient?

Doubts also tarnish the rest of the members of the Commission. It is true that Commissioners must have the approval of Parliament, to which they must set out their political program, and it is also true that the president of the Commission can reject them, but in principle it is the European Council, and therefore the governments of the Member States, who propose and appoint them.

With regard to the control mechanisms of the executive, it should be noted that the Commission's actions, particularly in the run-up to the drafting of legislative proposals, are not sufficiently transparent to allow Parliament to exercise real control over them (Mendes, J., 2011), The meetings of the Commission are not public and are held in camera. This difficulty of controlling their actions is increased by the existence of regulatory agencies, created by the Commission itself and charged with carrying out functions of a governmental nature, but which at the same time exist outside the institutional framework. In this way, when one considers the role of Parliament as the institution responsible for exercising control over the European executive, it becomes more clear that the system of parliamentary supervision and control over the executive is imperfect.

Of course, all these circumstances are reflected in the normative statements of the European institutions and are all the more pronounced if we bear in mind that most Community rules are adopted on the basis of the ordinary legislative procedure, which in practice involves a legislative procedure in which both institutions, Parliament and the Council, have a similar specific weight in the production of rules. For this reason, the problems identified as a result of the form of election of the European Parliament and its modus operandi is to also add the problem posed by the other legislative institution, that is, the Council, endowed with legitimacy indirectly, of the second degree which is provided tangentially by the parliaments of the member states to elect the heads of government.

In addition, the Council of the European Union is not precisely characterized by its transparency, but by the closed nature of its decisions, since this institution has not yet made public the agendas of meetings, the legislative proposals of the presidency, or the texts of all the legislative amendments proposed by the governments that make up the Council. As if that were not enough, it should be noted that the Council of the European Union functions through committees and working groups, which are composed of officials from the Member States, not even politicians with indirect legitimacy, even of the third degree. This means that this institution is very far from the citizens, and therefore the regulatory decisions in which its action is translated are also so. Parliament's control of the Council is not provided for in the Treaties, so the motion of censure can only be put before the Commission.

In practice, this means considering that, of the two European legislative institutions, only one of them has been elected by the citizens, although both have the same legislative potential. Therefore, in the process of legitimizing European standards, only one of the intervening institutions finds the direct legitimacy to justify its participation in the process of normative production and this with the limits pointed out above. It is for this reason that the instruments of participatory democracy are necessary, since they have to complement the shortcomings that the system of representative democracy presents within the European Union at the level of Parliament.

NATIONAL PARLIAMENTS AS A SOURCE OF DEMOCRATIC LEGITIMACY FOR EUROPEAN STANDARDS

As we have indicated above, the democratic legitimacy of the European Union is indirect, which does not mean that it was originally insufficient, given the limited initial powers of the European Communities. However, its lack of legitimacy has become progressively evident as the process of European integration progressed. Resorting to the rhetorical figure of metaphor, we can imagine European evolution as a universe in the process of expansion that increasingly holds greater powers and that increasingly moves away from the citizens and therefore from its original source of legitimacy. Following the simile cosmic, the power of legitimacy, i.e. the link between the institutions and its citizens is becoming more and more weak as the European institutions to increase their public power and legitimacy is revealed as insufficient and unable to justify and sustain the gradual transfer of sovereignty, which extends to areas not just economic, but also political and affects the core of the sovereign power. This emerging deficit of democratic legitimacy is being addressed in two ways:

- Directly involving citizens through their democratic participation.
- Involving their representatives at internal level, which will lead to the strengthening of the role of national parliaments in the production and implementation of Community decisions.

In sub-paragraphs later we will discuss the mechanisms of participatory democracy that have been articulated with the purpose aforesaid, but by the time it is our intention to carry out some short reflections on the role of national parliaments in the process of production rules, a process that has been strengthened over time, and it has been consolidated with two protocols, relating to the role of national parliaments in the European Union (protocol No. 1), and the protocol on the application of the principles of subsidiarity and proportionality (protocol No. 2) .

The first of these aims to encourage greater involvement of national parliaments in the activities of the European Union and to increase their capacity to intervene in draft Community legislation. To this end, two distinct aspects are distinguished: information to national parliaments and inter parliamentary cooperation. The second of the aforementioned protocols lays down the procedure for national parliaments to monitor the application of the principle of subsidiarity in the early stages of the regulatory procedure.

In a legal system such as the European one, where the rules result from competence titles shared with the states in most cases, and where the states directly, or their citizens, are the addressees of such rules, the need and importance for the national bodies that have direct legitimacy to participate in the processes of production of rules cannot be ignored. And in this sense, participation implies more than the mere issuance of reports or opinions to show their conformity or non-conformity with the rule and their adaptation to the principle of subsidiarity. In this context, participation should involve genuine consideration of the objections raised by national parliaments leading to a reconsideration of the content of the draft legislation. However, nothing is further from reality, since the early warning mechanism is devised so that their true legal effectiveness requires at least a third of the total votes allocated to the national parliaments of the member states, in that the Commission will be obliged to reconsider its proposal.

The intervention of the national parliaments is very limited and circumscribed to the so-called early warning mechanism, whose effectiveness is conditional on the obtaining of a particular representative percentage of votes capable of being bypassed by the Commission, unless the opinions contrary to the principle of subsidiarity are backed by the 55% of the members of the Board or by a majority of the votes cast in the European Parliament on the terms stated in the Protocol for the application of the principles of subsidiarity and proportionality.

It should also be borne in mind that the principle of representative democracy which governs the European Union also requires the cooperation of national parliaments on the basis of the principle of inter parliamentary cooperation, as laid down in Protocol No 1. However, despite the various mechanisms of cooperation articulated, the establishment of effective cooperation between the internal parliamentary institutions and the European Parliament is far from effective in normative terms, and there may even be a certain rivalry between them.

In these circumstances the question whether the deficit of democratic legitimacy of the European Union and its decisions can be understood remedied by the intervention of the legislative bodies internal question that doesn't fit but give a negative response, and to ensure the democratic legitimacy of the Union is necessary in gear that coordinates the national parliaments, the European Parliament works perfectly.

THE ELABORATION OF EUROPEAN STANDARDS UNDER THE SLOGAN OF PARTICIPATORY DEMOCRACY

As opposed to the forms of direct or representative democracy, there is a third form that is participatory democracy, which, although it implies a direct participation in public affairs, does not imply the exercise of political power, nor does it constitute a manifestation of popular sovereignty, but a recognition of the capacity for collaboration that gives citizens the opportunity to provide the public institutions sufficient to ensure the success and opportunity (the efficiency and effectiveness if we use the terminology used for the concept of smart legislation) of the rule.

The forms of participatory democracy involve the collaboration of citizens with the public authorities, but not at any time replacing the decision-making capacity of the competent institutions, but simply as a complement to their action. In short, it is a question of complementing representative democracy with the mechanisms of participatory democracy. It cannot be denied that there is currently a progressive lack of interest on the part of citizens in public affairs, as well as a certain discredit on the part of public authorities, perhaps motivated by the lack of rapprochement or harmony between the public sphere and the true interests and concerns of citizens. Hence the need, not only to respect the mechanisms of participatory democracy already established, but also to expand them, even recognizing the initiative of citizens to submit to the public authorities' proposals of a normative nature.

Since the legitimacy of the European Union is indirect, the objective of this type of participatory mechanisms is to overcome the democratic deficit by promoting formulas that involve bringing citizens closer to the institutions (Renda, A.,2006). And in this sense, it is important that attempts to address the deficit of democratic legitimacy are realized in mechanisms with incidence of participation in the process of production of law because the democratic participation in the process of development of European standards allows you to influence the fundamentals and legal bases that legitimize it after the execution, implementation, and management of European policies and the adoption of standards of implementation and enforcement of such policies. In short, what that means is that, since the development of regulations, directives and decisions determines the functioning of the European Union, the policy-making processes that are open to the participation emphasize the legitimacy of the actions, policies and actions to take after the European institutions to bring causes of these general rules which have been previously agreed with the citizens.

The mechanisms of participatory democracy in the context of production of European standards allow you to precisely "agree" with the citizens, the actions of a European institutions whose existence and origin has been mediated by the action of the states, and that therefore they are away from the citizens in terms of democratic legitimacy, as citizens only indirectly involved in the process of the creation of such institutions. And if the democratic legitimacy is measured in terms of the relationship between rulers and ruled, and also in terms of the effectiveness of the decisions taken, it is evident that citizen participation and the mechanisms of participatory democracy allow you to connect, in each case in which such participation operates, to the citizens with the rulers, and at the same time are more of a factor with an impact on the success, opportunity, and ultimately the effectiveness of the adopted standards. Thus, although the democratic legitimacy of the Union is in origin an indirect legitimacy, the mechanisms of participatory democracy, partly make it possible to remedy this fact by establishing a link between the citizens and the European institutions through collaboration between the two.

EFFECTS OF PARTICIPATORY DEMOCRACY MECHANISMS ON EUROPEAN LEGISLATION

The regulation of the previous mechanisms of participatory democracy underlies an attempt to strengthen the legitimacy of the decisions taken by the Community institutions. Participatory democracy at the level of the European Union is a way of giving legitimacy to it through its rules rather than through its institutions, as is the case with representative democracy and elections to the European Parliament. In fact, participatory democracy increases the connection between rulers and governed, since the decision-making process also includes the citizens who are made to participate in the normative process (Xanthaki, H.,2013). It is another matter to assess the participation mechanisms put in place so far, and whether they really give citizens a real capacity to intervene in the legislative results and to have their opinions taken into account, issues that we will analyze in detail when referring to each of these instruments.

In short, the mechanisms of participatory democracy applied to rule-making procedures imply that the citizens of the Union have instruments expressly provided for to convey their views and interests to the Community bodies, and that these mechanisms are different from those provided for in the constitutional systems of the Member States. Consultations, or citizen initiatives institutionalize public participation by citizens. And unlike the mechanisms of direct democracy, they do not imply an appeal to the electoral body as the holder of sovereignty, nor do they imply the exercise of public power by citizens (Van Gossum, P., Arts, B., & Verheyen, K., 2010). They merely involve their collaboration with the public authorities and an expression of their interests on certain issues of European competence. The mechanisms of participatory democracy directly connect the institutions of the European Union with the citizens without the need for the mediation of the states and legitimize their actions.

Unlike the rights of political content, such as the right of active and passive suffrage in local elections, which involve the exercise of popular sovereignty, and that therefore must be regulated and recognized by each of the member states, the mechanisms of participatory democracy at the European level have been developed by the European institutions. In this way, the Community institutions have been developing a status specific to the European citizen, different from that of nationals of the Member States, although derived from this status and which allows the citizen to establish a direct relationship with the institutions of the Union.

These instruments enhance the democratic nature of the Union and institutionalize citizen participation in regulatory processes (Gunningham, N., & Sinclair, D. 2017). The European Union has created a system of democratic legitimacy own and specific to have unique mechanisms that are independent of those provided for in each of the constitutional systems of the member states, thus giving greater participation of citizens in the process of adoption of standards. These mechanisms help to create a common public sphere, which can be an additional element in the process of cohesion and a sense of belonging to a community capable of catalyzing the process of political integration. It is true that the sense of common identity is far removed from reality, especially after the economic crisis and the decisions taken by the European Union, which have accentuated its unpopularity. From a theoretical point of view, however, there is no doubt that the institutionalization of the mechanisms of European participatory democracy is an attempt to assist citizens in the integration process (Volkova, A. V., Kulakova, T. A., & Aliaskarova, Z. A., 2017).

In any case, increasing the influence and participation of citizens in the normative process is not intended to replace the mechanisms of representative democracy, but simply to respond to the demands for greater openness and transparency of public actions in a democratic system that is in the process of evolution.

Citizen initiatives and public consultations are mechanisms of participatory democracy. Both refer to the participation and collaboration of citizens in regulatory production, either on their own initiative (in the first case), or following an appeal from the European institutions in the case of consultations. However, while citizens ' initiative has been expressly regulated in the Treaties, public consultations are not subject to detailed legal treatment in the Treaties, being regulated generically by the precepts recognizing the right of every citizen to participate in the democratic life of the Union.

It is curious to note how the Treaties have preferred to highlight the figure of the citizens ' initiative, much more pompous and striking, although of very limited real effectiveness, and to leave public consultations in the background, despite the fact that this mechanism is part of the legal tradition of most states, including Spain, which recognize it even at the constitutional level in some of its variants, such as the public hearing.

CONCLUSION

The mechanisms of participatory democracy are key to the consolidation of the concept of European citizenship, that is, to create a European awareness of belonging to a project and a common political community that helps to build the demo absent in the European Union.

The challenge is complicated, because we are talking about entrusting instruments of democratic participation and the rights resulting from them with the capacity to bring together a collectivity composed of different nationalities, with a different historical and cultural background. There are no precedents in history that can predict the outcome. What is true is that the European Union can only be sustained if it is legitimized in the eyes of its citizens. Citizenship must underpin the European edifice, but that citizenship must evolve. If the edifice is to be well established, European citizenship cannot be subject to nationality, nor can it be a concept derived from it, but must be a concept built on the basis of democratic rights that create awareness of belonging to a common political system and that transcend the idea of nationality of a member state.

The exercise of democracy through mechanisms of citizen participation that create collective rights of individual exercise is not only a way of bringing the European Union closer to its citizens and legitimizing it before them, but also a way of starting to build a concept of citizenship on a more solid basis than those provided by nationality. In this sense, the theme of the latest citizen initiatives presented reveals a certain awareness of belonging to a political community. The consolidation of citizenship through the exercise of European democracy is certainly a difficult path to take, but in my opinion, the only viable option to advance the process of European integration. The European edifice must cease to be based on the states, but must be based on their citizens. Only through the creation of a European demo can political integration become a reality, and in this sense, democracy is the way, because to paraphrase the Greeks (money for money), democracy also generates more democracy.

In short, we advocate the potential of the instruments of participatory democracy to make citizens participate in the European project and involve them in it on the basis of the idea that the EU must be legitimized by the citizens, that is, it must have their support and support. Participatory democracy is the most powerful instrument currently available to the European Union to consolidate the concept of European citizenship. Citizen participation increases the connection between rulers and ruled, as well as the effectiveness of the results obtained by the management of the European institutions, as in the decision-making process are also present citizens who were made partakers of the legislative procedure. The institutional design of a political system must allow a certain balance between the political capacity of the institutions and the citizens, and in this sense, the Treaty of Lisbon sought to democratize integration by promoting the participation of citizens in the construction of the European process. It cannot be said that through this system the European Union and its regulatory decisions will be able to gain direct legitimacy, but it can be said that through this system the European institutions are a little closer to legitimizing themselves before the European citizens, as well as creating a feeling of the European people that can be decisive in the evolution of the European Union.

Through the instruments of participatory democracy, the democratic nature of the European Union is increased and public participation in decision-making is institutionalized. The European Union has its own and specific system of democratic legitimation by having its own democratic mechanisms that are independent of those provided for in each of the constitutional systems of the Member States. These mechanisms are undoubtedly one more element in the process of cohesion and a sense of belonging to a community capable of catalyzing the process of political integration.

The concept of intelligent legislation refers to a type of rule "thought" (not improvised), both in terms of its substantive content, and the anticipation of its possible effects and that also establishes mechanisms to be evaluated and evaluated throughout its validity. In short, the term "smart legislation " promotes the controlled creation of "quality" regulations in a context of regulatory hyper regulation.

The quality of the standard is determined by the ultimate objective of that regulation, i.e. to adopt legislation that is as efficient and effective as possible for the implementation of Union policies, that is simple and clear, that avoids overregulation and administrative burdens for citizens, administrations and businesses (especially small and medium-sized enterprises) and that is designed to facilitate their transposition and application and to strengthen the competitiveness and sustainability of the Union economy . The concept of smart legislation therefore has an economic background, which seeks to rationalize the adoption of new rules and defends the idea that it is not a question of regulating more, but of doing so in the most effective way, considering the importance of maintaining an attractive and flexible economic environment for the private sector.

The term originates in the context of the European Union and other international organizations, such as the OECD. Regulatory policies based on this concept start from the idea that legislation must serve economic purposes, and therefore laws must be aimed at creating the right conditions to improve national economies and promote economic growth, or at least not hinder it. The aim is to adopt rules that promote and facilitate growth and competitiveness, adopting legislation that is appropriate to achieve these ends. Under these premises, laws serve economic objectives and must create a favorable regulatory environment for business development, especially for small and medium-sized enterprises. To achieve these premises, smart legislation applies democratic principles and resorts to citizen participation and participatory democracy mechanisms. In doing so, the European institutions aim to gather information that will make it possible to make informed decisions that are more effective in terms of opportunity, usually economic. Citizen participation is not an objective in itself, but a consequence of the implementation of the principles that govern this regulatory policy.

With this aim of refining the legislative technique and ensuring its substantive quality, regulatory production procedures are subject to a series of principles that are basically based on four pillars:

- Improved legislative planning, so that rule-making is not the result of improvisation, but follows a pre-conceived and pre-conceived normative road map. And this planning extends not only to the stage prior to the adoption of the standard, but also to a later stage, on the basis of which the obligation is established to periodically evaluate the application of the standards in force with a view to their revision, amendment, recast, or even repeal.
- The establishment of a series of principles of good regulation, such as necessity, proportionality, legal certainty, transparency, accessibility, simplicity, efficiency and effectiveness, which will constitute the parameters against which to evaluate both current and emerging legislation. Thus, the legislator will have to state in the preamble or statement of reasons of the rule how it satisfies these principles throughout its validity.
- Subjecting standards to impact assessments, both ex ante and ex post, and setting up bodies to assess the correctness of such assessments.
- The increase of the participation of the citizens, gathering your opinion from the earliest stages of production of the standard, and even prior to the existence of the proposed regulations (prior consultation), so that the institutions can know in advance the problems that have to be addressed, as well as the possible alternative solutions, regulatory or otherwise, that may arise to meet the intended goals.

REFERENCES

Baldwin, R. (2005). Is better regulation smarter regulation? *Public Law*, (Autumn), 485–511.

Bouza Garcia, L. (2015). *Participatory Democracy and Civil Society in the EU: Agenda-Setting and Institutionalization*. Palgrave Macmillan. doi:10.1057/9781137436849

Cuesta Lopez, V. (2010). The Lisbon Treaty's provisions on Democratic Principles: A legal framework for Participatory democracy. *European Public Law*, *16*(1), 123–136.

Espaliú Berdud, C. (2016). A reflection on the participation of individuals in the creation of European Law through the European Citizens´ Initiative and its scope in International Law. *Cuadernos Europeos de Deusto*, *54*, 181–202.

Gunningham, N., & Sinclair, D. (2017). Smart regulation. Regulatory theory: Foundations and applications, 133-148.

Karatzia, A. (2017). *European Citizens´ Initiative. General Court rules of the Commission's obligation to give reason for refusing to register proposals*. http://eulawanalysis.blogspot.com/2017/02/european-citizens-initiative-general.html

Mendes, J. (2011). Participation and the role of law after Lisbon Treaty: A legal vie on article 11 TUE. *Common Market Law Review*, *48*, 1849–1878.

Prieditis, M. (2002). Participation in the legislative process and new information technology. *Proceedings of the Fifth Congress of the European Association of Legislation in Athens, Nomos*, 99-118.

Radaelli, C. M. (2005). Diffusion without convergence: How political context shapes the adoption of regulatory impact assessment. *Journal of European Public Policy*, *12*(5), 924–943. doi:10.1080/13501760500161621

Renda, A. (2006). *Impact Assessment in the EU: The State of the Art and the Art of the State*. CEPS.

Van Gossum, P., Arts, B., & Verheyen, K. (2010). From "smart regulation" to "regulatory arrangements". *Policy Sciences*, *43*(3), 245–261. doi:10.100711077-010-9108-0

Volkova, A., Kulakova, T. A., & Aliaskarova, Z. A. (2017). "Smart Regulation" and Models of Civil Cooperation in the Context of Global Challenges. In Globalization and its Socio-Economic Consequences (pp. 2903-2910). Academic Press.

Xanthaki, H. (2013). European Union legislative quality after the Lisbon Treaty: The Challenges of smart regulation. *Statute Law Review*, *35*(1), 66–80. doi:10.1093lr/hmt005

Chapter 9
Non-Financial Reporting:
A Comparative Analysis With a Focus on the Combat Against Money Laundering – Evidence From Major European Banks

Catia Nunes
Nova School of Business and Economics, Portugal

Leonor Fernandes Ferreira
https://orcid.org/0000-0002-6544-7511
Nova School of Business and Economics, Portugal

ABSTRACT

This chapter analyzes disclosures on anti-money laundering made by the largest banks in Germany, France, and the United Kingdom. Directive 2014/95/EU has transformed the way banks with more than 500 employees disclose non-financial information. The findings show that between 2013 and 2017 compliance has increased, while understandability has remained the same. Despite the fact that a common regulation is applied, this research highlights differences and similarities in disclosures of non-financial information among banks. It adds an overview of the non-financial reporting in the banking sector to the literature, which is not limited to a specific country.

INTRODUCTION

As Rousseau once said, reports are meant to be "the eyes of the public" (Rousseau, 1772).

Since the financial crisis of 2008, the reputation of the financial sector has been damaged enormously. There has been a loss of trust on the part of stakeholders as well as gigantic financial losses. As Groenfeldt (2017) pointed out, trust is always central to the debate on the banking industry. Therefore, it is an indispensable area to study, even almost fourteen years after the crisis. This is specially to evaluate what they were willing to report at the time in order to increase transparency and understandability and what they must report nowadays in order to comply with the new regulation.

DOI: 10.4018/978-1-7998-9410-0.ch009

The annual report has been an important source of information about the financial position of a company, its economic performance, and changes in its financial position. The general purpose of financial reporting is to provide financial information about the entity to potential investors, lenders, and other creditors (IASB, 2008). The annual reports include financial information and non-financial information. Financial information had played a key role in companies' annual reports. However, non-financial information has been gaining attention, interest, and importance (Accountancy Europe, 2017)[1]. A possible reason is the growing demand not only from shareholders but also from other stakeholders for more information about non-financial data to better comprehend the company and to deal with expectations regarding the future. This information relates to environmental and social issues, as well as risk management and corporate governance issues among others, therefore facilitating the task of informed choices (EY, 2016).

Given the inadequacy of capacity of value creation, the Global Integrated Report (GRI) was published, and the European Union (EU) issued Directive No. 2014/95/EU about the disclosure of non-financial information. The Directive emphasizes the need for improving company disclosure by adding non-financial information to which the various stakeholders give importance, to increase the understandability and comparability of the information disclosed among companies and across countries. Larger EU companies, which are those with more than 500 employees (the "larger companies"), must apply the 2014 EU Directive from the year 2017 onwards.

This paper is therefore timely. It looks at the disclosure of non-financial information by eight European banks from France, Germany and the United Kingdom which represents 62% of the total assets of the top 3 largest banks in Europe. It analyses the framework for disclosure, the compliance level interns of the EU regulation and some characteristics of the information disclosed by banks, namely understandability and comparability.

The study focuses on the largest European banks in Germany, France and the U.K., which are obliged to comply with the new regulation. Moreover, these banks need to comply with the new disclosure inserted into national legislation by the year 2018.

Begina (2016) and Venturelli, Caputo, Cosma, Leopizzi and Pizzi (2017) studied non-financial information in the context of Directive 2014/95/EU. Both studies focus on specific countries, respectively Italy and Greece, and both compare the *national vs international situation*. This paper is not limited to a specific country and adds to the literature by providing an overview of non-financial reporting in the banking sector in three additional countries: France, Germany, and the UK.

Directive 2014/95/EU[2] refers to the combat against money laundering as part of the non-financial information that has to be disclosed, providing protection not only against the misuse of national financial systems but also against criminal activity and, potentially, against terrorism (Goodman, 2015). In this respect, this information has an added value as a bulwark for the welfare of society. Hence, this research focuses on *anti-money laundering* disclosures, with the aim of evaluating compliance with the new EU regulation, as well as understandability in non-financial reporting and comparability. For this purpose, an analysis of annual reports was performed.

The paper proceeds as follows. Section two introduces the conceptual and regulatory framework for *anti-money laundering* disclosures and provides information about the regulation in force (EU Directive). Section three reviews the empirical literature about *anti-money laundering* while Section four outlines the research questions, methodology, sample, and data. Section five answers the research questions and discusses the results. Finally, Section six presents the main conclusions of the study, its limitations and gives suggestions for future research.

CONCEPTUAL AND REGULATORY FRAMEWORK

The term *Non-Financial Information* can simply be understood as information on everything that is not linked to finance. However, such a broad definition is open to various interpretations. According to the Federation of European Accountants (FEE), "non-Financial Information is all the information disclosed by the company that cannot be explained with a currency. In spite of the appellation, it can have financial consequences such as those regarding the profit and loss account or the market value of the company due to reputational damage" (FEE, 2016: 6).

Tarquinio and Posadas (2020) concluded that the meaning of non-financial information is still ambiguous as there is not a generally accepted definition of the term. They also concluded that the term can be used to refer to information about society and the environment in general, or it can be understood as coming under corporate social responsibility issues and intellectual capital information.

The major issue at stake with this concept is the value of the disclosed information. Therefore, making a distinction between non-financial and financial information is imperative. The former influences decisions taken, and informs providers of corporate finance. Insights on internal and external policies, values, management, and perspectives are within the scope of this definition (FEE, 2016).

After the 2008 financial crisis, *transparency* has become a hot topic within financial institutions. *Transparency* is defined as "the *availability* of information to the general public and *clarity* about government rules, regulations and decisions" (Asian Development Bank, 1995:11).

The consequence of the indisputable importance of transparency nowadays is more legislation about disclosure of information and the increase of investors' interest in non-financial data. Eccles, Krzus and Serafeim (2011) forecast that "market interest in non-financial data will increase exponentially as more companies disclose more non-financial information, more knowledge is developed by research and teaching programs in business schools, and more investors develop more sophisticated valuation models." (Eccles, Krzus, & Serafeim, 2011:127). The number of companies disclosing non-financial information has grown significantly, from 30 in the early 1990s to more than 6,000 in 2010 (Serafeim, 2014), demonstrating the growing importance that stakeholders are giving to this type of information.

However, there is a lack of comparability and understandability in non-financial disclosures (Burrit, 1999) and it has become clear over the years that voluntary transparency has reached its limit (EY, 2017).

Comparability and understandability are key concepts in the motivation for approval of regulations in this matter. According to IASB (2018:2.24-2.25), comparability is described as the characteristic that "enables users to identify similarities in, and differences among, items." and understandability assumes that "classifying, characterizing and presenting information clearly by and concisely makes it understandable." (IASB, 2018:2.34-2.36).

The purpose of the Directive is to overcome the lack of trust in the information disclosed on the part of stakeholders, consumers, and investors. Each member state has to approve national legislation in accordance with the Directive and they will adopt different approaches (options) when making changes to the legislation. Thus, there will be variables such as in the definition of a large undertaking, public interest entity, report topics and content, reporting framework, disclosure format and diversity reporting required, could differing from member state to member state (CSR Europe and GRI, 2017). Nevertheless, the United Kingdom imposed regulation regarding non-financial information in 2016 and France and Germany in 2017. All three countries have similarities and differences in the application, presentation, and framework they rely upon to follow the Directive. Appendix 1 summarises these aspects.

To enhance understandability and comparability in non-financial information disclosed by EU companies, the EU approved Directive 2014/95/EU. The new regulation is an amendment to Directive 2013/34/EU regarding disclosure of non-financial and diversity information by certain large undertakings and groups. According to Article No. 1 of the Directive, this new regulation applies to EU member states, more specifically to their Public-Interest Entities (PIE) with an average number of 500 employees during the financial year. These entities are required to disclose information in a consolidated non-financial statement on several non-financial matters, with the aim of providing an understanding of the undertaking's development, performance and position, and of the impact of its activities. The Directive gives a typology of non-financial information and helps to define its categories by stating that at a minimum amount of information must be provided on the following subjects: environmental matters, social and employee-related matters, respect for human rights, and anti-corruption and bribery matters. Besides a brief description of the company's business model for each of the above matters the following information disclosures are required: a description of policies, including due diligence processes implemented; outcomes of these policies; related risks and how the company manages them; non-financial key performance indicators relevant to the particular business (Directive 2014/95/EU: Article 19a).

As to the framework used to report non-financial information, the Directive is vague, mentioning that "undertakings (…) may rely on national frameworks, Union-based frameworks (…), or international frameworks (…)" (Directive Nr. 2014/95/EU). The frameworks for reporting and verifying non-financial data are the Global Reporting Initiative, Integrated Framework, Global Compact, AA 1000 and ISAE 3000 (Deloitte, 2015). That leaves room for wide interpretation resulting in several types of frameworks around non-financial information reporting. To overcome this problem, a possible solution would be to create a global framework for non-financial information reporting, with clear guidelines and structure, like IFRS[3] Standards (Accountancy Europe, 2017).

Some of the challenges of non-financial information are related to the age of a company: startups may be facing issues related with definition, data collection, governance, and organizational competence. On the other hand, companies which have already started the process may face challenges related to reporting complex supply chain disclosures (EY, 2016). Deloitte (2015) refers to benefits of reporting non-financial information, such as the reputation of the business, generation of positive exposure in the media, enhancement of relationships with customers, suppliers and stakeholders and improvement of efficiency and process management. Recent research has shown that transparency also leads to better performance as the disclosure of environmental, social, and governance (ESG) has positive economic effects: "higher levels of transparency reduce informational asymmetries between the firm and investors, thus mitigating perceived risk." (Cheng, Ioannou & Serafeim, 2014: 27). Moreover, investors can better assess the opportunities and risks of future investment if they are provided with an insight into the policies and performance of non-financial aspects of the business (EY, 2017).

LITERATURE REVIEW

Early empirical research about non-financial disclosures focused on companies' voluntary disclosure practices, as there was no regulation imposing disclosure of such information. Nonetheless, since mandatory non-financial disclosure emerged, studies have revealed that this change has forced companies to improve their performance as regards the environment (Delmas, Montes Sancho & Shimshack, 2010).

These facts lead us to ask about the potential externalities of voluntary versus mandatory disclosure, and whether more regulation means companies disclose more information and of a higher quality.

Corporate Social Responsibility (CSR) is clearly based on a voluntary approach (European Commission)[4]. Dincer (2011) reveals that this voluntary basis of CSR leads shareholders to recognize the positive impact of disclosing this type of information, given the usefulness of data disclosure to enhance reputation and to satisfy the stakeholders' needs for more corporate information. Additionally, several externalities from non-financial reporting could arise. Sial, Zheng, Khuong Nguyen and Usman (2018) concluded that firms with higher performing firms report more on CSR activities than lower performing ones. Similarly, Fijałkowska, Zyznarska-Dworczak and Garsztka (2018) found that higher performing banks located in Central and Eastern Europe were more financially efficient and more conscientious in reporting CSR activities, indicating that the larger the bank (in financial terms), the more it shows concern and openness to non-financial, corporate and social responsibility issues. De Groen et al. (2020) gathered survey responses from more than 200 companies as part of the ongoing implementation of the Directive. In two-thirds of the surveyed companies the obligation to publish a non-financial statement had contributed to changes in internal practices and procedures.

The empirical literature regarding the disclosure of non-financial information has shifted to the discussion on the benefits and consequences of mandatory versus voluntary reporting. Aksu and Espahbodi (2016) for example, analyzed whether the mandatory or voluntary regulation provides better quality of disclosure and found that mandatory implementation of IFRS had a positive impact on Transparency & Disclosure for a sample of companies listed on the Istanbul Stock Exchange. Carungu, Di Pietra and Molinari. (2021) studied 184 non-financial reports from a sample of 92 Italian companies that have been previously disclosing non-financial information on a voluntary basis. The research did a content analysis and concluded that "the quality of non-financial information does not increase when moving from a voluntary to a mandatory basis". In another study, Begina (2016) analyzed the quality and comprehensiveness of sustainability disclosure in the voluntary and mandatory cases. The author compares the level of non-financial disclosure of Greek banks with that of the largest European ones, using an "evaluation tool based on the reporting principles included in the GRI G4 Guidelines and conducted an evaluation of stand-alone CSR annual reports published by three Greek banks and the ten largest European banks" (Begina, 2016:4).

Shehata M. (2020) studied the relationship between the elements of non-financial information and the mechanisms to address the risks of management fraud. A positive relationship was found which had a statistical significance among the mechanisms of processing the risks of management fraud and all the elements of non-financial information related to the internal control system dealing with environmental issues. The importance of non-financial information in the field of control systems was highlighted, particularly in reducing the risks of management fraud (Shebata, 2020).

Venturelli, Caputo, Cosma, Leopizzi and Pizzi (2017) evaluated the information gap for Italian companies and the adjustments required by the new EU Directive No. 2014/95/EU. To analyze the diversity and level of non-financial disclosure, the authors used the Non-financial Information Score (NFIS), a model that records the required information as a percentage. In a similar research, Lourenço (2017) analyzed the gap of information from companies listed on Euronext Lisbon, combining content analyses with the NFIS. These two papers, though insightful and relevant, only provide comparative views between a specific country and the largest companies in Europe.

To the best of our knowledge, empirical research exclusively focusing on the largest banks within the European Union does not yet exist, despite the new regulation - the Directive 95/2014/EU. Thus, this paper intends to bridge this gap and contribute to a better understanding of the non-financial information scenario in the largest banks of the European Union.

RESEARCH DESIGN

The purpose of this paper is to get an overview of non-financial reporting by the largest banks in the European Union from France, Germany and the UK. More specifically, it aims to characterize banks' disclosures on the combat against money laundering, by major European banks. The paper also assesses how the largest European banks' compliance level from France, Germany and the UK evolved from 2013 to 2017. The paper analyses understandability and comparability of information disclosed by banks and offers an insight on the framework banks used to disclose non-financial information by using descriptive statistic based on non-parametric statistics.

The research questions addressed are as follows:

RQ1: *Which non-financial information required by the Directive are banks disclosing?*
RQ2: *What format do banks use to disclose the non-financial information?*
RQ3: *How much non-financial information are banks disclosing?*
RQ4: *To what extent are banks complying with the new regulation?*
RQ5: *Is non-financial information understandable?*

Sampling Criteria

The universe of this research is composed of the 50 largest banks in Europe (see Appendix 2). All of them have more than 500 employees[5], the lowest limit for the mandatory disclosure of non-financial information. Out of this universe, only banks with headquarters located in the European Union were selected, to make sure that they apply EU Directives and EU regulations. All the banks selected will have shares listed in a Stock Exchange within the EU, so that all of them must report in accordance with the IFRS set by IASB. Therefore, seven banks were excluded,[6] leaving the sample with the 43 largest banks in Europe. France, the United Kingdom and Germany were selected since these countries contain the highest number of banks on this list, representing 62% of the total assets of the top 43 largest banks in Europe. Moreover, based on size criteria, the three biggest banks from each country (proxied by total assets) were selected for analysis; six of these banks are among the ten largest European banksAccording to Eurostat (2017), these countries have the largest GDP in the European Union[7]

Larger European banks generally offer better access to information and have more investors. They often serve as a benchmark for smaller ones and have much more to lose in terms of financial and non-financial consequences, given the scale of capital and reputation. The total assets of the nine banks included in the sample amount to more than 69 per-cent of the total assets of banks based on the three selected countries (see Appendix 3 and Table 1). The research covers the five-year period ranging from 2013 to 2017. The first year was immediately prior to the approval of the Directive 2014/95/EU and the last year when France and Germany imposed regulation on non-financial information. Moreover, all of these banks were audited by one of the Big Four companies (Deloitte, EY, KPMG or PwC) in 2017, meaning that the audit would not differ much from bank to bank.

Table 1. Sample characteristics

Bank	Country	Average of Total Assets*	Average of Number of Employees**	Year Country Law: Applicable
HSBC Holdings	U.K.	2,522	246,772	2016
BNP Paribas SA	France	1,982	190,000	2017
Deutsche Bank AG	Germany	1,603	98,955	2017
Société Générale SA	France	1,307	134,960	2017
Barclays Plc	U.K.	1,214	128,709	2016
Lloyds Banking Group	U.K.	827.8	76,913	2016
Commerzbank AG	Germany	514.4	50,028	2017
DZ Bank Group	Germany	442	29,641	2017
* 2013-2017 in bn EUR	** 2013-2017			

Source: (The authors)

The evolution of disclosure of non-financial information by the largest European banks in France, Germany and the UK is also studied, as changes occurred during this period and the diversity of information disclosed can be seen. Future research should explore the evolution from 2017 onwards.

Data Collection

The consolidated annual reports of each bank from 2013 to 2017 were downloaded from the banks' websites, and data was later retrieved from the 45 annual reports. Data was hand-collected and a content analysis on the reports was carried out, resorting to the word-search program embedded in Adobe Reader to locate key words such as *non-financial risks*, *risk appetite*, *financial crime*, *anti-money laundering*, *corruption*, *anti-bribery*, *core-markets*, *sanctions*, *KPI (Key Performed Indicator)* and *business model.* Given the time-consuming analysis which consisted in the review of 16,996 pages from 45 reports, the word-search program revealed to be an appropriate, feasible and thepractical choice of program to give answer to the amount of review the enormous performed and to overcome the lack of consistency in the reports when providing information on non-financial related topics.

During the data collection task, some difficulties arose. Of the nine banks selected for analysis, one bank (Crédit Agricole) did not present the annual report in a separate pdf file readily available for downloading from the bank's website. On the contrary, the bank opted to have all the information inside the website itself and scattered throughout the website. This fact justifies the exclusion of this bank's reports from the sample, given the lack of comparability and consistency and the risk of not finding all the information disclosed.

The length of the annual reports is one aspect that interfered with the data collection, since they are sizeable documents containing a large amount of information. Moreover, the same topic is mentioned in a different perspective several times, making the analysis and evaluation scores more difficult. One example of differing denominations is the name of the committee in charge of non-financial subjects, such as anti-money laundering or corruption. Frequently, it has different names across different banks and, sometimes they are part of other committees: COMMERZBANK with the Group Operational Risk Committee; Lloyds Banking Group with Anti-Money Laundering Officer and deutsche bank with the Group Financial Crime Governance Committee. Nevertheless, Appendix 5 shows how the understandability level was studied.

Variables and Score Analysis

Data was collected for the information presented in Appendix 5. The non-financial information is analyzed on the following variables: Framework used (RQ 2), Comparability along timelines and between banks and the Quantity of non-financial information banks disclose (number of pages) in relation to the report as a whole (RQ 3), Level of compliance (RQ 4), and Level of understandability (RQ 5).

To analyze the level of compliance (RQ 4), a "Score Analysis" was designed. This score includes five assessment grids that cover the disclosure requirements stated in Directive 2014/95/EU, namely: a brief description of the group's business model (*g1*); a description of the policies pursued by the group in relation to those matters, including due diligence processes implemented (*g2*); the outcome of those policies (*g3*); non-financial risks (*g4*) and non-financial key performance indicators relevant to the particular business (*g5*).

For each grid, a compliance percentage is computed by assessing the presence of the above-mentioned items. (See Appendix 5). The score was computed as follows:

$$S = \frac{\sum_{i=1}^{5} d_i}{\sum_{i=1}^{5} r_i} \qquad [1]$$

S: Compliance level's score; *i:* Number of grids; *r:* Total number of points in each grid required by the Directive; *d:* Total number of points in each grid disclosed by each bank.

Points were assigned to measure the degree of completeness in the annual reports to match the level of compliance with the requirements in the Directive. A binary scale (where 0 indicates the absence of disclosure and 1 indicates full disclosure) is used to designate a specific score to each section and generating an overall level of compliance as a percentage (by dividing the point score for each section by the maximum possible score, which is 28 points). This methodology is based on Venturelli et al. (2017).

Regarding the level of understandability (RQ 5), the same type of analysis was used.

For this purpose, four parameters were assessed. The first parameter consists in assessing whether the report contains the level of information required by stakeholders, avoiding excessive and unnecessary details. The second parameter evaluates whether the report is reader-friendly, such as containing diagrams, a quick overview of the topic, tables, and a glossary. The third parameter checks whether the reports avoid the use of technical terms, of acronyms, jargon, or other terminology likely to be unfamiliar to stakeholders. Finally, the fourth parameter assesses whether the information in the report is presented in a format that allows users to see positive and negative trends in the performance on a year-to-year basis.

The Likert Scale was used because it is highly flexible, provides quantifiable data about what would otherwise be merely a subjective opinion. It is one of the best-known scales and, therefore, easy to read and interpret. However, this scale requires a great deal of decision-making, and it can take a long time to analyze data (Likert, 1932).

As for the understandability of data, a univariate and bivariate analysis was used. Univariate analysis helps to better understand the data and to provide an overview of non-financial information issued by banks. To deepen the knowledge and comprehension about the characteristics of non-financial disclosure by banks, a bivariate analysis was done (contingency tables, correlation analysis between the compliance level and the ratio between pages of non-financial information, and the total number of pages in the report and the size of the bank). As the research is based on a small sample, non-parametric statistical tests were conducted, such as the McNemar test, which uses contingency tables with a dichotomous trait.

RESULTS

Content of Non-Financial Information Disclosures (RQ1)

According to the five grids that focus on combating money laundering, corruption and bribery mentioned in section 4, the largest European banks generally disclose similar types of information over time regarding the business model (grid 1). Scores are generally high and remain steady with time.

Table 2. Scores per type of non-financial information

Type of Non-financial Information	2013	2014	2015	2016	2017	Standard Deviation
Brief description of the group's business model (Grid 1)	0.76	0.78	0.79	0.76	0.76	0.011
Description of policies pursued by the group (Grid 2)	0.25	0.42	0.46	0.56	0.60	0.124
Outcome of policies (Grid 3)	0.00	0.08	0.23	0.25	0.35	0.126
Non-financial risks (Grid 4)	0.97	0.97	1.00	1.00	1.00	0.015
KPI - Key performance indicators (Grid 5)	0.13	0.21	0.29	0.33	0.33	0.081

Source: (The authors)

In the case of money laundering, data on the policies pursued by each bank (grid 2) shows a clear tendency of moreinformation bank disclosures about money laundering, corruption and bribery. However, two banks are exceptions to this trend. One bank (DZ Bank) has not disclosed any information in this grid since 2013 and Lloyds Bank, despite having a high score, seems to have no intention of disclosing further information, given that its score and the format of information disclosure does not change over time. The policies themselves are quite similar, going from detection and preventative tools to mandatory training for employees, and additional policies on KYC (Know You Customer) and increasing compliance levels.

With regard to grid 3, in which a description of the policies pursued by the group is required, the general conclusion leads us to believe that banks do not disclose much information about the outcome of the anti-money laundering policies. Nevertheless, this type of disclosure has been increased in the case of five banks.

Regarding non-financial risks (grid 4) all the banks achieve a high score, leading to the conclusion that banks have already been voluntarily disclosing information on this issue since 2013. In this respect, regulation did not change during the period of analysis. All banks in this study had, in some way, mentioned operational risk and financial and non-financial consequences of this risk. Therefore, there is an embedded culture regarding operational risk within these banks.

Finally, the fifth grid, KPI about anti-bribery, anti-corruption, and anti-money laundering has the lowest score. It appears that banks are not concerned about the disclosure of this type of information, as the scores remain steady throughout the period. The KPI's grid comprises three topics: anti-money laundering KPI, risk indicators and the outcome achieved by anti-money laundering (AML) policies. None of banks mentioned AML KPI, but with time, the risk indicators' topic began to appear. Deutsche Bank and HSBC are the only banks that in recent times have introduced some of the outcomes achieved by anti-money laundering, corruption, and anti-bribery policies. However, these outcomes refer to specific issues in the past, such as those already solved or those in the process of being solved, like the FIFA sanctions scandal in HSBC.

This leads to the belief that there is room to improvement and for more transparency in the matter of corruption, given that the Directive does not highlight any topic in particular (environmental, social, employee matters and human rights), as was mentioned in section 2.

Notwithstanding, this is an ongoing banking and national issue. The United Kingdom imposed regulation on non-financial information in 2016, and France and Germany did so in 2017, providing ample space for improvement and changes in the forthcoming years, hence future research could cover the differences of the disclosures made by the banks before and after the imposition of the regulation.

As a consequence, andaccording to Venturelli, et al. (2017) an information gap remains in topics such as anti-money laundering and bribery and KPIs, although the implementation of the Directive should help to fill this gap in the coming years. In fact, from the year 2017 banks did start to disclose more information.

Framework USED in Disclosures (RQ2)

All banks in the sample disclosed non-financial information every year from 2013 to 2017. The difference is in how they did it. Such a divergence was expected since, as mentioned before, the Directive itself gives freedom in this aspect and several methodologies that can be used in reporting (for instance, the Global Reporting Initiative, the Integrated Framework, among others).

Different formats of presenting the information were observed. Seven out of eight banks disclosed non-financial information in an integrated report. One bank, Deutsche Bank, disclosed non-financial information in a separate report with all the non-financial information shown together. Each bank adopted the same framework throughout the years 2013 to 2017, indicating that banks have not changed the manner in which they disclose information. In general, the integrated report is an extensive document (with an average of 390 pages in 2017[8]) where banks disclose all financial and non-financial information of the bank itself. As a result, this type of report is harder to analyze than if it was separated into topics in shorter reports. Table 3 and Chart 1 show a tendency over the last five years to reduce the number of pages in the integrated annual report. HSBC was the bank that reduced the length of the report by the greatest number of pages, given the 127.94 standard deviation. On the other hand, Commerzbank was the one that kept the number of pages constant over time.

Common to these reports is the density of the text contained in each report and the option to use tables to display financial information. On the contrary, when it comes to non-financial information, the written word is largely the format used to communicate with stakeholders

Table 3. Average length of the Integrated Annual Report (Number of pages)

Year / Bank	2013	2014	2015	2016	2017	Min	Max	Mean	S Dev.
HSBC	598	488	502	286	274	274	598	429.6	127.94
BNP Paribas	455	508	540	551	580	455	580	526.8	42.66
Barclays Bank	436	348	356	380	328	328	436	369.6	37.15
Societe Generale	486	573	495	532	566	486	573	530.4	35.52
Lloyds Banking Group	396	348	311	348	278	278	396	336.2	39.69
DZ Bank Group	372	408	420	242	388	242	420	366	64.15
Commerzbank AG	345	347	351	324	316	316	351	336.6	13.92

Source: (The authors)

Figure 1. Average length of the Integrated Report (Number of pages)

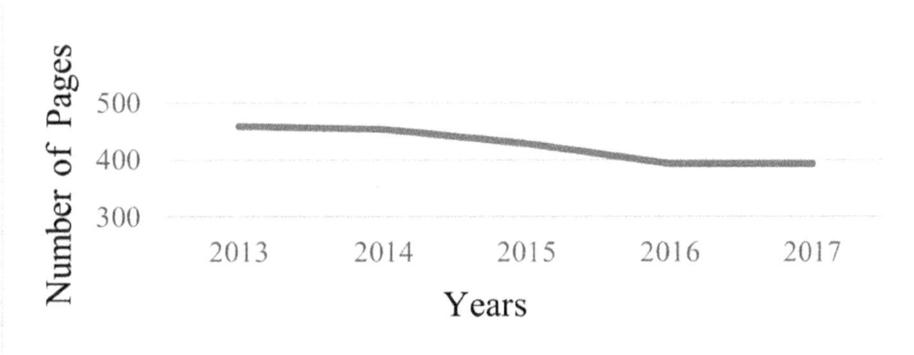

Extent of Non-Financial Information Disclosures (RQ3)

A ratio between non-financial information and total number of pages of the report was calculated.

The results show that the ratio has increased over time in two banks (HSBC and Deutsche Bank), but this is not true for the rest of the banks, as the ratio tends to be volatile. Although in 2015, after the Directive came into effect, banks did make an effort to increase this ratio compared to previous years, the change over time, was not that significant (Chart 2 and Table 4).

The findings below indicate that banks are still searching for the correct balance between non-financial information they disclose in relation to financial information. However, given that the total number of pages about non-financial information in the annual report does not change markedly over time (standard deviation is approximately 47 pages between 2013 and 2017), the change occurs not because banks are disclosing more non-financial information (numerator) but because banks are reducing the total number of pages of the annual report (denominator), taking into consideration that the standard deviation is approximately 226 pages in this period (see Appendix 7 and Appendix 8).

Another interesting finding is that there is no apparent correlation between time and the number of pages disclosing non-financial information. The effect of the new regulation did not seem to directly bear on either cutting or expanding number of pages in the reports in any given year.

Figure 2. Ratio of pages with non-financial information to number of total pages

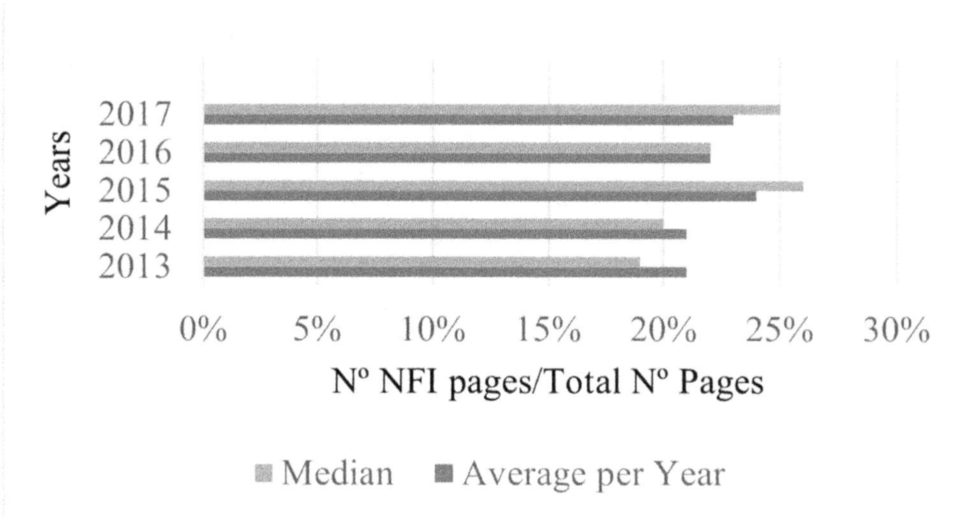

Table 4. Ratio between non-financial information and total number of pages

Year Bank	2013	2014	2015	2016	2017	Max	Min	Average
HSBC Holdings	15%	15%	27%	26%	28%	28%	15%	22%
BNP Paribas SA	31%	31%	36%	33%	32%	36%	31%	33%
Deutsche Bank AG	19%	16%	21%	20%	21%	21%	16%	19%
Barclays Plc	26%	30%	25%	24%	25%	30%	24%	26%
Société Générale SA	27%	24%	31%	26%	27%	31%	24%	27%
Lloyds Banking Group	14%	24%	28%	19%	26%	28%	14%	22%
DZ Bank Group	20%	14%	10%	14%	10%	20%	10%	14%
Commerzbank AG	14%	16%	17%	14%	16%	17%	14%	15%
Source: (The authors)								

Compliance Level (RQ4)

The deadline for adopting the Directive was 2017. So, as expected, the level of compliance increases over time, with 2017 having the highest percentage. Six out of eight banks show this trend, the exceptions being Lloyds Bank and HSBC, with minor differences, however. The median is the value that separates the 50% highest values of the sample from the 50% lowest values. In Table 5 the median per year shows the increase.

The median value supports the conclusion seen in the mean value (Chart 3). Over time, banks are disclosing more and more non-financial information in a way that leads them to have higher compliance levels.

Table 5. Compliance level

Year Bank	2013	2014	2015	2016	2017	Max	Min
HSBC Holdings	46%	64%	75%	93%	86%	93%	46%
BNP Paribas SA	39%	46%	46%	50%	54%	54%	39%
Deutsche Bank AG	57%	68%	75%	75%	79%	79%	57%
Barclays Plc	43%	50%	50%	46%	54%	54%	43%
Société Générale SA	46%	46%	61%	68%	75%	75%	46%
Lloyds Banking Group	57%	61%	71%	68%	68%	71%	57%
DZ Bank Group	46%	46%	46%	39%	46%	46%	39%
Commerzbank AG	54%	54%	54%	61%	61%	61%	54%

Source: (The authors)

Figure 3. Average compliance level (per year)

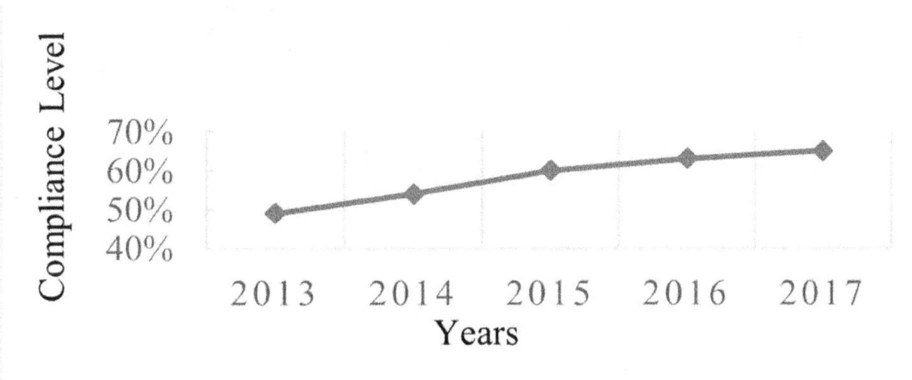

The correlation coefficient between the quantitative aspect of non-financial information disclosed, as seen in Table 4, and the level of compliance by year is -0.77 in 2013 and 0.43 in 2017. Nevertheless, a clear trend was not found, leading to the assumption that more does not necessarily mean better. The only exceptions to these results are HSBC and Lloyds Bank, which presented respectively a 0.86 and 0.73 correlation level across the five years. For a deeper overview of these variables, the McNemar's test was performed considering the mean. This test proved the statistical significance of these variables, as the result was approximately 0.89. Therefore, the correlation coefficient computed previously is also statistically significant.

In an attempt to understand whether the compliance level is associated to the size of the bank (proxied by total assets), a correlation index was computed between these two variables. For this purpose, 2017 was the year chosen to compare the compliance level to the bank size, and the average value of total assets (2013-2017) was also taken into account. The result was approximately 0.58, indicating that size has a moderately positive relation to the compliance level (the larger banks in the sample, comply more with the Directive).

Understandability Level (RQ5)

To measure the understandability of non-financial information disclosures (i.e., how clear, transparent, and understandable the reports are) an index of understandability was applied. It transpired that banks do not tend to drastically modify the structure and language of the reports over time, as shown by the steady level of understandability and the consistency of the wording and expressions used in the report itself. As an example, the committees, the explanation of the risks and type of risk generally do not change much over time. Should an explanation of a particular policy be required or needed, banks would normally reuse the text from the previous year (when available). The only exception is Deutsche Bank, which in 2017 made a big improvement in the index (See Table 6), mainly by the introduction of more diagrams, quick overviews and elimination of excessive details as opposed to previous years.

Table 6. Understandability level

Year Bank	2013	2014	2015	2016	2017	Average	Min	Max	Standard Deviaton
HSBC Holdings	60%	65%	65%	65%	65%	64%	60%	65%	2%
BNP Paribas SA	50%	50%	50%	50%	50%	50%	50%	50%	0%
Deutsche Bank AG	60%	65%	65%	65%	75%	66%	60%	75%	5%
Barclays Plc	70%	75%	70%	60%	60%	67%	60%	75%	6%
Société Générale SA	65%	65%	65%	65%	65%	65%	65%	65%	0%
Lloyds Banking Group	65%	65%	70%	70%	70%	68%	65%	70%	2%
DZ Bank Group	50%	50%	50%	50%	50%	50%	50%	50%	0%
Commerzbank AG	65%	65%	65%	60%	60%	63%	60%	65%	2%
Average per year	61%	63%	63%	61%	62%				
Median per year	63%	65%	65%	63%	63%				

Source: (The authors)

Barclays Bank and Commerzbank are the only banks that do not reach their highest level of understandability in 2017. In fact, their understandability level was better prior to 2015. This may be explained by the removal of diagrams and quick overviews from the banks' report, probably as a result of a change in the board in 2015 (chairman and group CEO in Barclays's case and CEO in Commerzbank's case), that could have culminated in a change of the staff in charge of preparing the annual reports.

CONCLUSION

This paper researched the reporting of non-financial information by the largest banks in Europe with a focus on the combat against money laundering, according to the regulatory framework approved by the Directive 2014/95/EU.

The study based on the disclosures in the annual reports reveals that banks in Germany, France and the United Kingdom increased the reporting of non-financial information over time and there is a trend for compliance levels to increase. However, there is room for improvement in areas of reporting such as anti-money laundering and bribery. Future research should explore these topics to a deeper level, such as by characterizing Key Performance Indicators or policies that are more concrete. As to how the information is disclosed, data showed that banks do not tend to change the format of disclosing. In fact, over time, there is a clear tendency to reduce the number of pages in the extended annual report and most of the banks chose to disclose non-financial information in an integrated report instead of a separate report. Furthermore, there is a moderately positive relation between the size of the bank and the extent of non-financial information disclosures, meaning that larger banks (proxied by total assets) tend to disclose more non-financial information.

This paper may provide a benchmark within the scope of the new regulation on non-financial information disclosure, which may be of use to smaller banks. For larger banks, this paper could help to fill the gap in non-financial information on the topic of anti-money laundering and bribery so that in upcoming years this information becomes more precise and detailed. Therefore, to the literature is added a study of a contemporary topic, not limited to a single country or region, and an overview of the non-financial reporting regarding anti-money laundering policies in the European Banking Sector, focusing on the largest economies in Europe (the United Kingdom, Germany and France).

There are some limitations such as the number of studied banks or the variety of countries explored. Hence, given that comparability across the years was an objective of this study, the sample of 40 annual reports might not be enough to have a fair overview of the actual situation throughout the whole of the European banking system.

The research on the topic studied in this paper could be furthered by looking at how banks will deal with disclosure practices in future years how these practices will develop over time. Including data on more European countries and more banks may allow comparisons not only in different periods of time, but also between countries or regions. A comparability analysis regarding the voluntary disclosure of non-financial information (before 2014) *versus* mandatory disclosure should be the next step, given that this could answer some questions concerning the actual impact of the Directive itself. All in all, hopefully this will lead to improvements in compliance, understandability, and transparency in the disclosure of non-financial information in the annual bank reports and contribute to combat money laundering practices.

ACKNOWLEDGMENT

This research received no specific grant from any funding agency in the public, commercial, or not-for-profit sectors. It is based on a work project written by Cátia Nunes under the supervision of Leonor Fernandes Ferreira. A preliminary version of this paper was presented to a jury in January 2019, as part of the requirements for the award of a master's degree in Management from NOVA – School of Business and Economics.

REFERENCES

Accountancy Europe. (2017). *Enhance the Coordination of Non-Financial Information Initiatives and Frameworks*. Author.

Aksu, M. H., & Espahbodi, H. (2016). The Impact of IFRS adoption and corporate governance principles on transparency and disclosure: The case of Borsa. Istanbul. *Emerging Markets Finance & Trade*, *52*(4), 1013–1028. doi:10.1080/1540496X.2014.998570

Asian Development Bank. (1995). *Governance: Sound development management*. Manila, Philippines: Asian Development Bank. https://www.adb.org/sites/default/files/institutional-document/32027/gov-policy.pdf

Begina, Ai. (2017). *Corporate and Social Responsibility Reporting in the Greek banking sector: a comparative approach* (Master diss.). International Hellenic University.

Burrit, R. (1999). Environmental Accounting and Reporting in Victoria, Australia. News Journal of the Asia Pacific Centre for Environmental Accountability, 5(3), 6-11.

Carungu, J., Di Pietra, R., & Molinari, M. (2021). Mandatory vs voluntary exercise on non-financial reporting: Does a normative/coercive isomorphism facilitate an increase in quality? *Meditari Accountancy Research*, *29*(3), 449–476. doi:10.1108/MEDAR-08-2019-0540

Cheng, B., Ioannou, I., & Serafeim, G. (2014). Corporate social responsibility and access to finance. *Strategic Management Journal*, *35*(1), 22–30. doi:10.1002mj.2131

CSR Europe GRI. (2017). *Member State Implementation of Directive 2014/95/EU: A comprehensive overview of how Member States are implementing the EU Directive on Non-financial and Diversity Information.* https://www.globalreporting.org/resourcelibrary/NFRpublication%20online_version.pdf

Delmas, M., Montes-Sancho, M. J., & Shimshack, J. P. (2010). Information disclosure policies: Evidence from the electricity industry. *Economic Inquiry*, *48*(2), 483–498. doi:10.1111/j.1465-7295.2009.00227.x

Deloitte. (2015). *Non-financial reporting*. https://www2.deloitte.com/content/dam/Deloitte/lv/Documents/strategy/Non-financial_reporting_2015.pdf

Dincer, B. (2011). do the shareholders really care about corporate social responsibility? *International Journal of Business and Social Science*, *2*(10), 71–76. doi:10.30845/ijbss

Directive 2014/95/EU: European Parliament and of the Council. (2014). *Official Journal of the European Union.*

Eccles, R., Krzus, M., & Serafeim, G. (2011). Market interest in nonfinancial information. *Journal of Applied Corporate Finance*, *23*(4), 113–127. doi:10.1111/j.1745-6622.2011.00357.x

Ernst & Young. (2016). *The road to reliable nonfinancial reporting*. Author.

Ernst & Young. (2017). *EU Directive: Disclosure of non-financial information and diversity information*. https://www.ey.com/Publication/vwLUAssets/Law_NFI_reporting/$FILE/Law%20NFI%20reporting.pdf

Eurostat. (2017). *Which Member States have the largest share of EU's GDP?* https://ec.europa.eu/eurostat/web/products-eurostat-news/-/DDN-20180511-1?inheritRedirect=true

FASB - Financial Accounting Standards Board. (1980). Qualitative characteristics of accounting information. *Statement of Financial Accounting Concepts*, *2*, 1–60.

FEE - Federation of European Accountants. (2016). *EU Directive on disclosure of nonfinancial and diversity information Achieving good quality and consistent reporting*. Author.

Fijałkowska, J., Zyznarska-Dworczak, B., & Garsztka, P. (2018). *Corporate social-environmental performance versus financial performance of banks in Central and Eastern European countries*. doi:10.3390/su10030772

Goodman, K. W. (2015). *Anti-money laundering: an often-overlooked cornerstone of effective compliance*. U.S. S.E.C. https://www.sec.gov/news/speech/anti-money-laundering-an-often-overlooked-cornerstone.html

Groenfeldt, T. (2017). Customer trust is the key for banks to build long-term relationships. *Forbes*. https://www.forbes.com/sites/tomgroenfeldt/2017/06/16/customer-trust-is-the-key-for-banks-to-build-long-term-relationships/#717e18fb3060

IASB. (2008). *Exposure draft on an improved conceptual framework for financial reporting: the objective of financial reporting and qualitative characteristics of decision-useful financial reporting information*. IASB.

Likert, R. (1932). A technique for the measurement of attitudes. *Archives of Psychology*, *22*(140), 1-55. https://legacy.voteview.com/pdf/Likert_1932.pdf

Lourenço, H. (2017). *A divulgação de informações não financeiras no âmbito da Diretiva 2014/95/UE: O atual grau de cumprimento das empresas cotadas em Portugal.* Tese de Mestrado. Universidade do Porto.

Member State Implementation of Directive 2014/95/EU: A Comprehensive Overview of How Member States are Implementing the EU Directive on Non-financial and Diversity Information. (2017).

Pieter de Groen, W., Alcidi, C., Simonelli, F., Campmas, A., Salvo, M., Musmeci, R., Oliinyk, I., & Tadi, S. (2020). *Study on the non-financial reporting directive*. https://www.ceps.eu/download/publication/?id=32991&pdf=EV0220277ENN.en_.pdf

PwC. (2017). *The non-financial reporting regulations What do they mean in practice?* PwC.

Rousseau, J.-J. (1772). *Considerations sur le gouvernement de Pologne*. Academic Press.

Serafeim, G. (2014). Turning a profit while doing good: aligning sustainability with corporate performance. *Governance Studies*, 3-9.

Shehata, M. (2020). Do non-financial information reduce the fraud risk management? *Journal of Behavioural Economics Finance Entrepreneurship Accounting and Transport*, *8*(1), 14–25. doi:10.12691/jbe-8-1-3

Sial, M., Zheng, C., Khuong, N., Khan, T., & Usman, M. (2018). Does firm performance influence corporate social responsibility reporting of Chinese listed companies? *Sustainability*, *10*(7), 112. doi:10.3390u10072217

Tarquinio, L., & Posadas, S. (2020). Exploring the term "non-financial information: An academics' view. *Meditari Accountancy*, *28*(5), 727–749. doi:10.1108/MEDAR-11-2019-0602

Venturelli, A., Caputo, F., Cosma, S., Leopizzi, R., & Pizzi, S. (2017). Directive 2014/95/EU: Are Italian Companies Already Compliant? *Sustainability, 9*(81-19).

ADDITIONAL READING

Abeydeera, S., Tregidga, H., & Kearins, K. (2016). Sustainability reporting – more global than local? *Meditari Accountancy Research*, *24*(4), 478–504. doi:10.1108/MEDAR-09-2015-0063

Quinn, J., & Connolly, B. (2017). The non-financial information directive: An assessment of its impact on corporate social responsibility. *European Company Law Journal*, *14*(1), 15–21.

Saravanamuthu, K. (2004). What is measured counts: Harmonized corporate reporting and sustainable economic development. *Critical Perspectives on Accounting*, *15*(3), 295–302. doi:10.1016/S1045-2354(03)00063-7

Sierra-Garcia, L., Garcia-Benau, M. A., & Bollas-Araya, H. B. (2018). Empirical analysis of non-financial reporting by Spanish companies. *Administrative Sciences*, *8*(3), 1–17. doi:10.3390/admsci8030029

Stubbs, W., & Higgins, C. (2015). Stakeholders' perspectives on the role of regulatory reform in integrated reporting. *Journal of Business Ethics*, *147*(3), 1–20.

Testarmata, S., Ciaburri, M., Fortuna, F., & Sergiacomi, S. (2020). Harmonization of Non-financial Reporting Regulation in Europe: A Study of the Transposition of the Directive 2014/95/EU. In S. Brunelli & E. Di Carlo (Eds.), *Accountability, Ethics and Sustainability of Organizations. Accounting, Finance, Sustainability, Governance & Fraud: Theory and Application*. Springer. doi:10.1007/978-3-030-31193-3_4

Thorell, P. & Whittington, G. (1994). The harmonization of accounting within the EU-problems, perspectives. and strategies. *European Accounting Review, 3*(2), 215–239. doi:10.1080/09638189400000019

Wagner, C. Z. (2018). Evolving norms of corporate social responsibility: Lessons from the European Union experience with non-financial reporting. *Tennessee Journal of Business Law*, *19*, 619–708.

Wang, K. T., & Li, D. (2016). Market reactions to the first-time disclosure of corporate social responsibility reports: Evidence from China. *Journal of Business Ethics*, *138*(4), 661–682. doi:10.100710551-015-2775-1

KEY TERMS AND DEFINITIONS

Anti-Money Laundering (AML): AML refers to the activities that financial institutions must comply with legal requirements to actively monitor customers transactions and report suspicious activities. Such activities are taken to prevent and combat money laundering, terrorism financing and other financial crimes AML includes procedures, regulations, and laws intended to prevent criminals from disguising illegally obtained funds as legitimate income. act. An AML compliance program should focus not only on the effectiveness of internal systems and controls developed to detect money laundering, but also on the risk assessment of customer activities.

Know Your Clients (KYC): KYC refers to the rules used by banks to verify the identity of their customers. Like AML, KYC is used by banks to secure financial institutions.

Mandatory Disclosures: Under Directive 2014/95/EU, large companies must publish information related to environmental matters, social matters and treatment of employees, respect for human rights, anti-corruption, and bribery. They also must disclose information about governance diversity regarding gender, age, professional and educational background of members in company boards.

Non-Financial Information (NFRD): NFRD is the Directive 2014/95/EU. This EU law set rules on mandatory reporting of non-financial and diversity information, such as disclose information on the way they operate and manage social and environmental challenges by public-interest entities (PIE). This Directive amends the Accounting Directive, that is Directive 2013/34/EU. EU rules on non-financial reporting currently apply to listed companies, banks, insurance companies, and large public-interest companies with more than 500 employees. NFRD can be used as a framework for stakeholders, such as investors, policy makers, and consumers, evaluate the non-financial performance of companies, and so it pushes companies to behave more responsible when doing business. The Corporate Sustainability Reporting Directive (CSRD) adapted in April 2021 by the European Commission would amend the NFRD.

Public Interest Entity (PIE): The definition of PIE is stated in the Directives 2013/34/EU on accounting (the Accounting Directive) and 2014/56/EU on statutory audits (the Audit Directive). PIE are (i) entities governed by the law of an European Union Member State whose transferable securities are admitted to trading on a regulated market of any Member State; (ii) credit institutions; (iii) insurance undertakings, or (iv) entities designated by Member States as public-interest entities, for instance undertakings that are of significant public relevance because of the nature of their business, their size or the number of their employees; incidentally, application of this fourth criterion varies significantly across Member States. On 17 June 2016, new rules on statutory audit became applicable across the EU. The new rules and key measures laid down in the EU Audit Reform apply exclusively to PIE defined as stated above.

ENDNOTES

1 Lately, this topic has been open to debate and is attracting the public interest, as evidenced by several recent events. Events such as the thirty-fifth session of the Intergovernmental Working Group of Experts on International Standards of Accounting and Reporting (ISAR) of the United Nations Conference on Trade and Development (UNCTAD) which has as one of the themes "Enhancing *comparability* of sustainability reporting" clearly demonstrate the interest in the topic. The "Non-Financial reporting workshop presentation: *Current state of affairs and future of information disclosure*" given by Deloitte Latvia in 2015 and the "12th Sustainability Reporting and Communications Summit 2018 - Impact of NFRD and TCFDs on the future of reporting", are events that also supports the relevance of this topic.

2 Non-financial information issues include environmental, social and employee, respect for human rights, anti-corruption, and bribery.

3 International Financial Reporting Standards issued by the International Accounting Standards Board (IASB).

4 https://ec.europa.eu/growth/industry/corporate-social-responsibility_en, accessed on December 8, 2018.

5 To make sure that every single bank has more than 500 employees, the number of employees was checked in the annual report of each bank.

6 Credit Suisse Group AG, UBS Group AG, PAO Sberbank of Russia, DNB ASA, Raiffeisen Gruppe Switzerland, JSCVTB Bank and Zurcher Kantonalbank were not included due to having their headquarters based outside the EU. The non-inclusion of Credit Suisse Group AG, Raiffeisen Gruppe Switzerland Zurcher Kantonalbank is also due to the fact that this bank adopted the U.S. GAAP in its financial reporting.

7 https://ec.europa.eu/eurostat/web/products-eurostat-news/-/DDN-20180511-1?inheritRedirect=true accessed on December 8, 2018.

8 Taking in consideration that Deutsche Bank did not enter in these computations since the bank has a separate report.

APPENDIX 1. IMPLEMENTATION OF DIRECTIVE 2014/95/ EU: GERMANY, FRANCE, AND THE U.K

Table 7.

	Germany	France	U.K.
Companies affected by the Directive	Over 500 employees Net turnover over EUR 40 million; *or* Balance sheet total > EUR 20 million Public Interest Entity: • Credit institutions • Insurance undertakings • Capital market-oriented companies in the legal form of a limited liability company or cooperative	Over 500 employees Net turnover over EUR 40 million; *or* Balance sheet total > EUR 20 million Public Interest Entities: • Listed companies • Credit institutions • Insurance providers 4. non-listed *sociétés anonymes* and non-listed investment funds shall comply if they have a net turnover over EUR 100 million).	Over 500 employees Public Interest Entities: • Listed entities • Credit institutions • Insurance undertakings
Information required	• Environmental performance • Social and employee issues • Human rights performance • Corruption and anti-bribery issues	• Environmental performance • Social and employee issues • Human rights performance • Corruption and anti-bribery issues.	• Environmental • Social and employee issues • Respect for human rights • Anti-corruption and bribery issues.
Mandatory disclosures	• A description of the undertaking's business model • Company policies relating to nonfinancial matters, and the outcomes of those policies • Principal risks related to non-financial matters and business activities • Any non-financial KPIs used	• A description of the undertaking's business model • Company policies relating to nonfinancial matters, and the outcomes of those policies • Principal risks related to non-financial matters and business activities • Any non-financial KPIs used	• A description of the undertaking's business model • Company policies relating to nonfinancial matters, and the outcomes of those policies • Principal risks related to non-financial matters and business activities • Any non-financial KPIs used
Presentation	• The management report, or • A separate non-financial report, within four months after the balance sheet date	• The annual report within eight months of the end of the financial year and made available on website for five years	• The strategic report
Compliance framework	• An international, EU based or national reporting framework • Comply and explain principle • Safe harbour principle • Diversity statement: applies to larger corporations with listed shares. • Auditor's involvement: presence of statement (applicable for the financial year starting on or after 1 January 2019) • Fines: up to, EUR 10 million, 5% of the total annual turnover of the company, twice the amount of the profits gained, or losses avoided because of the breach, whichever is the highest figure.	• An international, EU-based, or national reporting framework • Comply and explain principle • Diversity statement • Auditor's involvement: presence of statement and content is required if company has 500+ employees and has a turnover over EUR 100 million or balance sheet over EUR 100 million • Fines: no fine imposed unless an interested party asks for the disclosure of the non-financial information. if this is not available, a judge can impose financial penalties.	• An international, EU-based, or national reporting framework • Comply and explain principle • Safe harbour principle • Diversity statement • Auditor's involvement: presence of statement and consistency check of disclosures as part of the review of the management report • Fines: determined on a case-by case basis and imposed on the responsible individuals.
National Law	***Law-Bundesanzeiger No.20 vom 18.04.2017***	***Law-L. 225-102-1*** ***Law-R. 225-104/105/105-1/105-2***	***Law-Staturory Instruments 2016 No. 1245***

APPENDIX 2. UNIVERSE AND SAMPLE OF THE RESEARCH (continued)

Table 8.

Current Rank	Company (Ticker-exchange)	Headquarters	Accounting Principle	Total Assets (€B)
1	HSBC Holdings Plc (HS BA-LON)	U.K.	IFRS	2100,13
2	BNP Paribas SA (BNP-PAR)[1]	France	IFRS	1963,34
3	Crédit Agricole Group	France	IFRS	1763,17
4	Deutsche Bank AG (DBK-ETR)[2]	Germany	IFRS	1470,38
5	Banco Santander SA (SAN-MAD)[3]	Spain	IFRS	1446,15
6	Barclays Plc (BA RC-LON)	U.K.	IFRS	1275,62
7	Société Générale SA (GLE-PAR)	France	IFRS	1275,13
8	Groupe BPCE[4]	France	IFRS	1259,42
9	Lloyds Banking Group Plc (LLOY-LON)	U.K.	IFRS	914,14
10	ING Groep NV (INGA-MAS)	Netherlands	IFRS	846,22
11	UniCredit SpA (UCG-MIL)	Italy	IFRS	836,79
12	Royal Bank of Scotland Group Plc (RBS-LON)	U.K.	IFRS	830,78
13	Intesa Sanpaolo SpA (ISP-MIL)[5]	Italy	IFRS	801,01
14	Crédit Mutuel Group**	France	IFRS	793,45
17	Banco Bilbao Vizcaya Argentiria SA (BBVA-MAD)[6]	Spain	IFRS	671,02
18	Rabobank	Netherlands	IFRS	602,99
19	Nordea Bank AB (NDA SEK-OME)	Sweden	IFRS	581,61
20	Standard Chartered Plc (STAN-LON)	U.K.	IFRS	552,56
21	DZ Bank AG	Germany	IFRS	505,6
22	Danske Bank A/S (DANSKE-CSE)	Denmark	IFRS	475,39
23	Commerzbank AG (CBK-ETR)	Germany	IFRS	452,49
24	Cassa depositi e prestiti SpA	Italy	IFRS	419,53
26	ABN AMRO Group NV (ABN-AMS)[1]	Netherlands	IFRS	390,08
27	Caixa Bank SA (CABK-MAD)	Spain	IFRS	383,19
28	KBC Group NV (KBC-BRU)	Belgium	IFRS	292,34
29	Svenska Handelsbanken AB (SHB A-OME)	Sweden	IFRS	281,51
31	Nationwide Building Society (NBS-LON)	U.K.	IFRS	262,05
32	Skandinaviska Enskilda Banken AB (SEBA-OME)	Sweden	IFRS	260,41
33	Landesbank Baden-Wurttemberg	Germany	IFRS	238
34	La Banque Postale SA	France	IFRS	231,48
35	Swedbank AB (SWED A-OME)	Sweden	IFRS	225,11
36	Banco de Sabadell SA (SAB-MAD)	Spain	IFRS	221,35
37	BFA Sociedad Tenedora de Acciones SAL*[7]	Spain	IFRS	221,12
38	Erste Group Bank AG (EBS-WBO)	Austria	IFRS	220,66

continues on following page

Table 8. Continued

Current Rank	Company (Ticker-exchange)	Headquarters	Accounting Principle	Total Assets (€B)
39	Bayerische Landesbank	Germany	IFRS	214,52
41	Nykredit A/S	Denmark	IFRS	191,62
43	Dexia SA (DEXB-BRU)[8]	Belgium	IFRS	178,85
44	Belfius Banque SA	Belgium	IFRS	167,96
45	Norddeutsche Landesbank Girozentrale*	Germany	IFRS	165,22
46	Banco BPM SpA (BAMI-MIL)	Italy	IFRS	161,21
47	Landesbank Hessen-Thuringen Girozentrale	Germany	IFRS	158,35
49	Banca Monte dei Paschi di Siena SpA (BMPS-MIL)	Italy	IFRS	139,15
50	OP Financial Group	Finland	IFRS	137,24

APPENDIX 3. SIZE OF THE LARGEST BANKS IN FRANCE, THE U.K., AND GERMANY

Table 9.

Country selected	Total assets of each country from the 50 largest banks in Europe's List (€B)	Total assets of the three largest banks (€B)	Percentage of total Assets from the three largest banks
France	7,285.99	5,001.64	69%
U.K.	5,935.28	4,289.89	72%
Germany	3,204.56	2,428.47	76%
Total	16,425.83	11,720	71%

APPENDIX 4. RESULTS FROM THE DATA COLLECTED

Table 10a.

Bank	Country	Year	Disclosure Framework	Compliance Level	Understandability Level	Ratio between # pages with NFI and total #pages
HSBC Holdings Plc	U.K.	2013	Integrated Report	46%	60%	15%
HSBC Holdings Plc	U.K.	2014	Integrated Report	64%	65%	15%
HSBC Holdings Plc	U.K.	2015	Integrated Report	75%	65%	27%
HSBC Holdings Plc	U.K.	2016	Integrated Report	93%	65%	26%
HSBC Holdings Plc	U.K.	2017	Integrated Report	86%	65%	28%
BNP Paribas SA	France	2013	Integrated Report	39%	50%	31%
BNP Paribas SA	France	2014	Integrated Report	46%	50%	31%
BNP Paribas SA	France	2015	Integrated Report	46%	50%	36%
BNP Paribas SA	France	2016	Integrated Report	50%	50%	33%
BNP Paribas SA	France	2017	Integrated Report	54%	50%	32%
Deutsche Bank AG	Germany	2013	Separate Report	57%	60%	19%
Deutsche Bank AG	Germany	2014	Separate Report	68%	65%	16%
Deutsche Bank AG	Germany	2015	Separate Report	75%	65%	21%
Deutsche Bank AG	Germany	2016	Separate Report	75%	65%	20%
Deutsche Bank AG	Germany	2017	Separate Report	79%	75%	21%
Barclays Plc	U.K.	2013	Integrated Report	43%	70%	26%
Barclays Plc	U.K.	2014	Integrated Report	50%	75%	30%
Barclays Plc	U.K.	2015	Integrated Report	50%	70%	25%
Barclays Plc	U.K.	2016	Integrated Report	46%	60%	24%
Barclays Plc	U.K.	2017	Integrated Report	54%	60%	25%
Société Générale SA	France	2013	Integrated Report	46%	65%	27%
Société Générale SA	France	2014	Integrated Report	46%	65%	24%
Société Générale SA	France	2015	Integrated Report	61%	65%	31%
Société Générale SA	France	2016	Integrated Report	68%	65%	26%

(to be continued overleaf)

Table 10b.

Bank	Country	Year	Framework used to disclosure	Compliance Level	Understandability Level	Ratio between NFI and total #pages
Société Générale SA	France	2017	Integrated Report	75%	65%	27%
Lloyds Banking Group Plc	U.K.	2013	Integrated Report	57%	65%	14%
Lloyds Banking Group Plc	U.K.	2014	Integrated Report	61%	65%	24%
Lloyds Banking Group Plc	U.K.	2015	Integrated Report	71%	70%	28%
Lloyds Banking Group Plc	U.K.	2016	Integrated Report	68%	70%	19%
Lloyds Banking Group Plc	U.K.	2017	Integrated Report	68%	70%	26%
DZ Bank AG	Germany	2013	Integrated Report	46%	50%	20%
DZ Bank AG	Germany	2014	Integrated Report	46%	50%	14%
DZ Bank AG	Germany	2015	Integrated Report	46%	50%	10%
DZ Bank AG	Germany	2016	Integrated Report	39%	50%	14%
DZ Bank AG	Germany	2017	Integrated Report	46%	50%	10%
Commerzbank AG	Germany	2013	Integrated Report	54%	65%	14%
Commerzbank AG	Germany	2014	Integrated Report	54%	65%	16%
Commerzbank AG	Germany	2015	Integrated Report	54%	60%	17%
Commerzbank AG	Germany	2016	Integrated Report	61%	60%	14%
Commerzbank AG	Germany	2017	Integrated Report	61%	60%	16%

APPENDIX 5. GRIDS FOR COMPLIANCE LEVEL

Table 11.

Brief Description of the Group's Business Model	**Maximum of 9 points**
Clear description of the organization name	0/1
Clear description of the headquarters	0/1
Clear identification of the core markets	0/1
Clear identification of the number of employees	0/1
Clear identification of the profit	0/1
Diagram highlighting key elements of the business	0/1
Description of the macro environment	0/1
Description of the internal environment	0/1
Management declaration on the sustainability of the company	0/1
Description of the Policies Pursued by the Group in relation to those matters, including due diligence processes implemented	**Maximum of 6 points**
Identification of the current policies on AML	0/1
Identification of sanctions and concerns	0/1
Identification of processes to address AML policies	0/1
Reference to specific rules and standards on AML	0/1
Identification of subjects and committees responsible for decision-making on sustainability policy	0/1
Identification of short and long objectives	0/1
Outcome of Policies	**Maximum of 5 points**
First-time adoption or revision of sustainability policy	0/1
Comparison of the results against targets set	0/1
Comparison of the results with the previous period	0/1
Analysis of the evolution and impact of the results achieved	0/1
Entities affected by AML policy (within the bank)	0/1
Non-financial Risks	**Maximum of 4 points**
Explanation of the risks	0/1
Governance and structure	0/1
Role of the board and other company bodies in risk management	0/1
Business risk	0/1
Key Performance Indicators	**Maximum of 3 points**
Mention of any non-financial performance indicator on AML	0/1
Mention of risk indicators	0/1
Identification of the results and progress achieved by AML policies	0/1

APPENDIX 6. UNDERSTANDABILITY LEVEL

Table 12.

Topics	Score (*)
The report contains the level of information required by stakeholders but avoids excessive and unnecessary detail.	0/1/2/3/4/5
The report is reader-friendly: includes diagrams, quick overview about the topic, tables, and glossary.	0/1/2/3/4/5
The report avoids technical terms, acronyms, jargons, or other content likely to be unfamiliar to stakeholders.	0/1/2/3/4/5
The report presents the information in a format that allows users to see positive and negative performance trends on a year-to-year basis.	0/1/2/3/4/5
(*)1 = Strongly disagree; 2 = Disagree; 3 = Neutral; 4 = Agree; 5 = Strongly agree	

APPENDIX 7. NUMBER OF PAGES OF NON-FINANCIAL INFORMATION IN THE ANNUAL REPORT

Table 13.

Year Bank	2013	2014	2015	2016	2017	Total	Standard Deviation
Barclays Plc	114	103	88	92	81	478	12
BNP Paribas SA	142	156	197	184	183	862	20
Commerzbank AG	47	57	60	45	49	258	6
Deutsche Bank AG (*)	106	98	93	95	88	480	6
DZ Bank AG	75	56	44	33	38	246	15
HSBC Holdings Plc	88	75	138	75	78	454	24
Lloyds Banking Group Plc	56	82	87	66	72	363	11
Société Générale SA	132	135	152	138	153	710	9
Total	**654**	**664**	**766**	**633**	**654**	**3,371**	**47**

(*) Deutsche Bank has a Separate Report for non-financial information.

APPENDIX 8: LENGTH OF THE ANNUAL REPORT IN PAGES

Table 14.

Year Bank	2013	2014	2015	2016	2017	Total	Standard Deviation
Barclays Plc	436	348	356	380	328	1,848	37
BNP Paribas SA	455	508	540	551	580	2,634	43
Commerzbank AG	345	347	351	324	316	1,683	14
Deutsche Bank AG	572	610	448	478	412	2,520	75
DZ Bank AG	372	408	420	242	388	1,830	64
HSBC Holdings Plc	598	488	502	286	274	2,148	128
Lloyds Banking Group Plc	396	348	311	348	278	1,681	40
Société Générale SA	486	573	495	532	566	2,652	36
Total	**3,660**	**3,630**	**3,423**	**3,141**	**31,42**	**16,996**	**226**

Chapter 10
Changes in Personal Financial Behavior Amid the COVID-19 Pandemic in Brazil

Wendy Beatriz Witt Haddad Carraro
Federal University of Rio Grande do Sul, Brazil

Monise Pasetto Soster
Federal University of Rio Grande do Sul, Brazil

ABSTRACT

The study aims to identify changes in personal financial behavior in the midst of the COVID-19 pandemic. Data from 796 participants of the Extension Project 'Financial Education in Times of COVID-19', between April/2020 and February/2021, were analyzed. It is characterized as qualitative, quantitative, and descriptive research with a survey. The results indicate that 66% of respondents are women, 47% are married, 54% are between 25 and 44 years old, and 32% are financiers. Regarding financial behavior during the pandemic, 52% of the participants make a shopping list to go to the supermarket and search for cheaper brands, and 56% noticed an increase in expenses. Respondents changed their behavior and expanded their control over finances, starting to reflect more when buying. The research contributes, highlighting the importance of education and the use of money in facing financial crises. It enriches the literature regarding education and financial behavior in times of pandemic.

INTRODUCTION

In January 2020, there was already talk about covid-19 in some countries around the world, mainly in China. Shortly thereafter, the first suspected cases of the disease began in Brazil, which did not only affect the health of Brazilians, but also affected their finances. Some of the reflexes on the Brazilian economy were the increase in unemployment, fall in family income and GDP, mainly impacted by restrictions imposed on trade (Barros, 2020). With this economic scenario, many people needed to adapt or change their financial behavior (Apud, 2020), which is linked to psychological and social issues and, therefore,

DOI: 10.4018/978-1-7998-9410-0.ch010

does not always follow a rational logic. Individuals have their decisions influenced by several factors, such as: income, lifestyle, education, needs and desires (Marion, 2019). Therefore, with the scarcity of income, the population realized the need to change habits for a greater and better financial organization (Apud, 2020).

According to Toledo (2020), for individuals to make correct changes in their financial behavior, it is essential that they know the concepts of financial education and personal finance. For Forte (2020), this does not consist in just knowing mathematical formulas and control instruments, as it is based on concepts from psychology and behavioral economics "[...] and that is why accessing financial education is to provoke changes in behavior, through the reading of reality, life planning, prevention and individual and collective achievement" (Forte, 2020, p. 33).

Marques and Correia (2016) state that personal finance refers to the management of financial resources, individual or family, in search of a more financially calm future. Therefore, being financially well educated, knowing your personal finances and having a plan is important in facing unstable economic scenarios, especially when there is a loss of income by a provider in the family nucleus.

Given this context, the following problem question arises: what changes have occurred in personal financial behavior in the midst of the decovid-19 pandemic? To answer it, the study aims to identify changes in personal financial behavior in the midst of the covid-19 pandemic. This research seeks to identify these changes with the participants of the Extension Project 'Financial Education in Times of covid-19', which is part of the Extension Program 'Financial Education for all and for life'. The following steps are developed: description of the characteristics of the profile of research participants; identification of financial behavior during the pandemic; analysis of personal financial control practices identified by the participants; and description of changes in financial behavior exhibited by participants.

This study is justified as it presents at least three contributions. First, it highlights a relevant theme since the financial organization has always been important both individually and within the family. Second, with the covid-19 pandemic, this importance is even more evident, as the disease has not only affected the countries' health systems, but also their economies, increasing unemployment and causing many families to change their financial habits (TUON, 2020b). And third, it highlights the importance of the concepts of financial education, enriching the literature, which is still very scarce in relation to this topic.

BACKGROUND

Financial Education, Financial Behavior and Personal Financial Planning

Behrman et al. (2012) state that there is a strong correlation between financial education and the generation of wealth throughout the lives of individuals, since, in general, these individuals find it easier to maintain healthy financial habits, such as the construction of financial planning for control of your money. Forte (2020) highlights that financial education consists of more than a set of financial control tools, it is based on concepts from psychology and behavioral economics. For this reason, being financially educated can cause changes in the behavior of individuals, because it makes them better able to identify changes in the economic scenario and plan accordingly. Iris Souza, a volunteer with the Financial Education for Life Extension Project at the Federal University of Paraíba (UFPB), complements "[...] having financial education is having quality of life" (Carraro et al., 2021, p. 11).

Kaiser and Menkhoff (2017) indicate that financial education is directly linked to the financial behavior of individuals and classify it as the way people behave in relation to their finances (Kaiser & Menkhoff, 2017). Studies on financial behavior and behavioral finance emerged in the early 1970s, with Daniel Kahneman and Amos Tversky, who showed how these were affected by psychology and that decisions do not always follow a rational logic, but can be influenced by the environment (Marion, 2019). Atkinson and Messy (2012) complement by pointing out that financial behavior is linked to how individuals manage their financial resources, and this is affected by economic and social aspects such as: lifestyle, how they shop and how they evaluate your investments and debts. Toledo (2020) highlights that access to financial education is capable of contributing to changes in the population's financial behavior. Furthermore, Xiao et al. (2011) reinforce that individual with satisfactory financial behavior have less debt and develop more rational consumption behaviors.

In this context, financial planning, according to Cerbasi (2004, p. 31), is "[...] to establish and follow a precise, deliberate and directed strategy for the accumulation of goods and values that will form a person's assets and his family". Therefore, planning financially consists of having control of your money to achieve short, medium and long-term goals. With financial planning, individuals can better visualize and organize their finances to achieve individual or collective goals (Leal & Nascimento, 2012). It is important, not only for families to achieve their short-term goals, but also for them to build financial reserves and feel secure in the long term, especially in times of instability, such as job loss, for example Atkinson and Messy (2012) emphasize that short-term and long-term planning is necessary, as those who prioritize the realization of short-term desires are unlikely to be able to save to generate reserves or long-term investments. Marques and Correia (2016) highlight the importance of financial independence for the well-being of individuals and families, providing greater tranquility in financial decision-making, which even reflects in greater emotional stability.

According to behavior, knowledge and efforts to build an effective financial plan, Cerbasi (2004) classifies five financial profiles: saver, spender, uncontrolled, disconnected and financier. Savers are known to save money and lead a simpler lifestyle, with the aim of accumulating wealth. The spender is the opposite of the saver, individuals with this profile tend to spend more than they earn and only save money for short periods with the intention of short-term accomplishments (such as travel, buying a car or other desire). The uncontrolled do not have any control over their finances. The disconnected tries to get control of their finances but easily loses focus. And the financier is the most concerned about their finances, usually has a strict control of their income and expenses and thinks in the long term, saves their money thinking about living safely and comfortably in the future (Cerbasi, 2004).

After carrying out a financial planning, defining personal and family goals and putting into practice the cash flow or family budget, it is important to evaluate investment opportunities for the money saved. However, before allocating the money for investments, it is important to create a financial reserve for emergencies (Cerbasi, 2012), which allows for covering unforeseen events. Having this emergency reserve built, which offers financial security to the family, it is necessary to think about how to invest the saved money. There is a great diversity of products available in the financial market with the objective of multiplying money, however, it is necessary to study them to know the one that best adapts to the expectations and personal characteristics of each investor, as there are risky investments (you can lose or earn money) different (Cerbasi, 2012).

Cerbasi (2004) states that the attitude of planning financially brings benefits to families in the long term, which are noticed mainly in times of crisis, when they have greater peace of mind to make decisions that affect the family budget. Griffin (2012, p. 153) adds that the budget helps in the efficient use of money, which makes it an important form of control. All these financial control tools must take into account short-, medium- and long-term objectives. Furthermore, it is important to assess the current financial situation as a whole, knowing the sources of income, the main expenses and the capacity to generate savings (Cherobim & Espejo, 2010).

BRAZILIAN ECONOMIC SCENARIO IN TIMES OF COVID 19

The current Brazilian scenario is one of instability and uncertainty, due to the covid-19 pandemic, which hit Brazil and the world in 2020, causing quarantine to be decreed worldwide, which directly reflected in the development of the economy and has presented negative consequences for many Brazilian families (Ferreira & Rita, 2020). Unlike the crises of 1929 and 2008, caused by financial market factors, this crisis is linked to health, which makes it more difficult to make predictions, given the uncertainties in the possible treatments for the disease. In addition to social isolation, which restricted the movement of people and the functioning of sectors of the economy, these insecurities have caused instability in the financial and commodity markets (Ferreira & Rita, 2020).

The impacts of the pandemic also reflected in the Gross Domestic Product (GDP) of Brazil, which showed relevant drops in the first and second quarters of 2020. According to data from the Brazilian Institute of Geography and Statistics (IBGE), GDP had its biggest drop, since 1996, decreasing 9.7% compared to the previous quarter. This retraction was caused by falls in industrial production and demand in the service sector, due to restrictions on circulation and interruption of various activities imposed by the federal, state and municipal governments (Barros, 2020).

Another factor that drove the fall in GDP was the reduction in household consumption, which represents 65% of this index, caused by the reduction in income (Barros, 2020). IBGE data indicate that the 2020 unemployment rates were 14.6% and 13.9% in the third and fourth semester of 2020, respectively (Campos, 2021). Inflation closed at 4.52% in 2020, the highest since 2016, having been driven by the rise in the prices of rice, soy derivatives and fuels (Barros, 2021). This scenario reflects on the financial behavior of individuals, as shown in a survey published by the e-investor portal, carried out with classes A, B and C (with internet access), by IBOPE Inteligência, in which about 89% of the participants had changes of financial habits during the pandemic, mainly by reducing expenses. Of these, 22% said they started to delay payment of some bills (Apud, 2020).

In order to mitigate the economic impacts caused by the pandemic in the country, the government enacted several laws, regulations and provisional measures. Among those that brought greater relief to the population is Law No. 13982, of April 2, 2020, which provides for emergency aid for self-employed and informal workers, meeting certain prerequisites (Tamanini et al., 2020). In addition to the measures instituted, banks, finance companies and creditors in general proposed other alternatives to alleviate the effects of the crisis, such as extending payment terms and renegotiating debts.

METHODOLOGICAL PROCEDURES

The research carried out is classified according to the following aspects: regarding the approach to the problem: qualiquantitative; as to its objectives: descriptive; and as for the procedures: a survey. As for the objectives, the research was developed in a descriptive way, as it was intended to analyze the changes in financial behavior caused by covid-19. As for the procedures, the study was developed as a survey obtained by the Extension Project 'Financial Education in times of covid-19' which is part of the 'Financial Education for All and For Life' Program, developed at the Faculty of Accounting Sciences of the University Federal of Rio Grande do Sul (UFRGS).

An online survey was carried out from April 21, 2020, to February 24, 2021, obtaining 796 responses. It was intended to collect relevant information for studies related to the financial behavior of the population in the midst of covid-19. The questions analyzed for this study are found in Appendix A. Data were analyzed using Excel software. And, to verify the association between knowledge about finances and variables of the research participant's profile, and between the impact of social isolation and the respondents' financial style, the Chi-Square test was applied, respecting their assumptions. In all analyses, a $p<0.05$ was considered significant. All analyzes were created using SPSS 23.0 software.

ANALYSIS OF RESULTS

In this section, the profile of the respondents is presented; the description of your financial behavior; its main financial control practices; and the most relevant changes they had in their financial behavior, all in the midst of the covid-19 pandemic.

PROFILE OF PARTICIPANTS

From the responses obtained, the profile of the 796 survey respondents was identified: most are female (66%); 34% are male; 84% are residents of the southern region of Brazil; 7% from the Northeast region; 7% from other regions; and 1% resides outside the country. As for marital status, 47% are married; 44%, single; 8%, divorced; and 2%, others. Regarding the age group, most of them are between 19 and 34 years old (47%); 22% are between 35 and 44 years old; 16%, aged 45 to 54 years; 12%, aged 55 or over; and 3%, aged up to 18 years. Table 1 displays all this data.

Table 1. Profile of respondents

Região Brasil	% Participantes	Gênero	% Participantes	Faixa etária	% Participantes	Estado civil	% Participantes
Sul	84%	Fem.	66%	Até 18 anos	3%	Casado	47%
Nordeste	7%	Masc.	34%	Entre 19 e 24 anos	15%	Solteiro	44%
Sudeste	5%			Entre 25 e 34 anos	32%	Divorciado	8%
Centro-Oeste	2%			Entre 35 e 44 anos	22%	Viúvo	1%
Moro fora do Brasil	1%			Entre 45 e 54 anos	16%	Outros	1%
Norte	1%			Entre 55 e 64 anos	10%		
				65 anos ou mais	2%		

Regarding education, around 48% of the sample has completed or incomplete higher education; 44%, specialization, masters or doctorate; 7%, complete or incomplete high school. Regarding the respondents' main activity, 31% responded that they work in the public sector, 28% in the private sector, 11% are students, 10% are self-employed and 22% marked the option others. Regarding income, 44% reported having an income higher than R$3,000 to R$9,000; 22%, below R$3,000; 19%, from R$12,000 to R$15,000; and 15%, above R$15,000. From Table 2, it can be inferred that most respondents are from the Brazilian middle class and have a good level of education.

Table 2. Main activity

Education	% participants	Activity	% participants	Salary range up to	% participants
Specialization	28%	Private sector employee	31%	R$ 1.000,00	3%
Complete higher education.	24%	Public sector employee	28%	R$ 3.000.00	19%
incomplete higher education	24%	Student	11%	R$ 6.000,00	27%
Master's degree	11%	autonomous	10%	R$ 9.000,00	17%
Complete high school	7%	Businessperson	6%	R$ 12.000,00	11%
Doctorate degree	4%	Intern	4%	R$ 15.000,00	8%
Incomplete high school	2%	Unemployed	4%	Acima de R$ 15.000,00	15%
Others	2%	Others	8%		

Source: elaborated from research data (2021).

It is noteworthy that the survey was applied 100% online, so it is assumed that respondents had access to the internet. In addition to the questions regarding the socioeconomic profile, participants were asked to classify their financial style as proposed by Cerbasi (2014).

In this question: Which Financial style best suits you?, most respondents (32%) were classified as Financiers, who are those who maintain strict control of finances, with the objective of saving and accumulating wealth for the future (Cerbasi, 2004). Savers, 30% indicated, these, according to Cerbasi (2004), are the ones who know the importance of saving money and therefore do not mind restricting their spending in order to ensure a better future. The most worrying profiles represent 37% of respondents and are composed of Disconnected (19%) - the one who spends less than he earns and saves some money but does not maintain controls; by Uncontrolled (13%) - which has no control over their finances, tries to cut expenses, but without success; and by the Spender (5%) - who spends more than they earn, without any type of control (Cerbasi, 2004).

When asked: About your financial education, would you say that [...], the participants could select more than one alternative, and the one that scored the highest number of appointments was: "Searched for information", with 451 appointments. 56% marked this alternative, followed by "I received guidance from parents", with 271 appointments, and "I participated in training courses" with 190 appointments. However, 172 respondents said they had not been financially educated and 11 still have no interest in this subject, which is worrying, given the importance of financial education for a healthy financial future.

Atkinson and Messy (2012) claim that it is with financial education that individuals gain knowledge of important tools in financial control and thus have a greater ability to identify risks, opportunities and make safer decisions.

As for knowledge about personal finance, the question was: how do you rate your knowledge of personal finance?, the answers were scored on a scale from 1 to 5, where 1 represents "I don't have knowledge", and 5 indicates "I have solid knowledge". For Marques and Correia (2016, p. 3), "personal finance, in short, is the process of managing the financial resources of an individual or a family, aiming to build a financially calm life, both in the productive and in retirement." Thus, having knowledge about personal finances is critical for individuals' short- and long-term planning. A total of 38% classified as level 3; followed by 30%, level 4; and 16%, level 5. Therefore, 84% of respondents believe they have a good knowledge of personal finance, with a level between 3 and 5, 4% indicated level 1, and 12%, level 2, showing little or no knowledge.

In Table 3, knowledge about personal finance is related to the profile of respondents, and it appears that the profile of respondents who gave their grade from 3 to 5 for knowledge in personal finance, mostly, are of the Financier or financial style. Saver, have higher education (complete or incomplete) and are female. According to Cerbasi (2004), the Financier and Saver styles have similarities, as they know the importance of money and think about the future, fundamental concepts for personal finance (Atkinson & Messy, 2012).

Table 3. Knowledge about finances vs. profile, in number of responses

Variables	Level 1	Level 2	Level 3	Level 4	Level 5
	nº 34	nº 94	nº 302	nº 242	nº 124
Financialstyle					
Off	7	28	75	30	14
Financier	1	10	87	104	51
saver	2	24	83	84	47
Uncontrolled	17	22	46	14	6
Spender	7	10	11	10	6
Genre					
Feminine	31	77	211	148	56
Masculine	3	17	91	94	68
Education					
Complete primary education		2	3		
Incomplete high school	1	4	6	1	
Complete high school	2	16	20	13	2
incomplete higher education	8	22	89	48	23
Complete higher education.	23	50	184	180	99

Source: elaborated from research data (2021).

Briefly, based on Table 3, it is concluded that the profile of the participants was, in large part, female, with access to the internet, with good education, considering having good knowledge about personal finance and, somehow, having access financial education, whether on their own initiative, at school/ college, or by parents.

FINANCIAL BEHAVIORS DURING THE PANDEMIC

For the question: When shopping in the supermarket in isolation period, generally: 52% of respondents answered that they always make a shopping list; 54% reported that they search for the cheapest brands in the same supermarket. These actions help organize the budget and buy only what is necessary, avoiding waste, in addition to facilitating the search for products in stores (Carraro et al. 2021).

As for the perception of spending during the isolation period, most participants (56%) noticed an increase in spending on water, electricity and gas, while with telephone and internet most (71%) did not notice any change. Regarding the debts of those participants who have some type of financing or loan, around 40% of respondents renegotiated or intended to renegotiate their debts.

According to a survey by Ibope Inteligência, carried out with classes A, B and C (with internet access), there were changes in financial habits during the pandemic, mainly in relation to the reduction of expenses. Furthermore, participants stated that they needed to delay payment of some bills (Apud, 2020). According to the Consumer Indebtedness and Default Survey (PEIC), released in January 2021, household indebtedness worsened in the covid-19 pandemic, one of the causes raised by the survey is the population's lack of knowledge of the possibility of renegotiation (Confederation National of Goods and Tourism Trade – CNC, 2020). Table 4 shows that respondents had already renegotiated their debts or intended to renegotiate them, which leads to the inference that they already had knowledge about these opportunities.

Table 4. Perception of family expenses and debts in the isolation period in number of responses

Spending	reduced	I didn't notice change	Increased	
Drugstore	191	437	168	
Water/Light/Gas	100	243	453	
Telephone/Internet	85	569	142	
Debts	**I do not have**	**I didn't need to renegotiate**	**I intend to renegotiate**	**Renegotiated**
house financing	510	185	34	67
car financing	616	143	17	20
Loan	601	119	35	41
Payroll loan	623	123	29	21
Overdraft	581	162	32	21
Credit card	241	456	54	45

Source: elaborated from research data (2021).

In response to the impact of restrictions on domestic finances, Table 5 shows that the largest number of respondents said they had negative impacts (52%), while 31% responded that they had not noticed an impact so far, 13% had an impact positive and 5% were unable to respond. Relating financial style to the perception of impact, the respondents who found positive impacts are, in the majority, Financiers or Savers. In the words of Cerbasi (2004), the Financiers and Savers are concerned with saving money for the future, therefore, they were also the ones who most perceived negative impacts. This is explained by the fact that these profiles have greater control over their finances, which makes them more easily aware of positive and negative biases (Cerbasi, 2004).

Table 5. Impact of isolation on household finances, in number of responses

Variables	financial style				
	Off	Financier	Saver	Uncontrolled	Spender
Impact on domestic finances	nº 154	nº 253	nº 240	nº 105	nº 44
Had no impact so far	50	83	82	20	8
negatively impacted	74	122	121	71	22
Positively impacted	23	35	29	6	8
I do not know how to answer	7	13	8	8	6

Source: elaborated from research data (2021).

When asked: After this period of social restriction and isolation, in relation to their personal finances, the majority of the answers were: "I will continue doing the controls I always did and I will keep my reservations" (37%); followed by: "I will keep tighter control of my income and expenses and create a contingency reserve" (33%); in third, 13% responded: "I will just keep tighter control of my income and expenses". According to Apud (2020), the positive side of initiating greater control during the crisis is that individuals perceive the possibility of continuing these controls after the pandemic. However, about 18% of participants showed worrying behavior: 8% said they will maintain or create an emergency reserve; 5% answered "I don't know how to inform"; and 5% indicated not knowing whether to organize or not to carry out any control, as shown in Table 6.

Table 6. Impact of isolation on household finances, in number of responses

Attitude	% Participants
I will continue doing the controls I always did and will keep my reservations.	37%
I will keep tighter control of my income and expenses and create a contingency reserve.	33%
I will just keep tighter control of my income and expenses.	13%
I will just continue to maintain a contingency reserve as I have always done.	8%
I don't know how to inform.	5%
I don't know how to organize myself in relation to the control, I will not carry out any control	5%

Source: elaborated from research data (2021).

In summary, respondents became more careful when going to the supermarket, making shopping lists and comparing prices, as well as noticed increases in some household expenses such as water, electricity and gas, due to the fact that they spend more time at home. As for the impact of the restrictions on finances, the Financiers and Savers profiles were the ones that perceived the greatest impacts. Finally, respondents stated that, at the end of the restrictions imposed by the pandemic, they intend to continue controlling their finances.

MAIN FINANCIAL CONTROL PRACTICES MENTIONED BY PARTICIPANTS

285 answers were obtained for the question: What were your initiatives during the pandemic for better financial education or financial re-education?. Based on their analysis, it was found that the participants consider it important to control the financial budget, especially during the pandemic period. “The budget can help reveal efficient and inefficient uses of resources, which makes it an important management control.” (Griffin, 2012, p. 153). In Participant 56’s speech, there is a change in financial behavior with the use of the words “before” and “now”, and the importance of the budget: “Before this happened, I didn’t make a strict record of the family budget, now things are purchased and registered with more caution” (Participants 56).

The importance of the records is complemented by Participant 801, who cites the use of Excel software: “Recording of expenditures in Excel, recording of investments in Excel, I started a cycle of reading books” (Participant 801). Leal and Nascimento (2012) state that keeping a record of income and expenses is a fundamental tool for family financial control. Reinforced by the contribution given by participant 19: “Reinforcing the importance of financial control, guiding the family to follow what was established in our monthly expenses plan” (Participant 19).

Griffin (2012, p. 153) adds that, “for budgets to work, they must be realistic, reasonable and achievable, and based on a thorough analysis that includes a detailed assessment of financial needs, in order to make the goals be achieved. ”. In addition, financial planning is efficient when you know all our income and expenses and manage to keep them in balance, so that we always spend less than we earn, generating reserves and investments (Cerbasi, 2012). These concepts are reinforced by the responses of Participants 60 and 695.

Given that income has decreased a little, I am trying to reduce expenses in the same proportion to maintain how much I save per month and maintain my goal of owning a home (Participant 60).

Participant 60 demonstrates his change in behavior by stating that he is trying to reduce expenses in the same proportion as he realized the reduction in income.

Here at home, we are always questioning ourselves about whether what we want to buy is a desire or a need, so we can better define our priorities (Participant 695).

Participant 695 shows greater reflection on purchases. In addition to the importance of budget control, participants cite the importance of saving to achieve future goals. Financial control and planning not only intend to build up reserves to achieve personal goals, but seek to provide long-term financial security, so that in times of financial setbacks - such as what has been happening with the current covid-19 pandemic. - families are able to remain financially independent for a good period of time, as it is not known how long the pandemic will last and the restrictions caused by it (Barros, 2020). This is exemplified by respondents 27 and 729.

Any money left over, either because I can't leave the house or because I don't need to use it at all, I'm saving to reinforce the emergency reserve (and considering I can use it in the short term) (Participant 27).

I spend only what is necessary, whether in the supermarket or in stores. I control my monthly earnings and expenses and set goals to achieve, goals that serve as an incentive to save (Participant 729).

Participants also mention how important financial education is at this time, which can be seen in the statements of Participants 25 and 195:

I took courses, I follow Instagram's IGTVs that talk about financial education and are giving tips to get through this pandemic, I seek to improve my relationship with money (Participant 25).

I'm taking more online courses about this and also reviewing what is really necessary to acquire, I realized that the inevitable is food, besides, everything is a matter of choice, I think about wanting, deserving, needing, being able, having before buying something (Participant 195).

This interest of participants in seeking financial education in times of difficulty is evidenced in the study by Atkinson and Messy (2012), who show that, in times of crisis, individuals are more interested in financial education. These authors highlight the fact that it is at these times that people are most capable of changing their financial behavior.

CHANGES IN FINANCIAL BEHAVIOR MENTIONED BY PARTICIPANTS

For the question: Describe the positive financial impacts with the restrictions imposed by covid-19, if any, 313 responses were received, and most expressed that there was a reduction in transport and leisure expenses and greater awareness when making a purchase, rationalizing if it's really necessary. Respondents 199, 590, 612 and 677 have similarities in their answers: "I'm saving on everything I can. Assessing that he had unnecessary expenses before" (Participant 199). "Study, reflection and need for change." (Participant 590). "Awareness of unnecessary spending and cutting services that increased the value." (Participant 612). "Changing habits reduced the amount of water and electricity expenses by 50%." (Participant 677).

These respondents changed their behavior in relation to finances, seeking education and greater control of spending. In addition to the similarity of responses, they classified themselves with an Uncontrolled financial style, which is one that has no control over their finances (Cerbasi, 2004). Highlighting participant 590's response: "Study, reflection and the need for change" (Participant 590). Respondents with a Financist style, who, according to Cerbasi (2004), usually have strict controls in order to save, made these characteristics clear in their responses, for example, in the case of respondent 83:

I ended up spending a little less on food and hygiene products by buying them all together in one trip to the market, instead of going several times a month. In terms of food, it was also cheaper, since I used the company's VA less and I'm eating more at home; then I cook larger amounts of food and freeze what is left over (Participant 83).

Participants 158 and 777, financiers, also make clear the characteristic of caution with financial controls: "More caution with unnecessary expenses" (Participant 158). "Better control of expenses with supermarkets, water and electricity and better care for household food waste." (Participant 777). Participants 162, 203, 234, 588, self-declared disconnected, answered: "Better plan your expenses and think if I really need it" (Participant 162). "I had a great reduction in expenses and managed to balance the budget. Something I haven't been able to do for a long time." (Participant 203). "Less impulsive shopping with shopping trips." (Participant 234).

Savings on food for not eating out every day; economy with leisure (trips, tours, social events); savings with consumption (I have not made purchases that are not necessary) (Participant 588).

For Cerbasi (2004), the Offline style is one that generally spends less than it earns, but does not maintain any type of control and, if there is no money for a purchase, does not hesitate to divide it up in installments. It is noticed that there has been a change in relation to expenses and that the intention is to control them, as mentioned by participants 701 and 713: I did not make "new debts" and I tried to pay off all that I had/have (Participant 701). Create the habit of controlling expenses (Participant 713). Respondent 776, Spending style, also cites changes in relation to her expenses: "I was forced to review my expenses, change habits and plan for the future" (Participant 776). The Spender is the one who spends more than he earns and tends to indebted (Cerbasi, 2004). In the response of this participant 776, it can be seen that she is concerned with making financial planning and changing her behavior, thinking about the future.

"I started to invest and study financial education" (Participant 678). Participant 678 shows a change in behavior by using the word "I've started". A self-declared spender, he shows that he has started to seek more knowledge for his financial organization. According to Forte (2020, p. 33), "accessing financial education is to provoke changes in behavior, through reality reading, life planning, prevention and individual and collective achievement".

According to the answers analyzed above, many participants mentioned positive points, however, there were answers with negative perceptions, mainly related to the decrease in income. Some of the most recurrent negative aspects listed are the loss of one's own job, or that of a family member, or the drastic reduction in income. As mentioned by participants 242, 274, 537, 571. "I had to help pay the expenses of my parents who had decreased income." (Participant 242). "My self-employed husband is out of work, my salary was reduced by 25%, I lost food stamps and day care assistance." (Participant 274). "With all this I have all the bills overdue." (Participant 537). "Reduction of revenue." (Participant 571).

Since May 2020, Brazil has been experiencing high unemployment rates and difficulties in relocating to the job market. According to IBGE data, the 2020 unemployment rates were 14.6% and 13.9% in the third and fourth semesters, respectively (Campos, 2021). Another fact that had great influence was the increase in food prices, with a spike in the price of rice and soy derivatives. Inflation closed at 4.52% in 2020, the highest since 2016 (Barros, 2021). These aspects are verified in the answers of respondents

416 and 791: "Increase in the value of the item in the supermarket and not in the quantity" (Participant 416). "Inflation and price increase." (Participant, 791).

Based on the analysis of the positive and negative aspects mentioned by the participants, changes in some financial habits were evidenced. Regardless of the financial style, the main changes were related to greater control of income and expenses and the search for financial education and more knowledge on the subject.

CONCLUSION

The aim of this study was to identify which changes occurred in personal financial behavior in the midst of the covid-19 pandemic and, for this, data from the research 'Financial Education in Times of Covid-19' were analyzed. This was done online from April 2020 to February 2021 and got 796 responses. The results were presented in four sections: profile of respondents, financial behavior during the pandemic, main financial control practices, and changes in financial behavior mentioned by the participants.

Regarding the profile, most participants are female, represented by 66%; 47% are married; most are between 25 and 34 years old; 32% have a Financier financial style. As for financial behavior during the pandemic, respondents reported that they make a shopping list to go to the supermarket and search for cheaper brands. Furthermore, 56% noticed an increase in water, electricity and gas bills, and 40% renegotiated their debts. Regarding financial control practices, 285 responses were received, in which control practices and strict recording of the family budget, the use of software to monitor expenses (such as Excel) and the search for greater clarity about education were considered important. financial.

As for the changes in financial behavior, it was found that the self-declared Uncontrolled participants showed important changes in relation to their controls, with greater evaluation of expenses and reflection to reduce expenses. Financiers indicated greater concern with waste and consumption control. The Dismissed demonstrated control over impulsive purchases and created the habit of controlling expenses. Spenders responded that they had changed their habits in relation to financial control, for better planning for the future. In summary, regardless of the financial style, the participants showed changes in their financial behavior, always looking for more tranquility in relation to the uncertainties of the economy.

From these results, the importance given by the participants to financial education and financial control can be observed, in the midst of the covid-19 crisis. They demonstrated that they altered their financial behavior to adapt to the scenario imposed by the pandemic. However, some respondents did not change, and others only cited impacts and negative changes, such as loss of employment and income. Therefore, the objective of this study was achieved, as changes in financial behavior in the midst of the covid-19 pandemic were found. This research contributes by showing the importance of financial education in facing financial crises, as it is through it that individuals are able to analyze scenarios and adapt their routines using the correct tools. In addition, it enriches the national literature by presenting the relationship between financial education and financial behavior in the face of the pandemic.

Limitations of the research were the sample size and the fact that it was only applied online, restricting the profile of those surveyed. It is inferred that; therefore, the research did not reach the public most affected by the pandemic (low-income). It is suggested that, for future studies, the same participants are involved, but in a post-pandemic scenario. It is recommended that studies be carried out with broader samples and, if possible, that the research instrument be applied in physical and virtual environments.

REFERENCES

Apud, M. (2020). Ibope: 9 em cada 10 brasileiros mudaram hábitos financeiros na pandemia [Ibope: 9 out of 10 Brazilians have changed financial habits in the pandemic]. *Estadão*. https://einvestidor.estadao.com.br/educacao-financeira/ibope-brasileiros-habitos-financeiros-pandemia

Barros, A. (2020). PIB tem queda recorde de 9,7% no 2º trimestre, auge do isolamento social [GDP has a record drop of 9.7% in the 2nd quarter, the height of social isolation]. *Agência IBGE Notícias*. https://agenciadenoticias.ibge.gov.br/agencia-noticias/2012-agencia-de-noticias/noticias/28720-pib-tem-queda-recorde-de-9-7-no-2-trimestre-auge-do-isolamento-social

Barros, A. (2021). Inflação acelera em dezembro e chega a 4,52% em 2020, a maior alta desde 2016 [Inflation accelerates in December and reaches 4.52% in 2020, the highest since 2016]. *Agência IBGE Notícias*. https://agenciadenoticias.ibge.gov.br/agencia-noticias/2012-agencia-de-noticias/noticias/29871-inflacao-acelera-em-dezembro-e-chega-a-4-52-em-2020-a-maior-alta-desde-2016

Campos, A. (2021). Desemprego recua para 13,9% no 4º trimestre, segundo o IBGE [Unemployment drops to 13.9% in the 4th quarter, according to IBGE.] *Agência Brasil*. https://agenciabrasil.ebc.com.br/economia/noticia/2021-02/desemprego-recua-para-139-no-4o-trimestre-segundo-ibge

Carraro, W., Pereira, M., & Soster, Monise. (2021). *Dicas para manter sua educação financeira em alta* [Tips to keep your financial education high]. Educação Financeira UFRGS [Financial Education UFRGS]. https://lume.ufrgs.br/handle/10183/217579

Cerbasi, G. (2004). *Casais inteligentes enriquecem juntos* [Smart couples get rich together]. Gente.

Cerbasi, G. (2012). *Como organizar sua vida financeira: inteligência financeira pessoal na prática* [How to organize your financial life: personal financial intelligence in practice]. Elsevier.

Cherobim, A., & Espejo, M. (Eds.). (2010). *Finanças pessoais: conhecer para enriquecer* [Personal finance: knowing to get rich]. Atlas.

Confederation National of Goods and Tourism Trade – CNC. (2020) *Pesquisa de endividamento e inadimplência do Consumidor* [Consumer indebtedness and default survey]. http://cnc.org.br/editorias/economia/pesquisas/pesquisa-de-endividamento-e-inadimplencia-do-consumidor-peic-janeiro-2

Ferreira, R. Jr, & Rita, L. (2020). Impactos da Covid-19 na Economia: Limites, desafios e políticas [Impacts of Covid-19 on the Economy: limits, challenges and policies]. *Cadernos de Prospecção*, *13*(2), 459–476.

Forte, C. (2020). O papel da AEF-Brasil na execução da estratégia nacional. In Estratégia Nacional de Educação Financeira (ENEF): em busca de um Brasil melhor [The role of AEF-Brasil in the execution of the national strategy]. Riemma.

Griffin, M. (2012). *Contabilidade e finanças* [Accounting and Finance. Fundamentals Series]. Série Fundamentos. Saraiva. https://app.minhabiblioteca.com.br/#/books/9788502180161/

Leal, C., & Nascimento, J. (2012). Planejamento financeiro pessoal [Personal financial planning]. *Revista de Ciências Gerenciais, 15*(22), 163-186. https://revista.pgsskroton.com/index.php/rcger/article/view/2101

Marion, A. (2019). *Coaching financeiro* [Financial Coaching]. Atlas.

Marques, E., & Correia, J. N. (2016). *Gestão financeira familiar - como as empresas fazem* [Family financial management - how companies do it]. Alta Books.

Tamanini, M., Zorzeto, G., & Neves, A. (2020). A pandemia do coronavírus e os grupos vulneráveis: uma análise quanto à eficácia dos atos do governo brasileiro [The coronavirus pandemic and vulnerable groups: an analysis of the effectiveness of the Brazilian government's actions]. *Revista da Seção Judiciária do Rio de Janeiro, 24*(49), 112-135. http://177.223.208.8/index.php/revistasjrj/article/view/463

Toledo, A. (2020). Educação financeira: por que precisamos dela? [Financial education: why do we need it?]. In Estratégia Nacional de Educação Financeira (ENEF): Em busca de um Brasil melhor. Riemma.

Chapter 11
Gender Visibility of Female Directors in the Local Licensed Commercial Banks in Sri Lanka

Kennedy Degaulle Gunawardana
https://orcid.org/0000-0003-3381-9058
University of Sri Jayewardenepura, Sri Lanka

ABSTRACT

The purpose of this chapter is to expose and portray the investigation of female occupational compositions, trends in designation portfolio and female visibility in the board of directors and corporate management in local licensed commercials banks in Sri Lanka. The chapter identifies that most local licensed commercial banks in Sri Lanka still have not adopted dynamic corporate practices as they claim. The majority occupied secretarial and legal positions at the board level and deputy general manager positions at the corporate management level. All of them possessed higher educational/professional qualifications, longer number of years of experience, and many were members of third-party networks which would have felicitated their visibility towards career advancement. Further, the chapter identified that government banks were more rigid in occupying females into their apex levels than private licensed commercial banks.

INTRODUCTION

Women in Sri Lanka form approximately 57% of a total estimated population of 21 million. However, out of the total economically active population of 8.5 million persons, only 33.4% are women. Thus, almost 70% of the labour force constitutes economically inactive women. The Constitution of Sri Lanka deems that women are equal citizens entitled to all the privileges that men can avail themselves of. Article 12 of Constitution of Sri Lanka stipulates that all persons are equal before the law and are entitled to the equal protection of the law. Furthermore, no citizen shall be discriminated against on the grounds of race, religion, language, caste, sex, political opinion, place of birth or any such grounds. Nonetheless, constitutional guarantees notwithstanding, women in the workplace have not been provided with equal

DOI: 10.4018/978-1-7998-9410-0.ch011

opportunities to achieve leadership positions. While the only limiting factor should be the relevant levels of education, there seems to be a 'sticky floor syndrome' operating in many of the economically active areas of work. The main aim of the chapter is to verify the visibility of female representation in the Sri Lankan Corporate sector for the period of 2015 to 2018.

The main agriculture sectors of tea, rubber and coconut employ more women in labour positions as pluckers, tappers and coir workers respectively, while an insignificant number occupy management positions. The garment sector employs more women workers but fewer women in management positions. Sri Lanka's biggest foreign exchange earners are migrant workers among whom, women comprise a large percentage, mainly in the form of domestic labour. Table 1 represents Industry wise gender distribution.

Table 1. Gender distribution with industry

Major Industry Group	Sri Lanka (%)	Gender	
		Men	Women
Sri Lanka	*100%*	*100%*	*100%*
Agriculture	31.5	29.5	35.3
Industry	26.2	26.7	25.1
Services	42.4	43.8	39.5

(Source:2018 CBSL)

The highest percentage of women is in the service sector accounting for 39.5% of the labour force, while women in agriculture and industry account for 35.3% and 25.1%, respectively. Though statistics of women in management in Sri Lanka are not available, broad estimates would place the figure below 10% of the total cadre

According to ILO Women represented 48.5% of the global labour force in 2018 compared to 40.5% in 2008 (ILO, 2018). Although the numbers have grown, question remains as to whether it's a satisfying indication for women's visibility for career advancement. Women work 49 hours per week on average - same as male counterparts. More Women are relocating for career purposes compared to males. In most of the developing nations, remaining out the labour force might not be seen as an option but an obligation as per social norms. It's likely that women would opt to work if barriers are reduced. Total 51% of the Sri Lankan population currently comprise of women (CBSL, 2018). They are also the highest foreign exchange earners but mere 4% female decision makers exist. Although Sri Lankan women have reached greater altitudes locally/internationally, majority of women are still either being trapped in low-mobility occupations or being under-represented, which questions their visibility to reach zenith in their career ladder creating a clear gap to be considered.

LITERATURE REVIEW

To date research on females' career development largely focuses on inhibiting factors across a range of industrial sectors such as food marketing and sales (Mattis, 1995), hospitality (Purcell, 1996), banking (Parker et al., 1998) and retailing (Broadbridge, 2000). Such previous studies, have identified a range of inhibitors that may be grouped together broadly as employment practice and attitudinal barriers. Male-centred management styles and cultures (Alimo-Metcalfe, 1995; Traves et al., 1997; Broadbridge, 1999), and men's "power in higher management" (Parker et al., 1998, p. 156), together with networking patterns (Driscoll, 2004; Ogden et al., 2006) and the potential impact of work on parental responsibilities (Liff and Cameron, 1997), are examples of the former category. It is argued that views in much of the commercial sector are still "think female manager, think childless superwoman" (Liff and Ward, 2001, p. 32). A survey of middle managers in a UK bank, for example, finds that 56 per cent of women bank managers had no children compared to 14 per cent of male managers (Granleese, 2004). Similarly, in a study of 21 call centre managers, only three had children and over half were single (Belt, 2002). Negative gender stereotyping (Powell et al., 2002), females' tentativeness in career progression (Brown, 1990); and females' professional work-life balance sacrifices (Kellerman, 2003) or priorities (Hakim, 2004) are examples of the attitudinal category of barriers. Less is known however, about active enablers of females' career progression in the UK (Maxwell and Ogden, 2006).

Consequently, the aim of this paper is to investigate the Visibility of Female Directors in the Local Licensed Commercial Banks in Sri Lanka. This builds on work done recently by Ogden et al. (2006) which studies gender balance in the UK financial services industry. In his study First, enablers of female managers' career development are discussed from theoretical perspectives in terms of managerial gender composition, management styles, and impression management. Second, the findings from seven case studies, informed by 148 research participants, are presented as regards managerial gender composition, management styles, and organizational culture and impression management. As these cases have a number of female managers that is above their industry average, key implications on the nature and significance of enablers of the career development of female managers in financial services and retailing can be identified. In particular, these may help human resource managers and operational managers in finance and retail companies to recognize the extent to which these factors feature – currently or potentially – in the organizations and industries in which they operate. In addition to ILO 46.3% of global female participation was facilitated by the services sector in 2018. Also Culpan et al (2007) in their study quotes of professional financial service sector being female-dominant in general. In 2018, 45% of the labour force was comprised from the service sector in Sri Lanka. This industry has a fair amount of general female representation compared to other industries and therefore can be justified as the most suitable industry to conduct this investigation in a parametric concern.

2018 Department of Statistics indicates majority of underemployments/unemployed are females, while they are equally educated and literate as males. This definitely points out the disparity of stereotypes when it comes to occupational consideration. Therefore, analyzing the profiles of females who have broken the glass-ceiling effect has become a mandatory consideration which represents the possibility of shedding career barriers for female managers.

Female's visibility in upper echelons of management continues to be disproportional in comparative to men. Cai & Kleiner (1999) argues women are separated from positions with significant organizational power due to traditionally management being viewed a male occupation.

However, Glass Ceiling theory defined as "an invisible barrier which prevents women climbing the corporate ladder" Ryan et al. (2007); Thomas & Bierema (2004) explains real-world reasons. The assignment of the specific term 'Glass-ceiling' in 1980's exploited this issue in to the root-causes. According to Elmuti et al. (2003) generally these artificial obstacles are created by individual or organizational prejudices. The nature of these barriers is explained by Akande (1996) as "subtle that it is transparent, yet so strong that it prevents women moving up the organizational hierarchy".

Table 2. Types of Glass Ceiling

Type of Glass Ceiling	Description
Bamboo Ceiling	Excluding Asian-Americans from managerial roles
Concrete Ceiling	Types of strong barriers minority woman experience
Expatriate Glass Ceiling	After breaking through the 1st barrier, the next woman face
Glass Elevator	Rapid Promotion of men over woman especially in female dominant fields
Glass Cliff	Females being promoted to risky positions where the risk of failure is high
Sticky Floor	Woman trapped in low mobility/wage jobs
Sticky Ladder	Capable woman struggling to reach the top but stuck in the middle

According to literature there are different forms of Glass-ceiling as discussed as above; Elmuti et al. (2003) reveals more Women gave-up their careers because of marriage/children. Simpson & Altman (2000) 45% of younger-men were persuading their career life while having children compared to only 26% of younger-women. Broadbridge (2008) reports 50% perseverance rate of dual-roles as a barrier for women. They also felt that women still had more concern for the home and family. Labour statistic report (2008) states that among economically inactive females 53% was due to engagement in housework.

The glass ceiling remains a contemporary issue, with many surveys and reports being undertaken internationally (see Catalyst, 2007; Singh and Vinnicombe, 2004; EOWA, 2003). The constituent elements however appear to be entrenched, and among the key barriers consistently identified, which negatively affect women's advancement in corporations are the lack of transparency around promotion policies; work-family conflict; the old boys' network and the lack of senior or visibly successful female role models (see Cross and Linehan, 2006; Catalyst, 2007; Wirth, 2004). The operation of promotion policies in many organizations is an area which often has an unclear set of criteria attached. This allows for subjectivity rather than objectivity to enter the process, with senior management allowed considerable scope for discretion (Coughlan, 2002). Often, promotion criteria are focused on a preconceived notion of who should perform the role rather than the qualifications required. In an Irish study conducted by Cross and Linehan (2006) the findings suggest that promotional policies have a significant negative impact on the advancement of women to senior management positions. All of the interviewees in that study indicated a lack of transparency surrounding promotion procedures within their organizations, and Barriers to the executive suite 105 believed this to be directly responsible for the low numbers of women at senior management level. According to Flanders (1994) employers often believe that women are less committed to work and less able to undertake a full-time career than men, due to their biological make-up, rather than ability. When promotion opportunities arise and an employer is given the choice between a man and a woman with equal qualifications, the woman, he posits, is frequently viewed as the greater risk. Working in male dominated hierarchies is seen to reduce women's advancement levels as male managerial hierarchies are more likely to promote men for managerial positions, as men feel more comfortable with other men than they do with women (Tharenou, 2005). As Rosener (1995, p. 26) highlights, the "one-best model" of management and leadership leads to the thinking that anyone who is different "can only be worse". Thus, if the "standard of quality is male, then being a woman means being less good" (Rosener, 1995, p. 33). Another area which has generated much research in the search for answers to the dearth of senior female managers is the arena of work-family conflict. One of the major issues facing women is that women are still viewed by society and self as the primary care giver in relation to their domestic responsibilities. Studies on the division of non-market labour highlight that women retain a disproportionate responsibility for both domestic work and childcare, even in situations where both partners are working (see CSO, 2006; Gunter and Stambach, 2005; Rees, 1992). Although media reports indicate that men have increased their participation in domestic duties, the burden of responsibility is still carried by women. Schoon et al.'s (2006) findings highlight the persistence of the role of gender on individual career development. The dual roles of mother and labour-force participant appear to be both interdependent and in conflict. While, for men, the roles of father, and employee, are more independent, thus easier to combine. They posit that this suggests that the realities of working women's lives have not been accommodated for in the institutional arrangements. The importance of networking for female career progression is also a contemporary research theme (Bierema, 2005; Forret and Dougherty, 2001; Linehan, 2001). Studies have found that informal networks help men advance in management more than women, especially in mid-late career (Tharenou, 2005). In this regard, the "men's club" or the "old boys network" consistently emerges in the literature as a significant barrier encountered by women (see Cross and Linehan, 2006; Burke et al., 1995). The existence of the "old boys network" is often viewed as a primary reason why women are ignored and indeed discouraged from seeking top management positions. "Old boys" networks are effectively closed to females, as women have great difficulty in accessing and utilizing their possible potential (Ehrich, 1994), something which works to their disadvantage in the career advancement race.

Research on women on boards of directors is a growing field of study with great relevance for corporate, social and national policy (Bilimoria, 2000). Prior research on female directors has focused mostly on human capital variables such as education, experience in leadership positions, age and race (Singh et al., 2008; Johnson et al., 2013), generally seeking to understand gender differences in demographic characteristics and the effects of board diversity on board and firm performance. Though previous descriptive studies in the corporate governance literature point to the importance of network connections for board appointments (Lorschand and MacIver, 1989; Bilimoria and Huse, 1997; LeBlanc and Gillies, 2005), few studies have undertaken large sample quantitative examination of the role of professional networks on the relative probability of women being appointed to corporate boards as compared with men (Johnson et al., 2013). In the current study, we directly address this knowledge gap by investigating how human capital and professional networks contribute to women's as compared with men's odds of corporate board membership

Human Capital

A number of human capital characteristics are important for corporate board directors. In their role of monitors of corporate activity and advisors to the top management team, directors need to possess business acumen and/or knowledge of their field of expertise. Business acumen tends to be indicated by a combination of educational qualifications and professional experience. Age is frequently considered a proxy for the maturity to understand and participate in collaborative behaviors in the boardroom. The combination of director age, educational qualifications, profession and previous professional experience (particularly experience on other boards of directors) is commonly referred to as human capital (Withers et al., 2012). We have ample evidence, anecdotal and empirical, of the type of human capital that is desirable in corporate directors. For a long time, boards were considered a fiefdom of Caucasian men from established business families with degrees from Ivy League universities (Mills, 1956; Useem, 1984;

Branson, 2006). As ownership of companies transferred from family groups to shareholders and investment firms, board composition transformed to include members of top management teams, including from large investors. Changes in regulation and public sentiment led to the inclusion of community organizers and government officials. The evolution of the business environment transformed the composition of boards from fiefdoms to professional cliques. When the call for diversity in the boardroom became persistent and more powerful, women, particularly those in leadership positions, were invited to join in the boardroom. Early research found that female directors usually came from non-business occupations such as community service or philanthropic non-profit organizations, past government positions or educational institutions, often those that catered to women (Harrigan, 1981; Kesner, 1988). Two decades later, Hillman et al. (2002) found that not much had changed with respect to the "type" of women who were being invited to corporate boards – women on corporate boards still tend to be from non-CEO backgrounds, such as being experts in law or banking and from government or educational institutions.

METHODOLOGY

This paper used both annual report survey method and direct interview method to collect data for identify the female visibility level in listed annual reports in Sri Lanka. Since the investigation is regarding visibility levels and profiles of executive level female representation, it's important to identify a field where female participation is considerable in order to derive better generalization. Therefore, the commercial banking industry was selected due to female participation and authorized means to disclosed information etc. The below figure illustrates the Composition of the Licensed Commercial Banking Sector in SL under the Banking Act No. 30 of 1988

Figure 1.

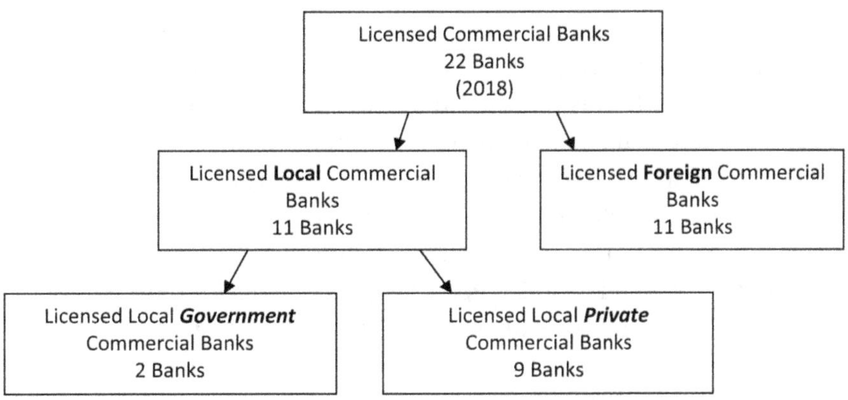

FINDINGS

To understand trends about the Apex-level female composition in local licensed commercial banks over the years, the following analysis has been done using data published in annual reports of local licensed commercial banks during the time span from 2015-2018.

The results clearly illustrate that female visibility in the local licensed commercial banks remains at a very low level. In a country with ethnic diversity, the local licensed commercial banking sector also follows same trends. Majority of females in this investigation were Buddhists reflecting on country's highest practiced religion. But it was surprising to see many Christians representing females in this level. There was only a single Muslim/Tamil female representative 2018.

Table 3. Comparison of % female directors over the years

Name of the Bank	2015		%	2016		%	2017		%	2018		%
	M	F		M	F		M	F		M	F	
Bank of Ceylon	06	02	25	06	01	14	06	01	14	04	01	20
Peoples Bank	08	01	11	08	01	11	07	01	12	06	-	00
Commercial Bank	08	01	11	09	02	18	11	02	15	11	02	15
DFCC Bank	07	02	22	07	02	22	06	04	40	07	03	30
Pan Asia Bank	12	01	07	12	01	07	04	01	20	08	01	11
Union Bank	08	-	00	11	02	15	10	02	16	09	02	18
Hatton National Bank	10	03	23	10	03	23	10	03	23	12	02	14
Nations Trust Bank	09	03	25	10	03	23	11	03	21	11	03	21
Sampath Bank	09	04	30	08	04	33	07	04	36	10	03	23
Seylan Bank	09	01	10	09	02	18	08	02	20	09	02	18
NDB	07	04	36	08	04	33	08	03	27	09	03	25
HDFC	08	01	11	08	01	11	08	01	11	07	02	22

The Average % of female directors in Govt. licensed commercial banks remain very minimum than the average of the industry. People's Bank having zero female directors have caused this situation. As the bank comprising the highest number of employees in the whole industry, this doesn't create a positive impression on gender diversity/HRM. This indicates that there's absolute low power in the Board level for females towards decision making and equal female representation in the Govt. licensed commercial banks in SL. This is a very sad situation for government institutes, since the equal-right/opportunities laws of SL are directly being breached in such circumstances. This confirms the weaknesses on the practical implication of equal-rights guaranteed under the 1978 constitution and the amended citizenship act in 2003. On the other hand, Pvt. Commercial banks report a higher average of female directors, exceeding the sector average. This might reflect on the extent of acceptance gender diversity management in Pvt. Banks.

According to the table above it's clear that in almost all these banks female representation is very low. Throughout the years none of the banks comprised 50% female representation at the board level. In 2017, DFCC Vardhana Bank reported the highest 40% of female representation. This is higher than the overall female representation in these banks during that year.

Comparing data published over the years in Annual reports, Sampath Bank comprises 1/3 female representation in the board-level. In government banks like people's bank did not have any significant female representation. Era of diverse workforce and especially majority of the Sri Lankan population being females, it's pathetic for a so called dynamic organization to not have generic representation at the Apex-level of their organizations. In such context, it's critical to state that they don't represent all employees equally in their organization. People's Bank and Union bank had zero female board representatives in 2015 and 2017 respectively.

DFCC Vardhana bank reported highest %female board representation during 2018, 2017, 2015 and HNB reported fixed %female board representation from 2015 to 2017. Both these banks reported fixed % of female directors than the average female visibility in the sample. Although PAB didn't report the highest % of female directors; their % of female directors remained below than the average visibility in all other years, except in 2017. However, these three banks can be considered as the most diversity workforce seeking banks in the local licensed commercial banks in SL. Percentage of Female representation in the board has increased over the years in many banks. But in NDB, BOC and Seylan bank it has reduced over the years. In 2016 and 2017 there were 2 banks with no female directors. Average visibility of females in the BOD remains low as 10.25% over the years. According to information presented, for an industry which comprised more female composition, there are very few voices representing them at the top of the hierarchy.

Female participation and visibility-levels can vary according to Government/private sector they are involved in. Specially developing nation situations are different from the developed nations. Organizational hierarchies in developing nation government organizations are more autocratic and rigid in nature. This acts as an obvious barrier for female's career advancement.

Despite this condition, when analyzing employee levels in the local licensed commercial banks, Government banks individually represents the highest numbers of People's Bank 8,401 and BOC 7,912 which accompanies the highest number of employees in the local licensed banking sector. Private commercial banks individually represent lower numbers of employees as shown in the graph below. (Excluding NDB since employee numbers aren't reported)

Figure 2.

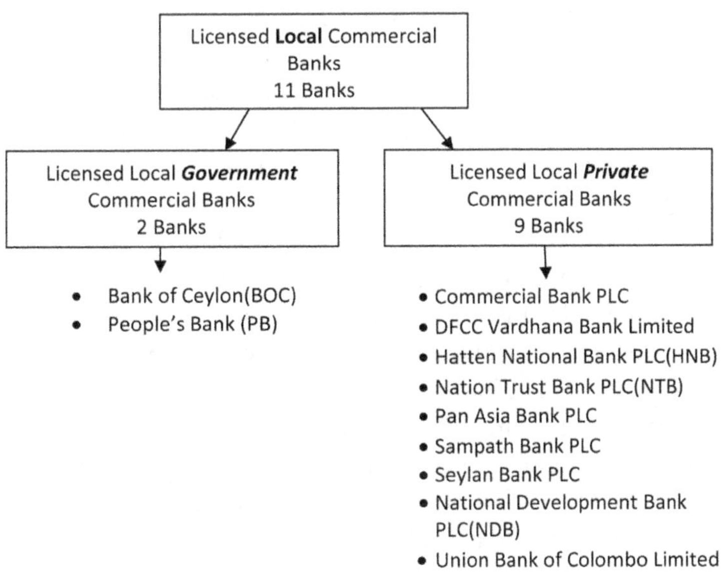

Figure 3. Employees in local licensed commercial banks in SL-2018

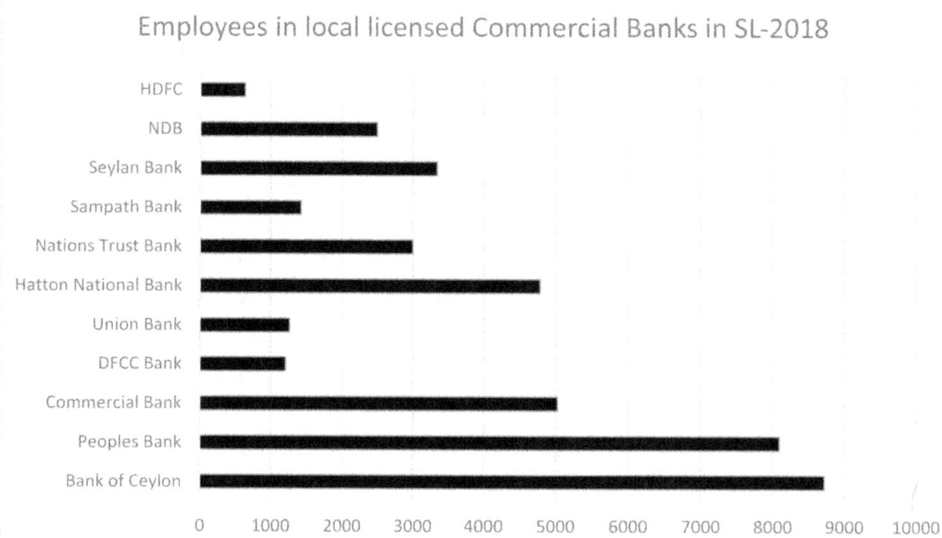

According to above mentioned information it's clear that, although the government sector comprises of 18% from the total local licensed commercial banks, it accounts to a powerful labour force of 16, 313. Remaining 72% banks total to only 17,399(excluding NDB). This is only 1,086(Excluding NDB) more employees than the government sector. This indicates the strength of the government sector as a labour force.

In such investigation it's interesting to analyses female gender composition in the Local licensed commercial banks. Yet, Gender composition/female employee information disclosure practices in these banks are limited. Operating in the twenty-first century, corporate practices have evolved from restricted formalities to more flexible and information rich environments. In many global and international banks, it's rather a norm to publish gender composition under the HR area of their Annual Reports. This specifies the quality of organizational practices. However, in the local context this situation varies drastically. Almost all banks have revealed BOD and most of the C.M information.

Yet, it's sad to note that only PAB, Commercial bank and HNB had disclosed gender composition. Therefore, numbers of female employees were only collectable from these 3 banks. This restricted the investigation from collecting information about the overall sector female composition through secondary data. Seylan Bank, BOC, NTB, People's Banks had disclosed information on Age/designation/education compositions/training etc under HR but not gender composition. Some banks didn't even have a section for HR and has limited information to HR and remuneration committee. Sampath bank in 2002 won the Top award for HRM Services, but even in 2008 they didn't have a special area allocated for HR in their Annual report. Does this poor information on important HR aspects reflect that Local licensed commercial banks are reluctant towards implementing gender diversity management practices? Or is it that these banks still have not embedded international corporate practices but claims on providing superior services? Surely the levels of disparity indicate that there's a disconnection at some point in these organizations. However, with the available information on the Apex level female representation this objective can be analyzed as follows.

Table 4. Comparison of % female corporate managers over the years

Bank	2015	2016	2017	2018
	%	%	%	%
Bank of Ceylon	20	17	17	25
People's Bank	16	15	15	16
Commercial Bank of Ceylon PLC.	0	7	6	10
DFCC Vardhana Bank PLC.	36	25	35	30
Pan Asia Banking PLC.	22	14	14	14
Union Bank of Colombo Ltd.	-	0	0	0
Hatton National Bank PLC.	30	25	22	20
Nations Trust Bank PLC.	-	20	18	18
Sampath Bank PLC.	0	0	0	7
Seylan Bank PLC.	-	6	16	16
National Development Bank PLC.	9	9	13	15
Average	**12%**	**13%**	**14%**	**16%**

The graph below is a summary of female representation at board level, C.M level and collective total compared against the Average % in the local licensed commercial banking industry.

Figure 4. Female directors and corporate managers by sector-2018

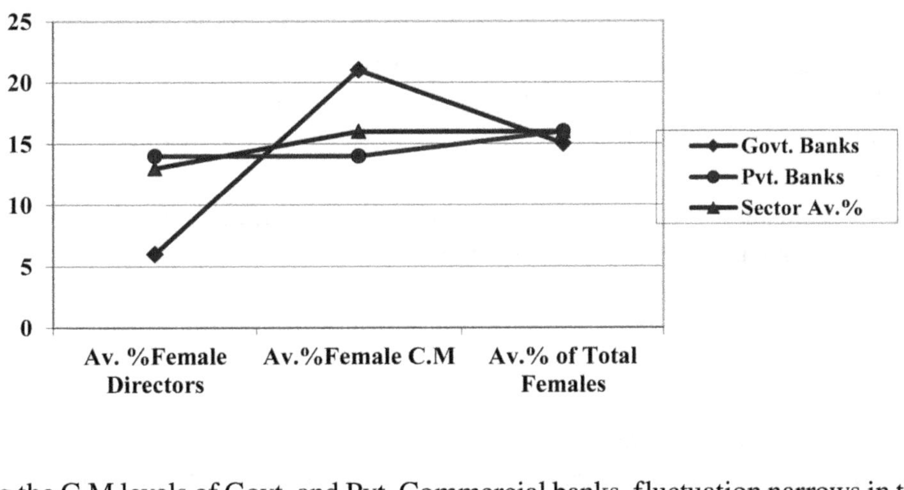

Considering the C.M levels of Govt. and Pvt. Commercial banks, fluctuation narrows in the negative side, represented by the Pvt. Banks and fluctuation widens towards the positive side, represented by the Govt. Banks. In this level govt. banks are more likely to occupy female corporate managers. However, since the absolute number of govt. banks remain very low, this representation might not be accurate 100%. But a general idea could be made. At this level, the higher rigidity of including females into the Govt. banks has reduced. Union bank did not account any female corporate managers from the Pvt.

Sector and Sampath bank accounted for only 7% female corporate managers. This would have affected the overall Pvt. Sector average to decline below average of the total sector.

In general, overall female visibility in local licensed commercial banks accounts only to 16% from the total positions available. This isn't a fair representation for the total females in the local licensed commercial banks. However, Pvt. Banks represents a total of 16% females at the apex level which is equal to the industry average. Govt. banks represent only 15% of females which is below the industry average. But the fluctuation with the average remains low.

In comparison, Pvt. Sector female representation at the Apex doesn't contain intense variations around the industry average. Therefore, the female visibility in Pvt. Banks encompasses a positive nature. But since the Govt. sector contains higher variations, the visibility of females depends on the management level.

Female visibility at the Director level comprise very low and at the corporate-level moderate in Govt. commercial banks. This proves the common trends of government institutions being male dominant at the Apex levels and encloses strong perceptions on female's eligibility towards performing corporate responsibilities. The glass-ceiling effect shows clear signs of existence in Govt. commercial banks.

Based on above information, one could argue that as the total number of employees increase in local licensed banks, there's an adverse effect on female representation at the very top-level. (E.g.: People's Bank, Commercial Bank, Sampath bank). Banks with lesser number of employees symbolize comparatively higher female representation at the top level. (E.g.: PAB, DFCC Vardhana Bank). But this trend changes as;

- Union bank has low number of total employees and lower number of female representation at BOD/C.M
- HNB has higher number of Total Employees and Higher number of female representation at BOD/C.M

Therefore, further research need to be done to justify the above argument. Information abstracted from the Annual reports will only not be adequate for conclusion.

By failing to inquire the success factors behind these thriving executive women, the real value of Human Resource could be under-estimated and future female-managers might not visualize their potentials due to unclear career guidance/real-life influences. This will under-utilize the talent-pool; Optimal usage of Human Resource in dynamic organizations will therefore be uncertain causing Costs in short/long-term.

CONCLUSION

In conclusion, this investigation report identifies female participation in the Board of Directors and Corporate Management levels from 2005-2018. Female participation at the Corporate Management is comparatively high than in the board level, proves that the intense of the glass-ceiling effect increases as females climb the career-ladder.

However, existing female directors and Corporate Management based on designations, education, experience and third party networks. This proves that all existing females' occupational profiles are rather attractive and strong. Designations held by these women were rather surprising as they held most common secretariat positions and Legal, Operations positions which are specified as male-oriented. Almost all of them had higher educational attainments/professional qualifications. Many had extensive years of experiences and the highest visible females had network relationships which literature reveals as career facilitators for women. It was ratified that Government banks had rigid structures towards female progression and Private Banks were in line with the industry average although the local licensed commercial banking sector in general under-represent females at the Apex-level.

Finally, by relating these leap-holes in the Sri Lankan local licensed commercial banking sector, recommendations are presented to reduce barriers opposing from corporate climates that isolate females, Pipeline barriers and institutional rigidity that restrict career growth.

RECOMMENDATION

Centuries gone passed with the glass-ceiling effect prevailing in the corporate world, it isn't yet too late to adopt remedies which will help women break through the invisible barrier against their career-advancement. Today's dynamic organizations perceive gender diversity as important and manage their labour force effectively in order to achieve optimum results in an extra competitive environment. Although Glass-ceiling hasn't hit hard in SL since it's a developing country and small in nature, it's always better to take early precautions to prevent the symptoms revealed in previous chapters through the analysis of the objectives. Therefore, the following are an amalgamation based on recommendations given by the FGCC, USA and ILO report (2004) which is transformed to suit the local licensed commercial banks in SL.

It's important to understand that neither the female employee nor the bank can alone accomplish these recommendations. A solid integration is therefore mandatory. These recommendations links with the narrow areas identified in previous objectives and discusses means to increase visibility of female managers. Below figure explains Female director's perceptions on low representation at Apex-level in USA.

Considering the facts in the figure and weaknesses identified when analyzing the above objective, the following Recommendations are made.

Revise HR Policies

According to previous Objective it's clear that local licensed commercial banks are less serious in handling HR related areas. By not publishing relevant information in their Annual reports, these banks at times claim to triumph HR awards. Banks should address HRM meaningfully and effectively therefore. It's important to practice a complete unit than mere escape because in the long-run it will be beneficial. The poor functioning of Gender-diversity issues and HR policies in local licensed commercial banks should be addressed.

It's impossible to exclude females from workforces as per law and best CSR practices. Banks should not hire females just to fulfill these aims, but with the objective of occupying them as intellectual capital. It's vital to have a diverse workforce since disadvantages of one category could be complemented by another. By providing these conditions, employees will benefit efficient working environments which will increase their motivation. Banks will be benefited in-terms of utilizing talent available effectively and automatically it will result in higher performance rates.

Diversity training deflates myths regarding suitability and capability of women for executive careers. It's important to create a culture that respects a female for who she is and not discriminate her demographic features. ILO report (2004) states that Women should not be penalized financially for motherhood, such as the elimination of retirement plan forfeitures for career-breaks, the provision of unpaid leave of absence with a guarantee of a job on return to work etc. This could be ensured by enforcing anti-discrimination legislation at the workplace so that women can increase their visibility and have better access to different levels of the organizational hierarchy.

These processes should inculcate with banks business plans and strategies. Involvement of male managers in this process is highly important. Their corporation can be enhanced through rewards systems. This process will make females familiar in working in corporative male environments. Ultimately, mono-stereotyped environments will not only result team-orientation but also key business areas like profits/capital-investment/productivity/market-share and quality will be improved according to the FDCC.

Mentoring/Career Encouragement for Female Managers

Kram (1985 cited women in management review 2003, volume 18, p.237) defined mentor support as "the support provided by a more experienced, highly ranked and influential member of the organization to less experience individuals". In Sri Lankan context most of these occupational mentors are males, since the male workforce is high. This creates a barrier for women due to differences in career-advancement factors affecting genders and male mentors don't exactly tally with career encouragement/motivation factors of women. Thomas and Bierema (2004, p. 64) says that women's career patterns are different from men. Women encounters more career interruptions and unplanned career paths.

Therefore, female managers should be focused more towards learning from career profiles of these existing Apex-level female managers in designing their career paths by inserting values seen in experienced female's profiles. Banks need to build relationships among existing Apex-level females and potential female employees so that they can learn through experience. This will build instrumental traits (work under pressure/competitiveness/Independence/self-confidence, decision making) and Expressive traits (Helpfulness, understanding, gentle etc) which are perceived as successful traits. Succession planning for future female executive managers will produce more focused/experienced managers with developed relationships which will increase their efficiency. This will not only facilitate visibility of female managers but also increase the quality of the future management teams due to prior experience.

FGCC says many successful businesses have identified mentoring as a critical factor in career advancement due to its function as a means of networking, socializing/forming ties with influential corporate leaders and learn to manage dual roles. Also by felicitating these high-level female managers, future female managers will be encouraged due to awareness of the possibility of females' reach zenith. This will add to female employee's visibility and will increase career encouragement.

Networking

Literature reveals that men use these networks more strategically than females. But Sri Lankan women by nature are shy and reluctant to make initiatives. Potential Females should be encouraged and educated of the importance of involving in such networks. By analyzing current profiles of executive females it's clear that their visibility has increased due to their involvement in many other networks like Charter Accountants, chamber of commerce, Charted Secretaries, institute of Bankers etc as explained in the 3rd objective.

Metz (2003) states that affinity groups are important development tools for those seeking higher responsibilities in organizations. These groups build support within and outside the banks for women creating networks increasing mobility of information, latest practices/knowledge-sharing, career planning/strategizing, professional support/encouragement and increased visibility. By establishing such networks within banks, future female employees will also be benefited and comfort levels of facing diversity will increase as a result. Such networks will create a strong/competent women's work force in Banks if initiatives are taken correctively.

These networks will also help female employees find solutions for work related problems from learning through experiences of others. Also it will build their reputation with the female employee being exposed to more interaction, speak up at meeting and seeking for high profiled assignment which will ultimately increase the visibility of the female manager.

By incorporating to these circumstances banks could make full use of the rich talent this nation has to offer. Also these recommendations would improve bank working conditions with inclusive workplace cultures, policies, and programs that support women's advancement.

REFERENCES

Alimo-Metcalfe, B. (1995). An investigation of female and male constructs of leadership. *Women in Management Review*, *10*(2), 3–8. doi:10.1108/09649429510146901

Belt, V. (2002). A female ghetto? Women's careers in call centres. *Human Resource Management Journal*, *12*(4), 51–66. doi:10.1111/j.1748-8583.2002.tb00077.x

Bierema, L. (2005). Women's networks: A career development intervention or impediment? *Human Resource Development International*, *2*(2), 207–224. doi:10.1080/13678860500100517

Bilimoria, D. (2000). Building the business case for women corporate directors. In R. J. Burke & M. C. Mattis (Eds.), *Women on Corporate Boards of Directors: International Challenges and Opportunities* (pp. 25–40). Kluwer Academic. doi:10.1007/978-90-481-3401-4_3

Bilimoria, D., & Huse, M. (1997). A qualitative comparison of the boardroom experiences of US and Norwegian women corporate directors. *International Review of Women and Leadership*, *3*(2), 63–76.

Branson, D. M. (2006). *No Seat at the Table: How Corporate Governance and Law Keep Women Out of the Boardroom*. New York University Press.

Broadbridge, A. (1999). Retail managers: Stress and the work-family relationship. *International Journal of Retail & Distribution Management*, *27*(9), 374–382. doi:10.1108/09590559910292825

Broadbridge, A. (2000). Retail managers: Their work stressors and coping strategies. *Journal of Retailing and Consumer Sciences*.

Broadbridge, A. (2008). Senior careers in retailing: An exploration of male and female executives' career facilitators and barriers. *Gender in Management*, *23*(1), 11–35. doi:10.1108/17542410810849105

Brown, S. (1990). Management of education and social welfare in Scotland: The places of women. *Scottish Journal of Adult Education*, *9*(3), 3–13.

Burke, R. J., Rothstein, M. G., & Bristor, J. M. (1995). Interpersonal networks of managerial and professional women and men: Descriptive characteristics. *Women in Management Review*, *10*(1), 21–27. doi:10.1108/09649429510077458

Cai, Y., & Kleiner, H. (1999). Sex discrimination in Hiring: The Glass ceiling. *International Journal: Gender in Management*, *18*(2-4), 51-55.

Catalyst. (2007a). *2007 Catalyst Census of Women Board Directors*. New York: Catalyst.

Catalyst. (2007b). *2007 Catalyst Census of Women Corporate Officers and Top Earners*. New York: Catalyst.

Coughlan, P., & Coghlan, D. (2002). Action research for operations management. *International Journal of Operations & Production Management*, *22*(2), 220–240. doi:10.1108/01443570210417515

Cross, C., & Linehan, M. (2006). Advancing female careers in the high-tech sector: Empirical evidence from Ireland. *Women in Management Review*, *21*(1), 28–39. doi:10.1108/09649420610643394

Cross, C., & Linehan, M. (2006). Advancing female careers in the high-tech sector: Empirical evidence from Ireland. *Women in Management Review*, *21*(1), 28–39. doi:10.1108/09649420610643394

CSO. (2006). *Women and Men in Ireland*. Central Statistics Office, Stationery Office.

Culpan, O., Marzotto, T., & Demir, N. (2007). Foreign banks: Executive jobs for Turkish Women? *Women in Management Review*, *22*(8), 608–630. doi:10.1108/09649420710836308

Driscoll, M. (2004, May 2). The girls ganging up on the old boy network. *The Sunday Times News Review*, p. 9.

Ehrich, L. C. (1994). Mentoring and networking for women educators. *Women in Management Review*, *9*(3), 4–10. doi:10.1108/09649429410056281

Elmuti, D., Minnis, W., & Abebe, M. (2005). Does education have a role in developing leadership skills? *Management Decision*, *43*(7/8), 1018–1031. doi:10.1108/00251740510610017

Equal Opportunity in the Workplace Agency & Catalyst. (2003). *Australian Census of Women Executive Managers and Women Board Directors*. EOWA.

Flanders, M. L. (1994). *Breakthrough: The Career Woman's Guide to Shattering the Glass Ceiling*. Paul Chapman Publishing.

Forret, N. L., & Dougherty, T. W. (2001). Correlates of networking behavior for managerial and professional employees. *Group & Organization Management*, *26*(3), 283–311. doi:10.1177/1059601101263004

Granleese, J. (2004). Occupational pressures in banking: Gender differences. *Women in Management Review*, *19*(4), 219–225. doi:10.1108/09649420410541290

Gunter, R., & Stambach, A. (2005). Differences in men and women scientists' perceptions of workplace climate. *Journal of Women and Minorities in Science and Engineering*, *11*(1), 97–115. doi:10.1615/JWomenMinorScienEng.v11.i1.60

Hakim, C. (2004). *Key Issues in Women's Work: Female Diversity and the Polarisation of Women's Employment*. Glasshouse Press.

Harrigan, K. R. (1981). Numbers and positions of women elected to corporate boards. *Academy of Management Journal*, *24*(3), 619–625.

Hillman, A. J., Cannella, A. A. Jr, & Harris, I. C. (2002). Women and racial minorities in the boardroom: How do directors differ? *Journal of Management*, *28*(6), 747–763. doi:10.1177/014920630202800603

International Labour Office. (2004). *Breaking Through the Glass-ceiling*. International Labour Office.

Johnson, S. G., Schnatterly, K., & Hill, A. D. (2013). Board composition beyond independence: Social capital, human capital, and demographics. *Journal of Management*, *39*(1), 232–262. doi:10.1177/0149206312463938

Kellerman, B. (2003). You've come a long way baby – and you've got miles to go. In D. B. Rhode (Ed.), *The Difference 'Difference' Makes*. Stanford University Press.

Kesner, I. F. (1988). Directors' characteristics and committee membership: An investigation of type, occupation, tenure and gender. *Academy of Management Journal*, *31*(1), 66–84.

Leblanc, R., & Gillies, J. (2005). *Inside the Boardroom: How Boards Really Work and the Coming Revolution in Corporate Governance*. John Wiley & Sons.

Liff, S., & Ward, K. (2001). Distorted views through the glass ceiling: The construction of women's understandings of promotion and senior management positions. *Gender, Work and Organization*, *8*(1), 19–36. doi:10.1111/1468-0432.00120

Linehan, M. (2001). Networking for female managers' career development: Empirical evidence. *Journal of Management Development*, *20*(10), 823–829. doi:10.1108/EUM0000000006237

Lorsch, J. W., & MacIver, E. (1989). *Pawns or Potentates: The Reality of America's Corporate Boards*. Harvard Business School Press.

Mattis, M. C. (1995). Corporate initiatives for advancing women. *Women in Management Review*, *10*(7), 5–14. doi:10.1108/09649429510095872

Maxwell, G., & Ogden, S. (2006). Career development of female managers in retailing: Inhibitors and enablers. *Journal of Retailing and Consumer Services*, *13*(2), 111–120. doi:10.1016/j.jretconser.2005.08.010

Mills, C. W. (1956). *The Power Elite*. Oxford University Press.

Ogden, S. M., McTavish, D., & McKean, L. (2006). Clearing the way for gender balance in the management of the UK financial services industry: Enablers and barriers. *Women in Management Review*, *21*(1), 40–54. doi:10.1108/09649420610643402

Ogden, S. M., McTavish, D., & McKean, L. (2006). Clearing the way for gender balance in the management of the UK financial services industry: Enablers and barriers. *Women in Management Review*, *21*(1), 40–53. doi:10.1108/09649420610643402

Parker, L., Guthrie, J., & Gray, R. (1998). Accounting and management research: Passwords from the gatekeepers. *Accounting, Auditing & Accountability Journal*, *11*(4), 371–406.

Parker, L., Guthrie, J., & Gray, R. (1998). Accounting and management research: Passwords from the gatekeepers. *Accounting, Auditing & Accountability Journal*, *11*(4), 371–406. doi:10.1108/09513579810231420

Powell, G. N., Butterfield, D. A., & Parent, J. D. (2002). Gender and managerial stereotypes: Have the times changed? *Journal of Management, 28*(2), 177–193. doi:10.1177/014920630202800203

Purcell, J. (1996). Human Resource Bundles of Best Practise: A Utopian Cul-de-sac? Paper presented in the ESRC/BUIRA seminar series on the Contribution of HR Strategy to Business Performance.

Rayn, M. K., Haslam, S. A., & Postmes, T. (2007). Reaction to Glass cliff: Gender differences in the explanations for the precariousness of women's leadership positions. *Journal of Organizational Change Management, 20*(2), 182–197. doi:10.1108/09534810710724748

Rees, T. (1992). *Women and the Labour Market*. Routledge.

Rosener, J. B. (1995). *America's Competitive Secret: Women Managers*. Oxford University Press.

Rosener, J. B. (1995). *America's Competitive Secret: Utilizing Women as a Management Strategy*. Oxford University Press.

Schoon, I., Martin, P., & Ross, A. (2006). Career transitions in times of social change. His and her story. *Journal of Vocational Behavior, 70*(1), 78–96. doi:10.1016/j.jvb.2006.04.009

Singh, P. (2008). Empirical assessment of ISO 9000 related management practices and performance relationships. *International Journal of Production Economics, 113*(1), 40–59. doi:10.1016/j.ijpe.2007.02.047

Singh, V., Terjesen, S., & Vinnicombe, S. (2008). Newly appointed directors in the boardroom: How do women and men differ? *European Management Journal, 26*(1), 48–58. doi:10.1016/j.emj.2007.10.002

Singh, V., & Vinnicombe, S. (2004). Why so few women directors on UK corporate boards: Evidence and theoretical explanations. *Corporate Governance, 12*(4), 479–488. doi:10.1111/j.1467-8683.2004.00388.x

Tharenou, P. (2005). Women's advancement in management. In R. J. Burke & M. C. Mattis (Eds.), *Supporting Women's Career Advancement: Challenges and Opportunities*. Edward Elgar Publishing. doi:10.4337/9781845423452.00009

Traves, J., Brockbank, A., & Tomlinson, F. (1997). Careers of women managers in the retail industry. *Service Industries Journal, 17*(1), 133–154. doi:10.1080/02642069700000007

Useem, M. (1984). *The Inner Circle: Large Corporations and the Rise of Business Political Activity in the US and UK*. Oxford University Press.

Wirth, L. (2004). Breaking through the Glass Ceiling: Women in Management. International Labour Office.

Withers, M. C., Hillman, A. J., & Cannella, A. A. Jr. (2012). A multidisciplinary review of the director selection literature. *Journal of Management, 38*(1), 243–277. doi:10.1177/0149206311428671

APPENDIX

Table 5.

Name of the Bank
Bank of Ceylon (BOC)
Peoples Bank
Commercial Bank
Development Finance Corporation of Ceylon Bank (DFCC)
Pan Asia Bank (PAB)
Union Bank (UB)
Hatton National Bank (HNB)
Nations Trust Bank(NTB)
Sampath Bank (SB)
Seylan Bank (SY)
National Development Bank (NDB)
Housing Development Finance Corporation (HDFC)

Chapter 12
Integrated Reporting Compliance in Portuguese Companies

Teresa Eugénio
https://orcid.org/0000-0003-1040-3164
CARM, Polytechnic of Leiria, Portugal

Filipa Domingues
Polytechnic of Leiria, Portugal

Henrique Carvalho
Polytechnic of Leiria, Portugal

Ana Fialho
CEFAGE, University of Évora, Portugal

Graca Azevedo
https://orcid.org/0000-0002-6346-4035
University of Aveiro, Portugal & CICF, Portugal

ABSTRACT

Given the growing importance of reports with non-financial information, the aim of this study is to assess integrated reports (IR) published by Portuguese companies. The assessment was carried out considering the compliance (or not) with the current international standards for reporting. To achieve this goal, an analysis of 23 integrated reports published by seven Portuguese companies between 2016 and 2019 from different activity sectors were analyzed using the content analysis methodology and through the construction of a working tool based on the essential contents disclosed by the International Integrated Reporting Council (IIRC) and by the European Commission Recommendation. In all the IRs analyzed, several topics that are considered as essential for inclusion by IIRC and the European Commission - COM 2017/C 215/01 were not disclosed by companies. This reveals that the need persists to develop the contents to be included in the IRs. The level of compliance is low, which demands greater consideration regarding the effect of standards for reporting practices.

DOI: 10.4018/978-1-7998-9410-0.ch012

INTRODUCTION

As pointed out by Kumar et al. (2021) corporate sustainability is now considered imperative to contribute toward sustainable development by organizations all around the world. With the evolution of the corporate world, globalization, and worldwide competition, the information disclosed by companies to comply with legal obligations has become ever more important, as needs to make the comparability of that information easier. Both of these issues demand the standardization of procedures. Issues related to corporate social responsibility, clarity, and transparency of information are topics that are little explored in traditional accounting. Adams (2015) describes as main shortfalls in the information provided by companies the lack of strategic information, lack of disclosure of risks, goals, and forecasts of future performance, as well as the lack of proper descriptions of the relationship between financial and non-financial indicators.

Integrated reporting gained considerable attention in the last few years as an evolutionary approach to corporate communication (Camodeca et al. 2018). This kind of report emerges as an attempt to help overcome this shortcoming to the extent that it compiles information of financial and non-financial nature. The aim is to provide useful information that allows key stakeholders to make decisions in a more conscious and informed manner. According to Melloni et al. (2017), integrated reporting (IR) is an international attempt to link financial performance with a company's environmental, social, and governance performance. In 2010 the International Integrated Reporting Council (IIRC) emerged, identifying the need for financial reporting to keep up with the evolution occurring in companies, and creating a framework to help organizations produce their Integrated Reports according to guidelines for essential contents. Different authors call for attention to the importance of the integrated reports (Abhayawansa et al., 2019; Aceituno, 2013; Owen, 2013).

Leocádio (2017) mentions the importance of non-financial reporting as now recognized by numerous international organizations, notably including the United Nations Global Compact and the guidelines developed by the Global Reporting Initiative (GRI). The European Union recognizes the importance of this type of reporting through Directive 2014/95/EU of the European Parliament and of the Council of 22 October 2014 amending Directive 2013/34/EU, about the disclosure of non-financial information and information on diversity by certain large companies and groups. This Directive was published to increase the clarity of disclosure of non-financial information and environmental and social performance. In 2017 a Communication was also published by the European Commission (COM 2017/C 215/01) seeking to guide non-financial disclosure in the environmental, social, and governance domain. Due to the importance of this topic, the European Commission hosted a public meeting between February and June 2020 on the revision of the Non-Financial Reporting Directive (NFRD), to strengthen the foundations for sustainable investment.

The current study evaluates integrated reporting in Portugal and assesses whether it is in line with the essentials prescribed by IIRC and the European Commission - COM 2017/C 215/01. This topic has special importance nowadays, given the current debate on future guidelines for the reporting of non-financial information and the possibility of the publication of International Non-financial Reporting Standards by the IASB. Our analysis will identify the themes that are most often disclosed by Portuguese companies in their IRs and the issues that need to be further developed. The content analysis technique was used to examine the 23 Integrated Reports of the Portuguese companies that comprise the sample.

Since there are few studies on how Portuguese companies are using integrated reporting or empirical studies that demonstrate the reality of Portuguese companies *vis-à-vis* international guidelines, this study makes a valuable contribution to the literature.

Following this introduction, the integrated report is discussed, contextualizing some topics of the IR such as the objectives, laws, and regulations, and the content. Then we describe the methodology for sample selection and data treatment. This is followed by the presentation and discussion of the results and finally the conclusions, limitations, and avenues for future research.

BACKGROUND

Integrated Report: Concept, Objectives, and Advantages of Integrated Reporting

According to IIRC (2011), the integrated report aims to present an exposition of the company about its strategy, governance, performance, and prospects for the future. A reflection on the social, environmental, and commercial impact of the organization is also provided in this type of report. Through the integrated report, key stakeholders can obtain financial and non-financial information about the company, as well as a clear, concise, and substantiated representation of how the company creates and sustains value. According to Smith (2014), three pillars are revolutionizing internal operations and the way organizations interact with their stakeholders: (1) stakeholder engagement, (2) corporate governance, and (3) integrated reporting. Integrated reporting aims to group concise, objective, reliable, and relevant information about the company to its stakeholders. IIRC (2013) adds that the main purpose of integrated reporting is to enable concise communication about the company's strategy, governance, performance, and perspectives. Integrated reporting reflects a commitment to corporate social responsibility, transparency, and reliability (Leocádio, 2017). In the literature, we find mention of several advantages obtained by companies that publish integrated reports. The study by Zhou et al. (2017), based on companies in South Africa, demonstrated that companies that produce integrated reports enjoy benefits associated with the cost of capital. This advantage is directly related to the fact that the company shares non-financial information in its integrated report that reduces the risk of information gaps that can constrain the business. The conclusion of this study leads to the belief that the company's image in the market improves with the preparation and disclosure of the IR. Hoque (2017) identifies IR as a prominent tool for business organizations and adds that it is the most effective way to communicate with stakeholders and that it contributes to the improvement of strategy.

IIRC (2011) mentions some advantages of IR when compared to traditional reports, such as information adapted to what stakeholders want to know; more clarity and assertiveness in the non-financial information provided; increased confidence among the main stakeholders; better and more efficient management in the allocation of resources (regarding cost reduction and risk management); and greater commitment to investors, customers, suppliers, employees, local community, and other stakeholders.

Lopes (2018) also refers to the advantages and opportunities associated with the presentation of the IR by small and medium-sized enterprises, observing that it: provides evidence on the economic, environmental, and social impact of the company for internal and external stakeholders; supports managers in the preparation of budgets, planning, and implementation of actions; improves the company's image and credibility by demonstrating its sustainability projects; and aids in obtaining funds and attracting investments.

In addition to the advantages, Fernandes (2019) presents a danger that may be associated with this type of reporting: the possibility of companies reporting less reliable information. The sharing of flawed information, addressing only positive aspects and camouflaging the negative aspects, i.e., "green painting" or "greenwash". According to the author, to minimize this problem a cultural change is needed so that companies begin to give the correct value to sustainable development.

Directives, Laws, and Regulations

In the European Union the reporting of non-financial information, often considered by companies as integrated reporting or sustainability reporting, is addressed by Directive 2014/95/EU of the European Parliament and of the Council of 22 October 2014. According to this document, non-financial statements must be published by large undertakings that are public-interest entities and which, at the closing date of their balance sheet, exceed the criterion of an average number of 500 employees during the financial year.

To support companies covered by Directive 2014/95/EU (and therefore required to disclose non-financial information), the European Commission published guidance in July 2017 through COM 2017/C 215/01 (European Commission, 2017). This communication was prepared with Directive 2014/95/EU in mind. It refers to methodologies so that the non-financial information disclosed is relevant, useful, consistent, and comparable across companies and sectors. COM 2017/C 215/01 (European Commission, 2017) provides non-binding guidance and therefore does not create legal obligations. This communication presents the fundamental principles that the information made available should follow and the content of that information. For each of the points presented this communication provides examples of how disclosure should be made.

The integrated report appeared in Portugal as a result of the transposition of directive 2014/95/EU, giving rise to Decree-Law No. 89/2017 of 28 July. This decree-law applies to reports issued from 2017 and thereafter by public interest entities that at the balance sheet closing date have an average number of employees greater than 500. According to article 65 of the Commercial Companies Code, this report must be prepared and approved by the company's management bodies.

Content of the Integrated Report

Through the framework developed and the European Commission COM 2017/C 215/01, the IIRC provides guidelines on what the content of the integrated report should be. Furthermore, Directive 2014/95/EU states that the management report of companies should contain "sufficient information for an understanding of the development, performance, position, and impact of their activities, concerning at least environmental, social and employee-related issues, respect for human rights, and combating corruption and attempted bribery".

The basic principles released by both the European Commission and the IIRC underpin the preparation of information so that it is important and accurate and that stakeholders can trust it. Only if all disclosed information follows the same principles can it be compared. As for the essential contents, also described by the European Commission and IIRC, they concern essential information for an IR to be complete. If all integrated reports disclose information on each of these points, it is possible to compare information from different companies. Table 1 establishes a comparison between the basic principles of the two sources analyzed:

Table 1. Comparison of the main principles of IIRC and - COM 2017/C 215/01

IIRC	European Commission - COM 2017/C 215/01
Focus on strategy	Strategies and prospects
Directions for the future	
Information connectivity	Logical and coherent
	Rigorous, balanced, and understandable
Stakeholders' responsiveness and inclusion	Targetted to stakeholders
Conciseness, reliability, and appropriateness	Comprehensive, but concise
	Discloses relevant information (materials)

Source: Author, based on the basic principles of the IIRC (2011) and the European Commission COM 2017/C 215/01

As for the content that should be included in the integrated report, IIRC (2011) and the European Commission COM 2017/C 215/01 (European Commission, 2017), identify six questions that should be answered throughout the report. The communication made by the European Commission also suggests guidelines for companies on reporting systems, stating that "companies (...) can use national systems, Union systems, such as the Eco-Management and Audit Scheme (EMAS), or international systems, such as the United Nations Global Compact".

The description of the diversity policy in the governing bodies is not part of the non-financial statements. However, listed companies are to disclose a note on corporate governance. Thus, the European Commission directs that this point be addressed by identifying the diversity of age, gender, qualifications, and professional background.

In Table 2 it can be seen that the essential contents of the IIRC have in common with the Communication of the European Commission only the requirement of information on the business model adopted.

Table 2. Comparison of essential contents

IIRC (2011)	European Commission
Content elements	**Content elements**
Overall view of the organization and business model	Business model
	Policies and due diligence
	Results
	Main risks and their management
	KPIs
Performance	
Strategic goals and strategies to achieve them	
Governance and compensation	
Operational context	
Future prospects	
	Thematic features

Source: Author, based on the basic principles of the IIRC (2011) and the European Commission COM 2017/C 215/01

METHODOLOGY

This study aims to analyze in detail the content of the integrated reports of Portuguese companies to determine the topics that are covered most and if they comply with the guidelines of the IIRC proposal and the European Commission Communication COM 2017/C 215/01 (essential contents).

Sample

The sample was obtained from the GRI database (with reference to 16 April 2019) with an indication of the companies, worldwide, that published sustainability reports between 1999 and 2019. For the period 2016-2019 a total of 38 Portuguese companies were found, which during these 4 years published a sustainability report or integrated report. The sample comprised all the Portuguese companies that in the name of the report had the expression: "Integrated Report", to investigate compliance with the Integrated Reporting guidelines. The final sample is composed of 23 integrated reports, published by 7 companies between 2016 and 2019 (Table 3). The period under analysis includes the first year following the publication of the IIRC guidelines and includes the year of publication of the European Commission Communication (2017) and two subsequent periods.

Table 3. Companies in the sample

	Reports			
Company	**2016**	**2017**	**2018**	**2019**
BONDALTI	n/a	Integrated Report	Integrated Report	Integrated Report
Brisa	Integrated Report	Integrated Report	Integrated Report	Integrated Report
CTT	n/a	n/a	Integrated Report	Integrated Report
Galp Energia	n/a	n/a	Integrated Report	Integrated Management Report
José de Mello Saúde	Integrated Report	Integrated Report	Integrated Report	Integrated Report
Garrigues	Integrated Report	Integrated Report	Integrated Report	Integrated Report
SATA	Integrated Report	Integrated Report	Integrated Report	Integrated Report

Source: Author

Treatment and Analysis of Data

We used content analysis and developed a tool using Excel software. Content analysis is a method used to describe and interpret the content of documents and/or texts to create qualitative or quantitative descriptions to achieve an understanding that goes beyond common reading. Content analysis is widely used in studies in the area of Social and Environmental Accounting (De Villiers and van Staden, 2006; O'Dwyer and Unerman, 2016; Melloni et al., 2016; Adhikariparajul et al., 2019; Mata et al., 2018). According to Hsieh and Shannon (2005), there are three possible approaches to content analysis: conventional, targeted, and summative. In the present study, the summative approach was chosen since it involves counting and comparisons, followed by interpretation.

We assessed the compliance with the disclosure of essential contents for IR according to IIRC (2011) and the European Commission document COM 2017/C 215/01, as described in the Appendix.

For the working instrument, a scale of 0 to 3 was considered, taking into account the absence or absence of information, the sense of that information, and the detail, as described in Table 4.

Table 4. Scale in the assessment of information in Integrated Reports

Scale	Information provided
0	No reference is made, although it may be important.
1	No reference is made to anything important.
2	Disclosed in poor detail.
3	Disclosed in detail and completely.

Source: Author

PRESENTATION AND DISCUSSION OF RESULTS

A summary presentation is made of each of the companies selected in the sample and a summary table is presented, per company, with the average rating assigned for each of the points of the essential contents according to IIRC and the European Commission - COM 2017/C 215/01, respectively.

Bondalti

Bondalti Capital, S.A. is 100% owned by José de Mello Capital, S.A. It is a corporate center of companies based in Portugal and Spain that is mainly engaged in the chemical industry. This company published its first IR in 2017. The reports analyzed (2017 to 2019) were prepared following the Integrated Reporting guidelines of the IIRC and GRI Standards in the essential agreement option. It should be noted that for this company only three IRs were analyzed since the first was published in 2017. Table 5 presents summarized information on Bondalti's disclosure.

It can be seen that the main themes disclosed refer to the overview of the organization and business model. The main themes are the description of the business environment through organograms with the production process; description of the business organization (also through organograms) with the organic and functional structure; strategic objectives mentioned in the topic of "Strategic Priorities" in 2017, "Forecasts 2019 and Future Objectives" in 2018, and "Ensuring the future" in 2019; and how the company generates value through a scheme of effort, result, and value created. Aspects such as the company's mission are also assessed here along with the main activities and the attitude toward risk.

Regarding the least addressed topics, four are identified: governance and remuneration, performance, future perspectives, and results. It is thus concluded that Bondalti does not disclose information about the decision-making process, executive remuneration, how it performed against the objectives set in earlier years, the preparation necessary for changes at the operational level in the future, and others.

The main change seen in 2018 has to do with risk management. "Risk Management was another of the important topics worked on by Bondalti in 2018" - Bondalti, Integrated Report (2018). In this year a document dedicated to risk management was published in which the attitude toward risk is mentioned, examples of risks are given, and correspondence between the impact criteria of the Risk Management Process is established.

Regarding the measurement of the objectives achieved in 2017, nothing was mentioned. However, in 2018 there is a point in the report at which a comparison is made between the operating result and the actual quantity of sales against what had been budgeted. Despite this, the information seems insufficient given the operational volume in question.

The variation seen in the 2019 classification compared to that of 2018 is because in 2019 an organizational chart with the organic and functional structure of the company was not disclosed, which would have allowed the entire structure of the company to be identified, from the board of directors to the corporate and support areas. Regarding the future, several future objectives are presented throughout the text. However, no mention is made of how it would be expected to achieve them or when. About other information, in 2017 the most prominent remark refers to the company's participation in a study to assess the socio-economic impact of the Estarreja Chemical Complex at the regional and national levels. In 2018 a report on the Intellectual Capital of the company was disclosed, as there are several Bondalti employees involved in R&D and Innovation activities.

Table 5. Bondalti classification summary

	Essential content	**2017**	**2018**	**2018**	**Total (average)**
IIRC	Overall view of organization and business model	2.4	2.7	2.6	2.5
	Operational context	2.0	2.4	2.4	2.3
	Strategic goals and strategies to achieve them	0.0	2.0	2.0	1.3
	Governance and compensation	0.0	0.0	0.0	0.0
	Performance	0.0	0.0	0.0	0.0
	Future prospects	0.0	0.0	0.0	0.0
European Commission - COM 2017/C 215/01	Business model	2.7	2.7	2.2	2.5
	Policies and due diligence	0.0	1.0	1.0	0.7
	Results	0.0	0.0	0.0	0.0
	Main risks and how to manage them	0.0	1.7	1.7	1.1
	Key Performance Indicators	2.0	2.0	2.0	2.0
	Thematic features	15	1.5	1.5	1.5
	Total	**10.5**	**15.9**	**15.4**	**13.9**

Source: Author

Brisa

The company Brisa Autoestradas de Portugal operates in several business areas of which we highlight road concessions, support services to road concessions, car inspections, technology, and mobility services. The IR includes all these segments and there is an individual analysis of each of them. The first year that Brisa issued its IR was 2016, with the aim of "improving the perception of all stakeholders" - Brisa, Integrated Report (2016). Table 6 summarizes the ratings – average ratings attributed for each point.

The main disclosures refer to the general vision of the organization and business model, including topics such as the company's mission – "To provide efficient mobility for people" - Brisa, Integrated Report (2017 to 2019); main activities as well as products and services that are presented with the title "Brisa Profile"; a form of value creation; and attitude toward risk. However, the points that received the highest rating were those related to thematic aspects, in which themes such as environmental and social issues were addressed, highlighting that the company provides information on each of the themes; issues related to corruption and bribery, where actions developed to respond to occurrences related to these issues are presented; and human rights issues.

Regarding the less addressed topics, four are identified: strategic goals and strategies to achieve them, performance, future prospects, and results. Therefore, we conclude that Brisa does not disclose information on how it measures the goals achieved, how it performed about the goals set in earlier years, the preparation necessary for changes at the operational level in the future, and others.

When comparing the four years analyzed, we see that the information and structure of the 2016 and 2017 IRs are very similar. The same did not happen in 2018 and 2019 since the structure was changed, as well as its content. The main variation between 2017 and 2018 concerns the critical factors assessed in the first point (Overview of the organization and business model) of the summary table. In 2018 these critical factors no longer appear, but in 2019 reference is again made to these issues. Also in 2019, the company included a table concerning the markets it operated in. In previous years, this information was scattered throughout the text.

Table 6. Brisa classification summary

	Essential contents	2016	2017	2018	2019	Total (average)
IIRC	Overall view of organization and business model	2.8	2.8	2.5	2.8	2.7
	Operational context	2.0	2.0	2.0	2.0	2.0
	Strategic goals and strategies to achieve them	0.0	0.0	0.0	0.0	0.0
	Governance and compensation	1.0	1.0	1.0	1.0	1.0
	Performance	0.0	0.0	0.0	0.0	0.0
	Future prospects	0.0	0.0	0.0	0.0	0.0
European Commission - COM 2017/C 215/01	Business model	2.3	2.3	2.3	2.5	2.4
	Policies and due diligence	1.0	1.0	1.0	1.0	1.0
	Results	0.0	0.0	0.0	0.0	0.0
	Main risks and how to manage them	3.0	3.0	3.0	3.0	3.0
	Key Performance Indicators	2.0	2.0	2.0	2.0	2.0
	Thematic features	2.8	2.8	2.8	2.8	2.8
	Total	**16.8**	**16.8**	**16.6**	**17.0**	**16.8**

Source: Author

During 2016 and 2017, although there was no topic related to the critical business factors, the following reference was made: "the virtuous relationships and positive externalities that a robust intellectual capital can generate, are among the main critical success factors" - Brisa, Integrated Report (2017). In the year 2018, no mention was made of this point. As for 2019, the reference made is as follows: "Brisa promotes the development of its business according to the most demanding standards of innovation - technological, services, processes, among others -, as a critical success factor" - Brisa, Integrated Report (2019).

Also in Brisa's IR, we verify that there is no evidence of the measurement of the objectives achieved. Regarding the future, several objectives are presented in the text. However, no mention is made of how it would be expected to achieve them or when.

Regarding the publication of other information, Brisa refers to "Highlights and Events of the Year" and "Awards and External Recognitions". Of note is the identification of anti-corruption policies in the four years, as well as the mention of confirmed incidents and actions taken related to corruption.

In conclusion, the integrated reports disclosed by Brisa have undergone positive changes, i.e. more information has been disclosed. The study by Adhikariparajul et al. (2019) is based on UK educational institutions and concludes that the level of IR disclosure has increased from 47% to 70%. Although Brisa is from a different industry, the increase in IR disclosure by the company studied may indicate that they are adopting an integrated thinking approach when reporting their activities.

CTT - Correios de Portugal

CTT - Correios de Portugal, SA is a company with a history of almost 500 years whose objective is to "bring the Portuguese and the world closer together". In 2019 CTT's business included four main activities: postal services, express and parcel delivery, banking, and financial services. Note that the first IR disclosed by CTT was in 2018, precluding an analysis of 2016 or 2017. Table 7 presents the summary rating, in which the average of the evaluation given for each essential content point is found.

By examining the "Total (average)" column in Table 7, we conclude that the main aspects disclosed in CTT's IR are related to the organization's overview and business model, in which topics such as the way value is generated are identified and presented through a scheme of inputs, business strategy, and outputs; identification of the markets in which the company operates is presented in graph form; strategic objectives are disclosed by business; trends that may affect the future are included in a section on future prospects; attitude to risk, in which the entire process is identified; and others.

In CTT's IR, as in Brisa's, the maximum score was attributed to the thematic aspects point. In this topic, aspects related to environmental and social issues are evaluated, highlighting that the company provides information on each of the themes; issues related to corruption and bribery; and human rights.

Regarding the least-disclosed topics, three are identified: performance, future prospects, and results. Therefore, we conclude that CTT does not disclose information on how it performed about the goals set in earlier years, the preparation necessary for changes at the operational level in the future, the results of the policies adopted, and others.

In CTT's IR there was no reference to the company's mission. Despite this, throughout the report mention is made of the slogan "Move us ...". Regarding the description of the business environment, no production scheme or anything similar was found. However, since this is a service company, this information was not considered relevant.

Table 7. CTT Classification summary

	Essential contents	2018	2019	Total (average)
IIRC	Overall view of organization and business model	2.4	2.3	2.3
	Operational context	2.0	2.0	2.0
	Strategic goals and strategies to achieve them	2.0	2.0	2.0
	Governance and compensation	1.0	1.0	1.0
	Performance	0.0	0.0	0.0
	Future prospects	0.0	0.0	0.0
European Commission - COM 2017/C 215/01	Business model	2.5	2.2	2.3
	Policies and due diligence	1.0	1.0	1.0
	Results	0.0	0.0	0.0
	Main risks and how to manage them	3.0	3.0	3.0
	Key Performance Indicators	2.0	2.0	2.0
	Thematic features	3.0	3.0	3.0
	Total	**18.9**	**18.5**	**18.7**

Source: Author

Again, about performance and prospects, the conclusions are incomplete. About performance, some indicators are presented, but they are not related to the objectives and strategy of the company. The future prospects are presented, but neither the operational assumptions nor the short, medium, and long-term management is explained.

Comparing the two years under analysis, the greatest difference can be seen in the point regarding the "Description of the business organization". In 2018 there was an organizational chart that described the relationship of the CEO to all other departments. There was an improvement in the point "Trends and factors that may affect the activity in the future", as in 2019 a section with future prospects was included.

As to other information, the references to awards and external acknowledgments, sustainability indicators, and sustainability commitments stand out. Since CTT's IR is one of those with the best rating, we may conclude that it is one of those that provides more information and therefore more quality. The more information disclosed, the more informed are the decisions made and the lower the risk.

Galp Energia

Galp Energia, SA, is a company in the energy sector that operates in three areas: Exploration and Production; Refining and Distribution; and Gas and Power. According to their IR, the aim is to "provide a global, transparent rigorous vision of the processes through which we create economic, environmental and social value, as well as to disclose our results for the period and our corporate governance practices." - Galp, Integrated Report (2018).

The main topics disclosed are related to the business model, in which aspects such as the description of the business environment through a map that explains the production process from extraction to arrival at the final client are identified; the business organization is described in the section on the governance model; strategic objectives are mentioned in the topic "Our strategy"; how the company generates value and the markets in which it is inserted; and trends and factors that may affect the activity in the future are disclosed in the section "How to plan the future". The point regarding risk management obtained the second-highest score. In this point, sub-points such as the identification of risk, how they affect the business model, and mitigation measures are classified. All these points are addressed in the chapter on risk management.

Table 8. GALP Energia classification summary

	Essential contents	2018	2019	Total (average)
IIRC	Overall view of organization and business model	2.1	2.1	2.1
	Operational context	2.0	2.0	2.0
	Strategic goals and strategies to achieve them	2.0	2.0	2.0
	Governance and compensation	1.0	1.0	1.0
	Performance	0.0	0.0	0.0
	Future prospects	0.0	0.0	0.0
European Commission - COM 2017/C 215/01	Business model	2.8	3.0	2.9
	Policies and due diligence	1.0	1.0	1.0
	Results	0.0	0.0	0.0
	Main risks and how to manage them	3.0	2.0	2.5
	Key Performance Indicators	2.0	2.0	2.0
	Thematic features	2.3	2.3	2.3
	Total	**18.2**	**17.4**	**17.8**

Source: Author

Once again, the least disclosed topics are performance, future prospects, and results. Galp does not disclose information on KPIs of performance about its strategic objectives; the Company's impact on the resources and relationships on which it depends; short-term actions, how to achieve them and associated uncertainties; a useful, fair, and balanced vision of the results of its policies, and others.

About Galp's IR, as in CTT's, no reference is made to the company's mission. The Company's business areas, as mentioned, are exploration and production, refining and distribution, and gas and power.

The rating attributed to the 2019 report shows a slight decrease because in this year no risk mitigation measures were disclosed. Nevertheless, in that year a section on "How to plan for the future" was disclosed that slightly raises the score. In the remaining topics, the reports are very similar.

Regarding the commercial aspects, an organization chart is presented with the identification of the main clients (families, companies, and cities). However, no reference is made in terms of applicable laws and regulations as mentioned with IIRC.

Contrary to what was found in previous IR analyses, Galp dedicates none of its IR content to environmental or social issues. Mention is made of the subject in a section on non-financial information, but only the policies in force are shared and not the attitude of the company toward these thematic aspects. Despite this, the company has a link on its website with information on the approach to sustainability.

The information provided in the items concerning the company's performance and future prospects is the most incomplete. As far as performance is concerned, there is a section regarding the performance achieved but no comparison is made with the intended goals. As for future prospects, none of the subtopics identified by the IIRC are addressed in this report.

It should be noted that the company presents a section on the commitment to long-term sustainability. According to the study by Zhou et al. (2017) on listed companies in South Africa, there is a negative relationship associated with investors' earnings forecast error. As it is a listed company, GALP will have every interest in keeping its investors correctly informed.

José Mello e Saúde

José Mello e Saúde is the company that owns CUF private hospitals, which on 26 June 2020 changed its name to CUF, SA. The José Mello e Saúde Group is an economic group with activities in the chemical industry, health, infrastructures, and mobility. It should be noted that this group of companies holds other companies already reported on above such as Bondalti and Brisa. In 2016 the company published its first IR "according to the structure proposed by the International Integrated Reporting Council (IIRC) to present the company's economic, social and environmental performance and financial results" - José Mello e Saúde, Integrated Report (2016).

Table 9. José Mello e Saúde classification summary

	Essential contents	2016	2017	2018	2019	Total (average)
IIRC	Overall view of organization and business model	2.5	2.5	2.5	2.5	2.5
	Operational context	2.4	2.4	2.4	2.4	2.4
	Strategic goals and strategies to achieve them	3.0	3.0	3.0	3.0	3.0
	Governance and compensation	0.0	0.0	1.0	0.0	0.3
	Performance	0.0	0.0	0.0	0.0	0.0
	Future prospects	0.0	0.0	0.0	0.0	0.0
European Commission - COM 2017/C 215/01	Business model	2.2	2.2	2.2	2.2	2.2
	Policies and due diligence	1.0	1.0	1.0	1.0	1.0
	Results	0.0	0.0	0.0	0.0	0.0
	Main risks and how to manage them	3.0	3.0	2.7	2.7	2.8
	Key Performance Indicators	2.0	2.0	2.0	2.0	2.0
	Thematic features	1.5	1.5	1.5	2.0	1.6
	Total	**17.5**	**17.5**	**18.2**	**17.7**	**17.8**

Source: Author

As seen in Table 9, the main topics disclosed refer to the general vision of the organization and business model and the main risks and their management. Regarding the first point, the mission, main activities, products, and services are disclosed. Regarding the value creation sub-point, a business model is presented. Regarding risk and its management, a section is dedicated solely to this topic.

The least disclosed topics are similar to those identified so far: performance, future prospects, and results. The company does not disclose information on how it performed about the objectives established in earlier years, the preparation necessary for changes at the operational level in the future, the results of the policies adopted, and others.

About the point on "Governance and remuneration", the change in 2018 relates to the fact that a point was disclosed on the variable part of the remuneration that depends on "the assessment of the achievement of collective, annual and multi-year performance objectives, reviewed annually" (José Mello Saúde, Integrated Report (2018)). However, in 2019 there was no longer any mention of aspects related to variable remuneration according to performance.

As for the point concerning Risk Management, the evolution was negative, that is, in the 2016 and 2017 IRs, a table was presented indicating the risk and describing the mitigation strategy. In 2018 and 2019 the way of presenting this information was changed and the types of risks were presented and described in the section dedicated to risk management, but no direct reference was made to mitigation measures.

It should be noted that this is the first company reporting a point related to "Strategic objectives and strategies to achieve them", with a rating of 3, since the published IR identifies a point where the strategy and achievements in the year are described. This point also identifies the goals for the following year. In 2019, there is an evolution regarding bribery and corruption, as an anti-corruption policy was published.

Once again, in this IR no reference is made to aspects related to performance, policies, due diligence, or results. Also, in the point related to future prospects, the classification was 0. Despite this, the company presents a point with goals for the following year but does not fulfill the sub-topics proposed by the IIRC.

Garrigues

Garrigues is a Spanish law firm but provides services internationally. It acts in numerous areas of activity such as banking, industrial property, venture capital, sports, and entertainment. Garrigues' integrated report 2019 is the 14th consecutive one published by the company, although it was previously called "Social Responsibility Memoir".

In our study of all the IRs herein, we found that at the end of each of the reports there is a verification of the Sustainability Report by AENOR. In other words, although this report is entitled "integrated report", the information contained refers to a Sustainability Report in accordance with the GRI Standards. It should be noted that when a company follows the GRI guidelines, it is not necessarily following the essential contents defined by the IRRC and COM 2017/C 215/01. This situation translates in limitation to the study derived from the selected sample, i.e., the selection was made through the title of the publications, i.e., "Integrated Report". In this case, the title of the publication is adequate to the study, but the guidelines followed, are not. Table 10, presents the summary classification attributed to these IRs.

The most disclosed information is related to performance indicators and thematic aspects. For performance indicators, information and key financial indicators are described that allow comparability between companies. Regarding thematic aspects, two sections of information are provided on natural capital and social and relational capital.

Table 10. Garrigues classification summary

	Essential contents	2016	2017	2018	2019	Total (average)
IIRC	Overall view of organization and business model	1.2	1.3	1.3	0.9	1.2
	Operational context	1.6	1.6	1.6	1.6	1.6
	Strategic goals and strategies to achieve them	0.0	0.0	0.0	0.0	0.0
	Governance and compensation	0.0	0.0	0.0	0.0	0.0
	Performance	0.0	0.0	0.0	0.0	0.0
	Future prospects	0.0	0.0	0.0	0.0	0.0
European Commission - COM 2017/C 215/01	Business model	0.7	1.0	1.3	1.3	1.1
	Policies and due diligence	0.0	0.0	0.0	0.0	0.0
	Results	0.0	0.0	0.0	0.0	0.0
	Main risks and how to manage them	0.0	0.0	0.0	0.0	0.0
	Key Performance Indicators	2.0	2.0	2.0	2.0	2.0
	Thematic features	2.5	2.5	2.5	2.5	2.5
	Total	**8.0**	**8.4**	**8.7**	**8.3**	**8.3**

Source: Author

In Table 10 we see that many topics were left unaddressed in this company's IR. Between 2016 and 2017, the improvement is seen in the point related to "Description of the business organization". In the second year, the corporate organization is presented, in text, i.e. the CEO is identified as well as those who support him, i.e., the Partners' Council and the Professional Practice Committee. Comparing 2017 and 2018, a difference is found in the disclosure of information on the "Business model", more specifically in the point "Strategic objectives". The IRs of 2018 and 2019 present "Objectives of the next financial years" although they do not explain how to achieve the objectives mentioned. In 2019 the only difference relates to the disclosure of the company's mission. From 2016 to 2018 it had identified the mission as "To provide the best advice to our clients wherever they are doing business by being one of the world leaders in presenting legal and tax services." - Garrigues, Integrated Report (2018).

The point regarding corruption and bribery has a rating of 2 because only one mention is made of a pact (United Nations Global Compact in the year 2002) that initiated the communication of principles in the areas of human rights, labor, environmental issues, and anti-corruption. Despite this, the "Internal rules on the prevention of money laundering and the financing of terrorism" are mentioned. To obtain a higher rating, a mention should have been made in this report of actions in response to the occurrence of situations of corruption or bribery.

In the four years of analysis, the field relating to "Strategic objectives" was rated 0. This situation is because in the reports a mention is made of the objectives for the following years, but the strategy for achieving them is not identified. There is a similar situation with the point "Measurement of achieved objectives", for which the rating was 0. However, milestones of the year under analysis are identified, but with no relation to the objectives defined in the previous year.

About the point "Key performance indicators", this company's classification was similar to those previously analyzed. However, the information made available in this report deals with non-financial indicators, such as the number of medical appointments, nursing, and others.

SATA Internacional - Azores Airlines

SATA Internacional - Azores Airlines is a company that operates flights outside the Azores, resulting from the transformation of Oceanair, acquired by SATA Air Açores in 1995. This company operates in four business areas: air transportation, aircraft handling, aerodrome management, and tour operating. The four reports under analysis were prepared based on IIRC guidelines, with 2019 being the eighth integrated report published by SATA Internacional

Table 11 shows that the topics on which SATA provides the most information are strategic goals and strategies for achieving them, and the main risks and their respective management. Both items were given the highest scores. Regarding the first point, ways of measuring the goals achieved were presented, showing through graphs and tables whether or not the goals had been achieved. Regarding risk management, the company reserves a section to identify risks, how they affect the business model, and risk mitigation measures.

Table 11. SATA classification summary

	Essential contents	2016	2017	2018	2019	Total (average)
IIRC	Overall view of organization and business model	2.4	2.4	2.4	2.5	2.4
	Operational context	2.0	2.0	2.0	2.0	2.0
	Strategic goals and strategies to achieve them	3.0	3.0	3.0	3.0	3.0
	Governance and compensation	1.7	1.7	1.7	1.7	1.7
	Performance	1.3	1.3	1.3	1.3	1.3
	Future prospects	0.7	0.0	0.0	0.0	0.2
European Commission - COM 2017/C 215/01	Business model	1.7	1.5	1.7	2.6	1.9
	Policies and due diligence	1.0	1.0	1.0	1.0	1.0
	Results	0.0	0.0	0.0	0.0	0.0
	Main risks and how to manage them	3.0	3.0	3.0	3.0	3.0
	Key Performance Indicators	2.0	2.0	2.0	2.0	2.0
	Thematic features	2.0	2.5	2.5	2.5	2.4
	Total	**20.6**	**20.3**	**20.5**	**21.5**	**20.7**

Source: Author

The least addressed points are related to future prospects, wherein only one of the years (2016) are the challenges for the following year and the results identified, and no information is presented on the useful, fair, and balanced view of the results of its policies.

When comparing the first two years, the variation is due to the classification attributed in the point "Future Prospects" and "Business Model". In 2017 no information was presented on "Short-term Actions, how to achieve them and associated uncertainties" and the organizational chart with the "Description of the business organization" was less complete than the one presented in the previous year. Despite this, in 2017 the company added "Plan for the Prevention of Risks and Corruption and Related Infringements covering all operations of Group companies." - SATA, Integrated Report (2017, 2018 and 2019).

In the year 2016, the "Challenges 2017" and how to address them are disclosed at various points in the IR. Despite this, the score on this sub-point was only 3, as uncertainties are not identified.

This difference is mitigated in 2018 by the fact that in the year 2017 an organizational chart with the "Description of the business organization" was not presented. This decrease is more visible in the overall ranking table because the summary table is averaged. In 2019 the score increases due to the disclosure of information on the point "Business Model", more specifically, in the sub-point "Trends and factors that may affect the activity in the future". In this year, a point was included on "Regulatory trends that will affect SATA in the future". - SATA, Integrated Report (2019).

The fact that SATA is the only company under analysis that presents any information on the "Leadership/Decision-making Process" stands out, mentioning that the shareholder is the one who makes decisions based on opinions on the economic and financial situation.

In addition to this point, also in the points "KPI's of performance about its strategic objectives" and "Measurement of objectives achieved" the rating was higher than that seen in previous analyses. In the four reports under analysis, the strategic objectives are identified and it is disclosed through graphs and tables showing the company achieved its objectives. However, it should be noted that no mention is made of the "non-achievement of objectives".

About the points of anti-corruption policy and human rights, the information provided has little detail, mentioning only from 2017 on the plan for the prevention of risks and corruption and respecting the fundamental principles of human rights.

As seen in Table 13 we were able to identify the five companies providing the most information in the integrated reports (and the year), by the standards outlined. Table 14 identifies the five companies that provide the least information and the respective years.

In general terms, SATA's report stands out from all those analyzed due to the disclosure of information on short-term actions to achieve future goals. In comparison with the other companies, this report also stands out due to the disclosure of information on the following sub-points:

- "Leadership/Decision-making process" - the following is mentioned on this point: "The preparation of opinions on SATA's economic and financial situation, provided by its supervisory body, contributes to informed decision-making by the shareholder." - SATA Integrated Report (2016 to 2019). Despite being brief and simple information, there is some concern about the disclosure of the manner and/or basis on which decisions are made. Thus, the rating attributed was not the maximum. However, this was the only company that obtained a score greater than 0.
- "Remuneration of executives and those responsible for governance is linked to performance" - about this point it is stated that the remuneration of members of the Board of Directors is variable according to the responsibilities taken on and the performance of the governing body. Furthermore, within the section "Governance" available in all the years analyzed, the remuneration of the members of the General Meeting, the Supervisory Board, and the Statutory Auditor is presented.

- "KPI's of performance about its strategic goals" and "Conclusion of how the company performed about its goals" - once again, SATA was the only company under analysis to obtain a score in the "Performance" item, due to the disclosure of maps, graphs, and texts identifying the strategic goals and whether it had achieved them. For example, one of the strategic goals was to reduce operating costs. Thus, in the years analyzed it is possible to verify whether or not this happened by inspecting the graphs of the last three years. Despite this, and as already mentioned throughout the text of the SATA IR analysis, in 2017 this goal was not achieved, but this is not stated clearly. We may be facing a phenomenon called greenwash.
- "Short-term actions, how to achieve them and associated uncertainties" - the information provided on this point is what makes the 2016 IR different from the others of this company. The rating on this point was not the maximum, but it was the only company that disclosed any information on this point. To do so, it identified the challenges for the year 2017 and how to address them. To achieve the maximum score, it should have mentioned some uncertainties, risks, or external factors that could jeopardize the achievement of the objectives.
- A paragraph on "Regulatory trends that will affect SATA in the future" was included in 2019 - SATA Integrated Report (2019).

CTT's 2018 IR stands out from the other reports analyzed in the following aspects:

- "Issues concerning respect for human rights" - indicators of "hours of training on policies and procedures concerning human rights aspects" and "Percentage and a total number of significant investment agreements and contracts that include human rights clauses" - CTT Integrated Report, 2018.
- "Combating corruption and bribery" - when referring to this point, the following indicators are disclosed: "Total number and percentage of operations submitted to risk assessments related to corruption and the significant risks identified"; "Communication and training on anti-corruption policies and procedures"; and "Confirmed cases of corruption and measures were taken" - CTT Integrated Report, 2018.
- "How it generates value" - about this point, a scheme of inputs, strategy, business model, and outputs is presented. To these outputs are still associated the main objectives per business.
- "Remuneration of executives and those responsible for governance is linked to performance" - about remuneration policy, this is defined annually by the Remuneration Committee. This remuneration policy takes into account several factors, such as: "The business objectives defined for this term of office, in the context of the Company's ongoing diversification strategy based on 4 business segments." - CTT Integrated Report, 2018.
- It also stands out negatively by not mentioning its mission.
- The IR for 2018 stands out because in this year a "Management Organization" is presented - CTT Integrated Report (2018), which allows the business organization to be described and understood. Nevertheless, in 2019 a chapter on future perspectives was presented, which increases the ranking in the sub-item "Trends and factors that may affect the activity in the future" compared to 2018.

Table 12. Classifications by the company and by year

		Bondalti			Brisa				CTT		GALP		José de Mello Saúde				Garrigues				Sata					
	Essential contents	**2017**	**2018**	**2019**	**2016**	**2017**	**2018**	**2019**	**2018**	**2019**	**2018**	**2019**	**2016**	**2017**	**2018**	**2019**	**2016**	**2017**	**2018**	**2019**	**2016**	**2017**	**2018**	**2019**	**Total (average)**	**Total (arithmetic sum)**
IIRC	Overall view of organization and business model	2	3	3	3	3	2	3	2	2	2	2	2	2	2	2	1	1	1	1	2	2	2	3	2.2	51.6
	Operational context	2	2	2	2	2	2	2	2	2	2	2	2	2	2	2	2	2	2	2	2	2	2	2	2.0	46.9
	Strategic goals and strategies to achieve them	0	2	2	0	0	0	0	2	2	2	2	3	3	3	3	0	0	0	0	3	3	3	3	1.6	36.0
	Governance and compensation	0	0	0	1	1	1	1	1	1	1	1	0	0	1	0	0	0	0	0	2	2	2	2	0.7	15.7
	Performance	0	0	0	0	0	0	0	0	0	0	0	0	0	0	0	0	0	0	0	1	1	1	1	0.2	5.0
	Future prospects	0	0	0	0	0	0	0	0	0	0	0	0	0	0	0	0	0	0	0	1	0	0	0	0.0	0.7
European Commission - COM 2017/C 215/01	Business model	3	3	2	2	2	2	3	3	2	3	3	2	2	2	2	1	1	1	1	2	2	2	3	2.1	47.9
	Policies and due diligence	0	1	1	1	1	1	1	1	1	1	1	1	1	1	1	0	0	0	0	1	1	1	1	0.8	18.0
	Results	0	0	0	0	0	0	0	0	0	0	0	0	0	0	0	0	0	0	0	0	0	0	0	0.0	0.0
	Main risks and how to manage them	0	2	2	3	3	3	3	3	3	3	2	3	3	3	3	0	0	0	0	3	3	3	3	2.2	49.7
	Key Performance Indicators	2	2	2	2	2	2	2	2	2	2	2	2	2	2	2	2	2	2	2	2	2	2	2	2.0	46.0
	Thematic features	2	2	2	3	3	3	3	3	3	2	2	2	2	2	2	3	3	3	3	2	3	3	3	2.3	52.0
	Total summary classification	**10.5**	**15.9**	**15.4**	**16.8**	**16.8**	**16.6**	**17.0**	**18.9**	**18.5**	**18.2**	**17.4**	**17.5**	**17.5**	**18.2**	**17.7**	**8.0**	**8.4**	**8.7**	**8.3**	**20.6**	**20.3**	**20.5**	**21.5**		
	Total complete classification	**55**	**69**	**65**	**74**	**74**	**72**	**76**	**76**	**73**	**73**	**71**	**69**	**69**	**71**	**70**	**36**	**38**	**40**	**37**	**77**	**76**	**77**	**81**		

Source: Author

Table 13. Top 5 companies by classification

Company	Year	Summary classification	Total classification
Sata	2019	21.5	81
Sata	2016	20.6	77
Sata	2018	20.5	77
Sata	2017	20.3	76
CTT	2018	18.9	76

Source: Author

Table 14. Bottom 5 companies by classification

Company	Year	Summary classification	Total classification
Garrigues	2016	8.0	36
Garrigues	2019	8.3	37
Garrigues	2017	8.4	38
Garrigues	2018	8.7	40
Bondalti	2017	10.5	55

Source: Author

Regarding the companies that disclose the least, Garrigues appears in every year analyzed. As already mentioned, this report is certified as a Sustainability Report. Although it is entitled to an integrated report, the information available in it is relative to a Sustainability Report. This may be a justification for the low rating obtained.

In the case of Bondalti, the information provided is reduced for the year identified in Table 5 and for the following year. In general, Bondalti's 2017 IR only discloses aspects related to the business model and some information about the operational context. It should be noted that in 2018 there are already some improvements, mainly in the disclosure of information related to risk management.

Table 15 shows the most and least disclosed topics.

Table 15. Summary of classifications by essential contents

	Essential contents	Total (average)	Total (arithmetic sum)
IIRC	Overall view of organization and business model	2.2	51.6
	Operational context	2.0	46.9
	Strategic goals and strategies to achieve them	1.6	36.0
	Governance and compensation	0.7	15.7
	Performance	0.2	5.0
	Future prospects	0.0	0.7
European Commission - COM 2017/C 215/01	Business model	2.1	47.9
	Policies and due diligence	0.8	18.0
	Results	0.0	0.0
	Main risks and how to manage them	2.2	49.7
	Key Performance Indicators	2.0	46.0
	Thematic features	2.3	52.0

Source: Author

The topic for which there is the greatest dissemination of information is the "Thematic Aspects", in which topics such as "Environmental Issues" and "Social Issues related to employees" are addressed. Also included in this topic are the sub-points "Issues related to the respect of human rights" and "Fight against corruption and bribery", which have lower scores compared to the other sub-topics.

The aspects that companies need to disclose more information on our "Future prospects" and "Results".

Table 16 identifies the IRs that were approved by entities outside the company, the entity that approved them, and the name of the approval given.

Table 16. Approval of RI by an external entity

<table>
<tr><th></th><th></th><th>Approved</th><th>Entity approving</th><th>Title of the approval document</th></tr>
<tr><td rowspan="3">Bondalti</td><td>2017</td><td rowspan="3">Yes</td><td rowspan="3">PWC</td><td rowspan="3">Relatório Independente de Garantia Limitada e Fiabilidade
Independent Limited Guarantee and Reliability Report</td></tr>
<tr><td>2018</td></tr>
<tr><td>2019</td></tr>
<tr><td rowspan="4">Brisa</td><td>2016</td><td rowspan="4">Yes</td><td rowspan="4">PWC</td><td rowspan="4">Relatório Independente de Garantia Limitada e Fiabilidade
Independent Limited Guarantee and Reliability Report</td></tr>
<tr><td>2017</td></tr>
<tr><td>2018</td></tr>
<tr><td>2019</td></tr>
<tr><td rowspan="2">CTT</td><td>2018</td><td rowspan="2">Yes</td><td rowspan="2">KPMG</td><td rowspan="2">Relatório Independente de Garantia Limitada e Fiabilidade
Independent Limited Guarantee and Reliability Report</td></tr>
<tr><td>2019</td></tr>
<tr><td rowspan="2">GALP</td><td>2018</td><td rowspan="2">Yes</td><td rowspan="2">PWC</td><td rowspan="2">Relatório Independente de Garantia Limitada e Fiabilidade
Independent Limited Guarantee and Reliability Report</td></tr>
<tr><td>2019</td></tr>
<tr><td rowspan="4">José de Mello Saúde</td><td>2016</td><td rowspan="4">N.A.</td><td rowspan="4">N.A.</td><td rowspan="4">N.A.</td></tr>
<tr><td>2017</td></tr>
<tr><td>2018</td></tr>
<tr><td>2019</td></tr>
<tr><td rowspan="4">Garrigues</td><td>2016</td><td rowspan="4">Yes</td><td rowspan="4">AENOR</td><td rowspan="4">Verificação do Relatório de Sustentabilidade
Verification of the Sustainability Report</td></tr>
<tr><td>2017</td></tr>
<tr><td>2018</td></tr>
<tr><td>2019</td></tr>
<tr><td rowspan="4">Sata</td><td>2016</td><td rowspan="4">N.A.</td><td rowspan="4">N.A.</td><td rowspan="4">N.A.</td></tr>
<tr><td>2017</td></tr>
<tr><td>2018</td></tr>
<tr><td>2019</td></tr>
</table>

Source: Author

Note that the approval through an independent report of limited assurance and reliability ensures compliance with the requirements of the GRI guidelines only, and does not guarantee compliance with the guidelines of the standards studied (IIRC and COM 2017/C 215/01). Thus, a certification by an external entity for integrated reports is called for. This certification may be based on the essential contents identified by IIRC and/or COM 2017/C 215/01. Another conclusion that can be drawn from an inspection of Table 20 is related to the fact that SATA, a company that occupies 4 of the Top 5 places in disclosure, does not present certification by an external entity in its IR.

Table 17 presents a summary of whether a sustainability report was published before the publication of the IR.

Table 17. Disclosure of sustainability report

	Bondalti	Brisa	CTT	GALP	Garrigues	José Mello e Saúde	SATA
2015	No	Yes	Yes	Yes	No	No	No
2016	No	No	Yes	No	No	No	No
2017	No	No	Yes	No	No	No	No
2018	No	No	No	No	No	Yes	No
2019	No	No	No	No	No	No	No

Source: Author

From Table 17 we conclude that both Brisa and CTT stopped publishing sustainability reports in the year they started publishing IRs. In this way, they achieve one of the objectives of integrated reporting, which is to synthesize company information in a single report. Garrigues, SATA, and Bondalti have never published a sustainability report. However, Garrigues has published a corporate social report since 2006. SATA and Bondalti did not publish a sustainability report between 2015 and 2019. In the case of GALP, the sustainability report was released in 2015 and thereafter the company released an IR only in 2018. Apart from this, on the company's website, there is nevertheless a tab dedicated to the sustainability topic. Finally, in the case of José Mello e Saúde, the sustainability report was disclosed in 2018 only. Once again, SATA which had never disclosed non-financial information stands out by occupying several places in the Top 5 of disclosures.

From the companies studied we conclude that Portuguese companies still have some way to go in disclosing the "Essential Contents" identified by IIRC and COM 2017/C 215/01. Arriving at the desired goal will require time, resources, knowledge, investment, and dedication.

In addition to the less-disclosed topics already identified, companies must also develop a relationship between financial and non-financial information. For example, most of the companies studied do not relate their results to the policies and objectives defined. Only with these relationships clearly stated will it be possible for stakeholders to make decisions about the quality of management policies, evaluate the strategy adopted to achieve objectives, identify ethical behavior regarding human rights, corruption, and bribery, and arrive at other essential conclusions about the company and its performance.

This is still a recent change of practice and not yet fully embraced by companies, but the framework defined by the IIRC provides very clear guidelines of the principles to be followed and the contents to be disclosed. The methodology disclosed by the European Commission is also clear and illustrates each of the principles with examples, which can be adapted to different realities and activity sectors. Nevertheless, there should also be an effort to encourage companies to obtain certification of IRs from external entities to avoid greenwash situations, thereby increasing trust and credibility in the information disclosed to readers.

CONCLUSION

The reports released by companies seek to convey information to all stakeholders. The IR is no exception, and this type of report adds to the effectiveness of the previous reporting on the strategy, governance, performance of the organization, and prospects for the future. With IRs it is possible to relate the financial information of a given organization to non-financial information. The IR emerged to compensate for the shortfalls in the disclosure of companies' non-financial information, such as risk management policies, business models, attitudes toward bribery, and corruption practices, among other topics. It is expected that the integrated report is also a reflection on the social, environmental, and commercial impact of the organization.

It is therefore fundamental that the information contained in the companies' IR meets the needs of investors and other stakeholders. Only in this way will it be possible for the potential added value of this type of report to be realized, and interest in producing IRs may even arise among companies that are not obliged to disclose this type of report.

According to studies made by KPMG in 2019, in Japan, there are already more than 500 companies that publish IRs, after having identified about 414 companies doing so in the previous year – revealing a growth rate of 21%. This growth demonstrates the beginning of recognition and appreciation of this type of reporting in Japan. The 2019 study discloses that 48% of the companies studied to explain the impact of their performance on their short, medium, and long-term strategies. The 2018 study concluded that 50% of the companies studied already revealed an inclination toward integrated thinking and disclosure of how the social and environmental results affect the economic value. Also, 66% of the companies displayed their way of creating value through a diagram.

From our analysis we conclude that companies in Portugal remain very focused on the past, disclosing little about their future prospects or relating their results achieved to their earlier-defined objectives. In short, the information made available in the integrated reports of Portuguese companies falls somewhat short of expectations regarding the development of integrated thinking and the relationship between financial and non-financial information.

The methodology that we used has some limitations related to the flexibility of the rating attribution process. In addition, the reports were analyzed by only one researcher. The way the sample was selected also has a drawback. By using the criterion that publications had to be called "integrated report" (without a revised analysis of the document), situations were identified such as that of Garrigues, in which the report is called an "integrated report", but is in fact a sustainability report, as verified by AENOR. For future work, a more detailed analysis of the topics identified by IIRC and COM 2017/C 215/01 could be carried out. That is, in the present study we identified only if the IR mentioned the topics identified as essential content, but not if all the sub-topics were mentioned.

ACKNOWLEDGMENT

This research is financed by National Funds of the FCT—Portuguese Foundation for Science and Technology within the projects UIDB/04928/2020; UIDB/04007/2020, and under the Scientific Employment Stimulus - Institutional Call CEECINST/00051/2018.

REFERENCES

Abhayawansa, S., Guthrie, J., & Bernardi, C. (2019). Intellectual capital accounting in the age of integrated reporting: A commentary. *Journal of Intellectual Capital*, *20*(1), 2–10. doi:10.1108/JIC-01-2019-223

Aceituno, J., Ariza, L., & Sánchez, I. (2013). Is integrated reporting determined by a country's legal system? An exploratory study. *Journal of Cleaner Production*, *44*, 45–55. doi:10.1016/j.jclepro.2012.12.006

Adams, C. (2015). The international integrated reporting council: A call to action. *Critical Perspectives on Accounting*, *27*, 23–28. doi:10.1016/j.cpa.2014.07.001

Adhikariparajul, M., Hassan, A., Fletcher, M., & Elamer, A. A. (2019). Integrated Reporting in UK Higher Education Institutions. *Sustainability Accounting. Management and Policy Journal*, *10*(5), 844–876. doi:10.1108/SAMPJ-03-2018-0093

Camodeca, R. E., Almici, A., & Sagliaschi, U. (2018). Sustainability Disclosure in Integrated Reporting: Does It Matter to Investors? A Cheap Talk Approach. *Sustainability*, *10*(12), 4393. doi:10.3390u10124393

De Villiers, C., & van Staden, C. J. (2006). Can less environmental disclosure have a legitimizing effect? Evidence from Africa. *Accounting, Organizations and Society*, *31*(8), 763–781. doi:10.1016/j.aos.2006.03.001

European Commission. (2017). Guidelines on non-financial reporting (methodology for reporting non-financial information). *Official Journal of the European Union. COM 2017/C 215/01.*

Fernandes, F. (2019). *Informação Financeira e Não Financeira* (Master Thesis). IPC.

Hoque, M. (2017). Why a Company Should Adopt Integrated Reporting? *International Journal of Economics and Financial Issues*, *7*(1), 241–248.

Hsieh, H., & Shannon, S. (2005). Three Approaches to Qualitative Content Analysis. *Qualitative Health Research*, *15*(9), 1277–1288. doi:10.1177/1049732305276687 PMID:16204405

International Integrated Reporting Council (IIRC). (2011). *Towards Integrated Reporting. Communicating Value in the 21st Century*. Available at https://integratedreporting.org/wp-content/uploads/2011/09/IR-Discussion-Paper-2011_spreads.pdf

KPMG. (2019). *Survey of Integrated Reporting in Japan 2019*. https://www.integratedreporting.org/wp-content/uploads/2020/05/jp-en-integrated-reporting2019.pdf

Kumar, K., Kumari, R., & Kumar, R. (2021). The state of corporate sustainability reporting in India: Evidence from environmentally sensitive industries. *Business and Society Review, 126*(4), 1–26. doi:10.1111/basr.12247

Leocádio, M. F. C. (2017). *Relato Integrado em Portugal* (Master Thesis). Economy Faculty of Oporto, Portugal.

Lopes, S. M. S. (2018). *Relato Integrado no Sector das Águas Minerais Naturais em Portugal: Relação ODS, Métricas e Estrutura* (Master Thesis). Polythetic Institute of Leiria, Portugal.

Mata, C., Fialho, A., & Eugénio, T. (2018). A decade of environmental accounting reporting: What we know? *Journal of Cleaner Production, 198*, 1198–1209. doi:10.1016/j.jclepro.2018.07.087

Melloni, G., Galio, A., & Perego, P. (2017). Saying more with less? Disclosure conciseness, completeness, and balance in Integrated Reports. *Journal of Accounting and Public Policy, 36*(3), 220–238. doi:10.1016/j.jaccpubpol.2017.03.001

O'Dwyer, B., & Unerman, J. (2016). Fostering rigor in accounting for social sustainability. *Accounting, Organizations and Society, 49*, 32–40. doi:10.1016/j.aos.2015.11.003

Owen, G. (2013). Integrated Reporting: A Review of Developments and their Implications for the Accounting Curriculum. *Accounting Education, 22*(4), 340–356. doi:10.1080/09639284.2013.817798

Smith, S. (2014). Integrated reporting, corporate governance, and the future of the accounting function. *International Journal of Business and Social Science, 5*(10), 58–63.

Zhou, S., Simnett, R., & Green, W. (2017). Does Integrated Reporting Matter to the Capital Market? *Abacus, 53*(1), 94–132. doi:10.1111/abac.12104

APPENDIX 1

Based on the evaluation of the information provided in the RI

Table 18.

	Essential contents	What to identify
IIRC	Overall view of the organization and business model	Mission
IIRC European Commission - COM 2017/C 215/01	Overall view of the organization and business model Operational context	Main activities
		Products and services
		Business model
		Way of creating value
		Critical factors
		Attitude regarding risk
		Commercial aspects
	Operational context Strategic goals and strategies to achieve them	Social aspects
		Environmental aspects
		Essential resources
		Essential relationships
		Main risks
		Main opportunities
		Measuring goals achieved
	Governance and compensation	Leadership/decision-making process
	Governance and compensation Performance	Leaders' decisions influence the organization's direction and strategy
		Compensation of executives and those responsible for governance are linked to performance
		Performance (KPIs) compared to strategic objectives
	Performance Future prospects	Impact of the business on the resources and relationships it depends on
		Internal and external factors that affect the achievement of the company's goals
		Assessment of how the company performed toward its goals
		What is the preparation for future operational level changes?
	Future prospects Business model	How does the company balance its short, medium, and long-term interests?
		Short-term actions, how does the company want to achieve its goals, and what are the uncertainties involved?
		How does the company create value?

continues on following page

Table 18. Continued

	Essential contents	What to identify
European Commission - COM 2017/C 215/01	Business model Policies and due diligence	Description of the business environment
		Description of the business organization
		The markets in which the company operates
		Strategic goals
		Trends and factors that could affect activity in the future
		Addresses non-financial aspects necessary to achieve goals
	Policies and due diligence Results	Risk management policies and their results
		A useful, fair, and balanced view of the results of the company's policies
	Main risks and how to manage them	Identification of risks
	Main risks and how to manage them Key Performance Indicators	How do risks affect the business model?
		Measures to mitigate the risks identified
		KPIs within the industry or value chain that allow for comparability
	Thematic features	Environmental issues
	Thematic features	Social issues regarding the employees
		Issues related to respect for human rights
		Combating corruption and bribery
		Others relevant in the sector/industry or value chain

Chapter 13
Sustainable Development Goals-Related Disclosures:
University of Minho's Webpage and Sustainability Reports

Sónia Monteiro
CICF, Polytechnic Institute of Cávado and Ave, Portugal

Verónica Ribeiro
CICF, Polytechnic Institute of Cávado and Ave, Portugal

Kátia Lemos
https://orcid.org/0000-0002-3961-5964
CICF, Polytechnic Institute of Cávado and Ave, Portugal

ABSTRACT

The aim of this chapter is to analyse the Sustainable Development Goals (SDGs)-related disclosures of University of Minho (UMinho), namely (1) how UMinho discloses information related to SDGs in its webpage; (2) the linkage to SDGs in its sustainability reporting, after the 2030 Agenda approval (in 2015); (3) whether sustainability reports' content was yet connected to SDGs themes in the previous period (2010-2014). This exploratory study used a content analysis of the information disclosed in UMinho's webpage and the stand-alone sustainability reports published by UMinho until nowadays. It considers five dimensions—environmental, economic, social, cultural, and institutional. Findings related to sustainability reports' content (2010-2014) reveals the connection to SDGs themes before the 2030 Agenda approval. The 2015 sustainability report published for the first time a formal alignment with the SDGs. Considering the SDGs-related disclosure analysis, the information is aligned with 9 of the 17 SDGs. The environmental dimension is the one that is linked to the largest number of SDGs.

DOI: 10.4018/978-1-7998-9410-0.ch013

INTRODUCTION

Public entities, such as Higher Education Institutions (HEIs), cannot remain apart to the reflection on Sustainable Development (SD) since they play an important role in the students' formation and transfer of knowledge of future professionals/ future citizens, who will have to democratically promote human rights and the common good. The HEIs should consider sustainability as a mechanism that, in addition to seeking academic success, should give significant contributions to the community within its area of influence, region, and country in general. This implies promoting sustainability as a strategic model in the mission, vision, and objective of the HEIs, carrying out activities of social commitment.

In 2015 the United Nations (UN) approved the 2030 Agenda, defining 17 Sustainable Development Goals (SDGs). The SDGs were designed to be universal and capable of being applied to all countries (Neary & Osborne, 2018), placing education at the heart of the strategy to promote SD (Annan-Diab & Molinari, 2017). In fact, the SDG4 - Quality education, goal 4.7. intends that, by 2030, all students acquire the knowledge and skills necessary to promote SD, inclusively, among others, through education for SD. Within the 2030 Agenda, educational responses are key to the vision of the SDGs. Education is both a goal in itself, and a means for attaining all the other SDGs.

The 2030 Agenda represents an opportunity for the improvement of higher education institutions' (HEIs) management, the relationship with stakeholders and the community, and the articulation and consolidation of education policies with the SDGs.

HEIs can and must provide an holistic approach to the 2030 Agenda and contribute to its progress, since the SDGs are a key aspect of their social responsibility (Vilalta et al., 2018). For that purpose, HEIs can use the following steps to start and deepen their engagement with the SDGs: (i) Map what they are already doing; (ii) Build internal capacity and ownership of the SDGs; (iii) Identify priorities, opportunities and gaps; (iv) Integrate, implement and embed the SDGs within HEIs strategies, policies and plans, and (v) Monitor, evaluate and communicate (report) their actions on the SDGs (SDSN, 2017).

Reporting can play an important role by informing the stakeholders who are attentive to the acts practiced by HEIs, analysing if they have sustainable behaviours. In order to do this, the dissemination of information about SD allows us to understand the concerns and strategies that guide decision making, contributing to the image they intend to pass on to the community in which they are inserted. Over the last few decades, reporting frameworks/guidelines have been developed to help assess and manage the current state of an organization's sustainability and communicate it to stakeholders. The guidelines/ standards developed by the Global Reporting Initiative (GRI) are singled out as the world's foremost internationally recognized and globally accepted sustainability reporting framework (KPMG, 2015). However, the GRI guidelines do not consider the specificities of HEIs, neither have a specific guidance on SDGs' disclosures. Reporting is a means of communicating the progress of HEI actions in support of the UN Agenda 2030 SDGs.

In SDGs Reporting, empirical research is scarce and mainly focused on the private sector (Schramade, 2017; Rosati & Faria, 2019a-b; Fonseca & Carvalho, 2019; Monteiro et al., 2020). Concerning the higher education, as far as we know, previous research on SDGs reporting approach is limited.Thus, this study represents a contribution to the existing literature and bridges the gap about the lack of empirical research on SDGs reporting.

Examining HEIs' disclosure on SDGs allows a deeper understanding of the concerns and strategies that guide decision-making as well as the messages that are conveyed to stakeholders. Our study provides insights into reporting's linkage with SDGs from a higher education perspective, focusing on an award-winning public Portuguese university – University of Minho (UMinho). Thus, the aim of this paper is to analyse the SDGs-related disclosures performed by University of Minho in its webpage and in its sustainability reports published since 2010. The specific objectives are to analyse: (a) how UMinho discloses information related to SDGs in its webpage; (b) the linkage to SDGs in its sustainability reporting, after the 2030 Agenda approval (in 2015); (c) whether sustainability reports' content was already connected to SDGs themes during the previous period (2010-2014) - before the publication of 2030 Agenda.

This work is structured in three sections. The first section, related to background, presents an empirical literature review regarding SDGs disclosure in HEIs. The research design is presented in the second section. The results of the study are presented and discussed in the last section. Finally, the solutions and recommendations, future research directions and the main conclusions are presented.

BACKGROUND

The 2030 Agenda places education at the heart of the strategy to promote SD. HEIs are in an unique position to lead the cross-sectoral implementation of the SDGs and advance the 2030 Agenda, providing an invaluable source of expertise in research and education on SDGs, in addition to being widely considered as neutral, trustful and influential players. The 2030 Agenda and the SDGs are a framework for collaboration by HEIs, in an holistic way, in all core activities (education, research, operations and community engagement) in pursuit of sustainability.

Dissemination and information on sustainability through reporting is a key element in the transparency and accountability of HEIs. The green book on Social Responsibility (SR) and HEIs, published in Portugal by ORSIES (2018), recommends that HEIs adopt SR Reporting Strategies, specifically the inclusion of the Social Responsibility theme in their institutional websites and the development of sustainability reports. By adequately using diverse communication channels, HEIs are able to mitigate risks associated with sustainability issues and obtain legitimation from their stakeholders. Thus, stakeholder and legitimacy theories are the main theoretical strands that address voluntary disclosure on sustainibility issues.

Regarding Higher Education, some studies seek to evaluate transparency and accountability in HEIs, through the analysis of information released on a voluntary basis (Ntim et al., 2017). In fact, HEIs are not required to elaborate mandatory reports on issues related to sustainability (Romolini et al., 2015). The literature review leads us to conclude that the level of sustainability reporting in HEIs is still very low. (Lozano, 2011, Lozano et al., 2013). Most of the reports published by HEIs are in Europe (54%) and America (32%), with the rest in Asia and Australia (Lozano et al., 2013). According to Martin and Moneva (2018) the lack of specific guidelines for universities could be the main reason for the later and lower level of development in comparison with other sectors.

The Global Reporting Initiative (GRI) methodology to prepare sustainability reports, as a well-structured indicator-based assessment tool, provides an opportunity to increase the transparency and consistency of any organization. The guidelines/standards developed by the GRI are singled out as the world's foremost internationally recognized and globally accepted sustainability reporting framework (KPMG, 2015)[1].

However, The GRI guidelines are not specifically designed to meet the needs of HEIs (Huber & Bassen, 2018) and there is no specific supplement to adapt the GRI methodology to higher education sector. So, HEIS are currently using their own, tailor-made versions (Yáñez et al. 2018).

In the opinion of Lozano *et al.* (2013), when preparing a sustainability report in an HEI, it is important to have an holistic perspective, addressing the different interrelationships between indicators, categories, and dimensions, as well as stakeholders throughout the higher education system.

The process of sustainability reporting provides ways to include SDGs-related disclosures, by offering organisations the possibility to transparently communicate their performance towards sustainable development to various stakeholders. Mapping of the SDGs Agenda represents an essential source of inspiration for improving the structure and presentation of sustainability reports. Six years have passed since the definition of the SDGs, under the 2030 Agenda, and HEIs play a key role in helping to achieve the SDGs. In this sense, it is important to align the organization's strategy and the organizational report with the SDGs, making known to stakeholders the current level of commitment to these goals.

The target 12.6 of Agenda 2030 encourages organizations to adopt sustainable practices and to integrate sustainability information into their reporting cycle. Thus, transparent and relevant reports on SDGs are important to communicate to stakeholders how HEI are meeting with their set goals.

Fleaca et al. (2018) conceptualized a model that allows guiding HEIs in the process of organizational culture change, aiming the alignment and operationalization of the HEI strategy with the SDGs. The proposed educational model can be customized and harmonized according to the different circumstances and priorities of the HEI. The last stage of the process is to report and communicate the progress made by HEIs in favour of SDGs, referring to the GRI guidelines as the benchmark to adopt.

Integrating the SDGs thematic into the HEIs reports represents, therefore, one of the challenges for HEIs. We share the opinion of Ceulemans et al. (2015), that HEIs should consider sustainability reports as a dynamic tool to plan the changes to be made in their actions in favour of sustainability, and not use them exclusively as a communication tool.

Even though different frameworks (from the GRI, the International Organization for Standardization and the United Nations Global Compact - UNGC) offer different ways to approach reporting on sustainability, none of them include specific guidance on SDGs disclosures. To bridge this gap, GRI provides a linkage document that shows the disclosures within the GRI Standards that can be used to report on specific SDGs and their targets. The latest version of the document 'Linking the SDGs and the GRI Standards' was updated in September 2020.

The need for a common set of performance indicators proves to be paramount for stakeholders to compare the contributions of companies to the achievement of SDGs. The frameworks already developed for SDGs reporting are still at a very embryonic stage (Amey & Whooley, 2018) and it is, therefore, necessary to foster their maturing.

In the last decade, there has been increasing interest in incorporating sustainability issues into the sphere of higher education. The stakeholder and legitimacy theories are usually used to underpin the theoretical framework of sustainability reporting studies. Stakeholder theory suggests that organisations should be managed in the interest and expectations of all their stakeholders (Martin & Moneva, 2018). In the case of HEIs, we can consider as main stakeholders the students, the local community, the academics, etc. Legitimacy Theory is considered one of the dominant theories in sustainability disclosure studies. This theory holds that entities seek to establish congruence between the social values associated with their activities and entities must act within the limits of what society identifies as sustainability acceptable behaviour in order to continue to operate successfully. Thus, according to this theory, HEIs published information to be deemed acceptable and legitimate by society (Dias, 2010).

The sustainability reporting in higher education is still very embryonic (Lozano, 2011; Ceulemans et al., 2015; Sassen & Azizi, 2017, Huber & Bassen, 2018; Sepasi et al., 2018a; Azizi et al., 2018; Son-Turan & Lambrechts, 2019; Vale et al., 2020) including in Portugal (Aleixo et al., 2016; Silva, 2017).

Concerning the analysis of HEIs' sustainability reports, many studies focus their analysis on specific case(s) of an institution (Yi *et al.,* 2017; Huerta-Riveros, & Gaete-Feres, 2017; Yáñez et al. 2018) or a region/country (Zorio-Grima et al. 2018; Sepasi et al, 2018a; Garde Sanchez, 2013; Fonseca *et al.,* 2011; Gamage & Sciulli, 2016; Sassen & Azizi, 2017-2018; Sassen *et al.*, 2018; Azizi *et. al.,* 2018; Son-Turan & Lambrechts, 2019). In a smaller number, there are comparative studies involving HEIs in an international geographical scope (Sepasi et al, 2018b; Romolini et al., 2015; Alonso-Almeida et al., 2015; Ceulemans et al., 2015; Larrán Jorge et al., 2018).

Some studies investigate the features of higher education institutions that disclose a sustainability report (Sassen et al, 2018; Martin & Moneva, 2018). In line with legitimacy and stakeholder theory, Sassen et al. (2018) provide evidence that the larger the size of the institution, the higher the probability of reporting. By other hand, Martin and Moneva (2018) found that research performance is identified as the best explanatory factor for the disclosure of universities. The findings of Ceulemans et al. (2015) showed that sustainability reporting has been predominantly driven by internal motivations.

All these studies are unanimous in concluding that sustainability reporting at HEIs is still in its early stages, pointing towards the slow and limited adoption of sustainability reports by HEIs worldwide, and the levels of disclosure of sustainability information are still rather low.

Nowadays, there is some debate on how SDGs should be assessed and reported by HEIs (Aleixo et al., 2020), and the empirical literature have started to explore the integration of SGDs in the reporting process.

In the scope of SDGs reporting, empirical research is few and recent, and most focus on companies. The results show a moderate reporting of SDGs by those companies in their reporting sustainability or integrated reporting (Schramade, 2017; Bebbington & Unerman, 2018; KPMG, 2018; Gunawan et al. 2019; Avrampou et al., 2019; Almontaser & Gerged, 2019; Rosati and Faria, 2019a-b; Hummel, 2019; Di Vaio &Varriale, 2020; Acuti et al. 2020; Wang et al., 2020; Tsalis et al. 2020), including in Portuguese companies (Fonseca & Carvalho 2019; Monteiro et al. 2020)

Concerning the higher education, we highlight the studies of Paletta et al. (2020) and Paletta & Bonoli, 2019) related to the University of Bologna. This University published the Report on the UN SDGs since 2016, defining the direct and indirect impact of the University's activities in their four dimensions—training, research, Third Mission and institution - in order to measure their contribution to the advancement of the 2030 Agenda.

The study of Caputo et al. (2021) review SDGs disclosures in 29 HEIs around the world. To understand the level of SDG disclosure the study was developed by associating a dichotomic variable to each GRI indicator, linked by the SDG Compass document to each Agenda 2030 goal. The authors concluded that the most popular SDG is SDG 4 Quality education, followed by SDG 10 Reduced inequalities and SDG 9 Industry, Innovation, and Infrastructure. The least disclosed goal was SDG 17 Partnerships for the goals.

De la Poza et al (2021) assess the level of reporting and alignment of SDG achievements with the overall the Times Higher Education (THE) ranking score. The results reveal that the overall ranking is related to the reporting of HEIs' SDG achievements. Moreover, the more positive actions related to health, education, industry, responsible consumption and production, climate action, and partnerships there were, the higher the position of HEIs in the general ranking was. The best-ranked universities are more committed to transferring knowledge to industry to satisfy its needs (SDG9), support strong institutions in their countries, and promote peace and justice (SDG16). Finally, SDG9 and SDG17 are the most relevant and constant SDGs when modeling the alignment of SDGs with HEIs' dimensions (teaching, research, citations, industry income, international outlook) and subjects (technological and social sciences and humanities).

In the Portuguese context, the studies of SD in HEIs are still scarce. Aleixo et al. (2016) sought to assess the practices of SD existing in Portuguese HEIs, through an analysis of the information disclosed on the websites. The authors concluded that Portuguese HEIs are at an early stage but incremental, and they already show some awareness on how they communicate their policies and strategies in favour of SD.

According to Aleixo et al. (2018a-b), Portuguese HEIs' initiatives to fostering sustainability take place in all core activities (research, education, campus operations, community engagement, institutional framework, on-campus experiences, and assessment and reporting) and in four dimensions: environmental, economic, social and institutional.

The study of Silva (2017), in turn, sought to identify and study the cases of sustainability report that exist or existed in Portuguese HEIs, contributing to understanding the main benefits that can encourage a greater sustainability report and the main barriers that the HEIs face in the dissemination of such practices.

Farinha et al. (2019) identify to what extent the integration of sustainability in the fourteen Portuguese universities was achieved, through their own strategic and activity plans and activity and sustainability reports. The results show that the movement made progress at the university level, with good examples and initiatives at several universities.

In summary, the reporting on sustainability by HEIs is not a common practice in most countries nor is it in Portugal. And it is even lower concerning SDGs-related disclosures. In the Portuguese higher education context, so far as we are aware, Monteiro et al. (forthcoming) present the first Portuguese approach to SDGs reporting, focusing on an award-winning public Portuguese university, analysing the evolution of UMinho's sustainability reports. Thus, the aim of this paper is to extend this previous study, analyse how this public Portuguese HEI has started to link its sustainability reporting to the SDGs after the 2030 Agenda approval (in 2015), and whether sustainability reports' content had already been connected to SDGs themes during the previous period.

The SDGs reporting is still in an embryonic stage in HEIs and the extent to which it has been adopted and its quality remains relatively unexamined. This study attempts to fill the research gap in comprehensiveness in SDGs-related disclosures in a recognized Portuguese university and help to better navigate the future trends in SDSs reporting practices of HEIs.

RESEARCH DESIGN

Sample and the Case in Study

In Portugal there is a total of 30 public HEIs: 15 Polytechnic institutions and 15 universities institutions. Most of the public HEIs are in center of Portugal (33,3%), followed by north (23,3%) (see table 1).

After a brief search for their websites in January 2021, we found that only 3 HEIs have published sustainability reports: University of Minho, Porto University and Polytechnic of Lisbon. The last two institutions published only one report in 2010 and 2011, respectively. University of Minho (Uminho) published its first sustainability report in 2012 (relating to 2010) and the following reports with an annual frequency, until 2016. In 2018 the social services of UMinho published its first sustainability report. Eventually, other Portuguese HEIs might prepare sustainability reports, but they were not available online for the researchers.

Table 1. Sample location

Location	Polytechnic	University	Total	%
North	4	3	7	23,3
Centre	6	4	10	33,3
Lisbon	2	4	6	20,0
Alentejo	3	1	4	13,3
Algarve	0	1	1	3,3
Azores	0	1	1	3,3
Madeira	0	1	1	3,3
Total	**15**	**15**	**30**	**100,0**

Thus, our study will focus on UMinho, because it is the only Portuguese HEI that publishes sustainability reports in a continuous way. Our study aims to verify if UMinho mentions the SDGs in their website and in their sustainability reports.

UMinho has been distinguished by the best result of a Portuguese institution in the "UI GreenMetric World University Rankings 2020" which evaluates the environmental sustainability indexes of academies around the world. UMinho has been a national leader since 2017. In 2020 Uminho stands out in the indicators of energy and climate change, waste, education and research.

UMinho was the first European university to align with the SDGs, as well as to be included in the International Sustainable Campus Network, having also joined the initiative United Nations Global Compact. This academy is a catalyst for its region, generating an annual positive economic impact of over 200 million euros, to which more than 5000 jobs are associated.

The UMinho is among the top 200 HEIs in the world in meeting the SDGs, according to the "2021 THE Impact Rankings". UMinho's growing commitment to sustainability is thus recognized in this ranking, which also highlights this academy in the top 50 in three of the 17 SDGs: SDG4- Quality Education (43rd position), SDG 1- No poverty (50th position) and SDG 9- Industry, Innovation and Infrastructure (50th position).

Methodology

There are six possible sources of evidence for case studies: documents, archival records, interviews, direct observation, participant observation, and physical artifacts (Yin,2003). Hence, qualitative content analysis might be an appropriate analysis and interpretation method for case study research (Kohlbacher & Florian, 2006).

This exploratory study used a qualitative content analysis methodology to analyse the information disclosed in UMinho's website and the stand-alone sustainability reports published by UMinho until nowadays. Content analysis of sustainability reports in higher education has been a widely used method in previous studies related to corporate social/environmental or sustainability reporting (Huerta-Riveros & Gaete-Feres, 2017; Garde Sánchez *et al.,* 2013; Sassen & Azizi, 2017-2018; Yi *et al.,* 2017; Azizi et. al., 2018; Larrán Jorge *et al.*, 2018).

The content analysis method consists in the codification of qualitative information into pre-defined categories (Krippendorff, 2013) that represent the different sustainable development dimensions, adjusted to the specificities of the HEIs. We consider the four dimensions used by UMinho in their sustainability reports: environmental, economic, social, cultural and we added an institutional dimension. This methodology does not allow to analyse the quality, or the amount of information disclosed, only to recognize whether there is some disclosure regarding each SDGs

In this way, a search for SDGs-related disclosures, in all UMinho's sustainability reports, was performed and afterwards classified into the 5 dimensions mentioned above. The 2015 sustainability report is the only one that presents an annex in which alignment with the SDGs is made. It should be mentioned that in 2018 the Social Services of UMinho published its first sustainability report, based on GRI guidelines. However, there is no linkage to SDGs issues.

RESULTS ANALYSIS AND DISCUSSION

After the UMinho webpage analysis, we have observed that there is not a separated tab on the website to disclose information related to SDGs. The mission/strategy is not aligned with the UN goals, and UMinho does not prioritize any SDGs on its website. We only found that some generic information about the SDGs is included in the tabs related to events and news.

Concerning the sustainability reports analysis, we found that the first UMinho report was published in September 2012 and had by reference the year 2010. Since then, annual reports have been produced, except for the years 2012 and 2013, the information of which has been compiled in a single report. The last report was published in December 2016, with reference to the year 2015, and is already in line with the United Nations SDGs[2]. UMinho has always adopted the GRI guidelines, and its sustainability reporting is at a fairly advanced level. In fact, 2015 report refers to the disclosures presented in the report itself and in the previous year.

These results supports the postulates of the legitimacy theory arguing that the main purpose of HEIs when they disclose information is not to be accountable but to persuade stakeholders to trust in the unersity and, as a consequence, sustainability reports usually emphasize the most favourable face of the university.

The 2015 report had an external verification, for the first time, by GRI and UMinho began its first approach to the SDGs in this report, presenting its contribution to the achievement of the goals. There is a clear concern to highlight the SDGs at a specific point of the report. UMinho states that it "cannot ignore the implications of integrating the SDGs into the institution and considers these goals critical to the future of humankind and to ensure the constant improvement of the quality of life of human beings. In fact, it is considered that Education, Research and Learning are at the centre of approaches to sustainable development" (page 14). Thus, in a specific annex (annex X) it is presented information that demonstrates the alignment with the 17 SDGs.

As mentioned before, we searched for SDGs-related disclosures in all UMinho's sustainability reports, considering the 5 dimensions stated previously. Table 2 summarizes the results of this analysis.

Table 2. SDGs-related disclosures in UMinho's sustainability reports

ENVIRONMENTAL DIMENSION			
SDGs		**Report periods**	**Content of Sustainability Reports**
6	**Clean water and sanitation**	2010; 2011; 2012_13;2014;2015	The principles of UMinho's sustainability policy are conservation and management of water for consumption and other purposes; as well as concerns about the quality of water sent to municipal systems.
7	**Affordable and clean energy**	2010; 2011; 2012_13;2014;2015	UMinho has been concerned about the energy efficiency of its buildings following energy standards that opt for cleaner, environmentally friendly energy sources.
12	**Responsible consumption and production**	2010; 2011; 2012_13;2014;2015	The main material resources used in UMinho's activities are paper and cartridges. The 2010 report was limited to quantifying the consumption of toner and paper by organic units. From then until 2015, UMinho reduced paper usage by 76% and the number of cartridges by 82%.
13	**Climate action**	2010; 2011; 2012_13;2014;2015	UMinho has been able to stabilize its greenhouse gas emissions due to the use of more environmentally friendly technologies such as renewable energy sources. It has been seeking to influence its stakeholders to adopt responsible conduct in mitigating negative impacts on the environment.
15	**Life on land**	2010; 2011; 2012_13;2014;2015	UMinho denotes concern with the collection of hazardous waste, its quantification and recycling. It has been separating and redirecting waste in need of treatment to an appropriate final destination. In 2015, the lowest value was found in the production of waste with special need for treatment.
SOCIAL DIMENSION			
5	**Gender equality**	2010; 2011; 2012_13;2014;2015	UMinho ensures gender equality and opportunity by guaranteeing any worker the right to engage in any activity, regardless of gender.
ECONOMIC DIMENSION			
8	**Decent work and economic growth**	2010; 2011; 2012_13;2014;2015	The distributed economic value demonstrates how the reporting entity generates wealth for its stakeholders. For UMinho, the Sustainable Development is transversal to all activities, either through the innovations that arise from the research activity, or the way in which it invests capital or how economic resources are treated. UMinho has an impact on the economy through the value it generates and circulates, both in the region and at the national level. It guarantees the best possible standard for general employment conditions and seeks to improve its offer in terms of services and benefits to employees (discounts for course attendance, sports practices and discounts for assistance to training courses and languages).
12	**Responsible consumption and production**	2010; 2011; 2012_13;2014;2015	UMinho refers to making purchases, not subject to public tender, within a radius of 100km from the university's poles. This distance was stipulated because it was accepted that from that point on purchases entail excessive transport costs and associated environmental impacts. UMinho's procurement practice prioritizes recycled products and services or that their origins are sustainable through "green" certification and minimize packaging waste or that are inserted in recovery programs.
CULTURAL DIMENSION			
11	**Sustainable cities and communities**	2010; 2011; 2012_13;2014;2015	It protects and safeguards cultural heritage by installing some of its organic units in historic buildings of public interest.
INSTITUTIONAL DIMENSION			
17	**Partnerships for the goals**	2012_13;2014;2015	UMinho has become a member of the United Nations Global Compact (UNGC) with the intention of developing business and societal strategies and policies through a commitment to 10 universal principles, committing itself to promoting the application of fundamental values with respect to human rights, labour, the environment and corruption.

According to 2015 sustainability report's attachment X, the 2014 and 2015 sustainability reports information is aligned with 9 of the 17 SDGs identified in Table 2: SDGs 5, 6, 7, 8, 11, 12, 13, 15 and 17. Having analyzed the previous reports, it was found that there was already information related to the same SDGs, except for SDG 17, whose associated information is only disclosed in the last 3 reports.

The environmental dimension is the one that presents the alignment with a greater number of SDGs (5 SDGs), followed by the economic dimension (2 SDGs). The social, cultural and institutional dimensions are only aligned with one SDG each.

The narrative used in the last two reports makes it easier to identify some of the SDGs' targets of the 2030 Agenda, even though the SDGs are not explicitly mentioned throughout the report.

We are aware that UMinho could have identified other SDGs, namely, among others, the SDG 4 - Quality education (being inherent to the HEIs' own activity and the fact that UM has an office for inclusion), the SDG 3 - Good health and well-being (fruit of the social action performed by HEIs) and SDG 16 - Peace, justice and strong institutions (specifically regarding combating corruption).

It was expected that additional sustainability reports have been published after 2016. The continuous communication through the sustainability report could serve not only as a reporting tool, but also to monitor performance, from an economic, environmental, social, and cultural point of view, allowing to link UMinho's contributions to the SDGs.

According to the results of interviews conducted by Silva (2017), UMinho was expected to continue to produce sustainability reports in the following years. However, strangely, until now, no further reports have been published. This may be, perhaps, related to the introduction of the GRI Standards in 2016, because they are considered more complex than the previous GRI guidelines, or because the University is eventually thinking about evolving to the <IR> Framework, as other HEIs (for example those in the United Kingdom).

This opinion finds some support in the fact that the language used in the latest sustainability report is much closer to the integrated reporting terminology than to the GRI sustainability reporting terminology. It is, therefore, pointed out that this type of report of the institution aims to "make public the integrated analysis (underline self-added) of the impacts of the University's activity, in environmental, social, economic and cultural terms, according to the best international practices "(p.3).

This study provides empirical evidence on how Uminho adopt and implement SDG disclosure as part of their sustainability reporting, contributing to a greater understanding of the role that the higher education can play in achieving the 2030 Agenda.

SOLUTIONS AND RECOMMENDATIONS

HEIs should integrate SDG actions into theirs strategic management and, consequently, reporting their SDG performance to promote sustainability and contribute to SD (De la Poza et al., 2021). This paper has practical implications, offering some insights to regulators in order to harmonize SDGs reporting at higher education, as well as to HEIS' managers that wish to follow these new trends. The development of a SDGs' reporting framework could assist HEIs on how to report on SDGs contributions and could be seen as a benchmarking tool to improve HEIs performance towards 2030 Agenda.

FUTURE RESEARCH DIRECTIONS

According to the results of interviews conducted by Silva (2017), UMinho was expected to continue to produce sustainability reports in the coming years. However, as mentioned above, to date Uminho has not published any other report[3], which can be explained by the greater complexity of the GRI standards compared to the previous guidelines.

Thus, in the future it will be interesting to advance in this exploratory study by conducting personal interviews as a research method to understand the reasons that motivate UMinho not to publish additional sustainability reports and how SDGs-related disclosures could be improved.

Our results are grounded on SDGs reporting from sustainability reports of Uminho which demonstrate the SDGs disclosures relevant to the university but do not evidence performance or efective practices towards SD. In view of the above, major challenges are placed on HEIs in identifying priority SDGs for their activities and stakeholders, incorporating them into their strategy, setting goals for their success, and measuring and reporting on their progress. Integrating the SDGs evaluation in the report implies showing to stakeholders their effective performance, resorting to the emerging non-financial reporting models. We defend that it is important to align the sustainability reports with the structure and scope of the SDGs and should, therefore, be a research area to be improved in the future.

CONCLUSION

The 2030 Agenda gives a new impetus towards aligning harmonization of sustainability reporting with the SDGs monitoring framework. This topic as been attracted the attention of researchers who have started to study the linkage of organizational disclosure with the SDGs (Paletta, et al., 2020; Monteiro et al., Forthcoming) often using webpages and sustainability reports as information sources. Following this line of research, the present study aimed to analyse the SDGs-related disclosures performed by UMinho in its website and also in their sustainability reports.

According to Bebbington and Unernan (2018), the SDGs have the potential to inform and advance research and practice on sustainability reporting. They state "academic investigation is needed to help understand where specific SDGs-related accounting initiatives lie on the continuum between pure rhetoric and meaningful action, and to inform the most effective use of the SDGs by a broad range of organizations in developing policies and practices that will contribute towards achievement of the SDGs" (2018:10).

This study used a content analysis methodology, based on the UMinho webpage and the Sustainability reports published since 2010. The university has always adopted the GRI guidelines and in the last published sustainability report (in 2015), UMinho has started linking its reporting with SDGs for the first time. Concerning the SDGs-related disclosure analysis, our findings indicate that sustainability reports' information is aligned with 9 of the 17 SDGs (SDGs 5, 6, 7, 8, 11, 12, 13, 15 and 17). These results confirms UMinho's commitment to the SDGs, not only through teaching, research and knowledge transfer, but also through the incorporation of these goals into its internal policies and practices. Considering the 5 defined dimensions, we found environmental dimension the one that is linked to the greatest number of SDGs.

ACKNOWLEDGMENT

This work is financed by national funds through FCT - Foundation for Science and Technology, I.P., within the scope of multi-annual funding UIDB/04043/2020.

REFERENCES

Acuti, D., Bellucci, M., & Manetti, G. (2020). Company disclosures concerning the resilience of cities from the Sustainable Development Goals (SDGs) perspective. *Cities (London, England)*, *99*, 102608. doi:10.1016/j.cities.2020.102608

Aleixo, A., Leal, S., & Azeiteiro, U. (2018b). Conceptualization of sustainable higher education institutions, roles, barriers, and challenges for sustainability: An exploratory study in Portugal. *Journal of Cleaner Production*, *172*, 1664–1673. doi:10.1016/j.jclepro.2016.11.010

Aleixo, A. M., Azeiteiro, U., & Leal, S. (2016). Toward Sustainability Through Higher Education: Sustainable Development Incorporation into Portuguese Higher Education Institutions. *Challenges in Higher Education for Sustainability*, 159–187.

Aleixo, A. M., Azeiteiro, U., & Leal, S. (2020). Are the sustainable development goals being implemented in the Portuguese higher education formative offer? *International Journal of Sustainability in Higher Education*, *21*(2), 336–352. doi:10.1108/IJSHE-04-2019-0150

Aleixo, A. M., Azeiteiro, U. M., & Leal, S. (2018a). The implementation of sustainability practices in Portuguese higher education institutions. *International Journal of Sustainability in Higher Education*, *19*(1), 146–178. doi:10.1108/IJSHE-02-2017-0016

Almontaser, T., & Gerged, A. (2019). (in press). Exploring corporate Sustainable Development Goals (SDGs) disclosure in a politically unstable environment: Evidence from the Libyan oil industry. *Sustainability, Accounting, Management, and Policy Journal*. Advance online publication. doi:10.1108/SAMPJ-04-2019-0136

Alonso-Almeida, M. del M., Marimon, F., Casani, F., & Rodriguez-Pomeda, J. (2015). Diffusion of sustainability reporting in universities: Current situation and future perspectives. *Journal of Cleaner Production*, *106*, 144–154. doi:10.1016/j.jclepro.2014.02.008

Amey, M. & Whooley. (2018, Dec.). Corporate reporting on the SDGs: mapping a sustainable future. *Advisor Perspectives*.

Annan-Diab, F., & Molinari, C. (2017). Interdisciplinarity: Practical approach to advancing education for sustainability and for the Sustainable Development Goals. *International Journal of Management Education*, *15*(2), 73–83. doi:10.1016/j.ijme.2017.03.006

Avrampou, A., Skouloudis, A., Iliopoulos, G., & Khan, N. (2019). Advancing the Sustainable Development Goals: Evidence from leading European banks. *Sustainable Development*, *27*(4), 743–757. doi:10.1002d.1938

Azizi, L., Bien, C., & Sassen, R. (2018). Recent trends in sustainability reporting by German universities. *Sustainability Management Forum, 26*, 65–85.

Bebbington, J., & Unerman, J. (2018). Achieving the United Nations Sustainable Development Goals: An enabling role for accounting research. *Accounting, Auditing & Accountability Journal*, *31*(1), 2–24. doi:10.1108/AAAJ-05-2017-2929

Caputo, F., Ligorio, L., & Pizzi, S. (2021). The Contribution of Higher Education Institutions to the SDGs—An Evaluation of Sustainability Reporting Practices. *Administrative Sciences*, *11*(3), 97. doi:10.3390/admsci11030097

Ceulemans, K., Lozano, R., & Alonso-Almeida, M. (2015). Sustainability reporting in higher education: Interconnecting the reporting process and organisational change management for sustainability. *Sustainability*, *7*(7), 8881–8903. doi:10.3390u7078881

DelaPoza, E., Merello, P., Barberá, A., & Celani, A. (2021). Universities' Reporting on SDGs: Using THE Impact Rankings to Model and Measure Their Contribution to Sustainability. *Sustainability*, *13*(4), 2038. doi:10.3390u13042038

Di Vaio, A., & Varriale, L. (2020). SDGs and airport sustainable performance: Evidence from Italy on organisational, accounting and reporting practices through financial and non-financial disclosure. *Journal of Cleaner Production*, *249*(10), 119431. doi:10.1016/j.jclepro.2019.119431

Dias, A. (2010). Responsabilidade Social Corporativa – Teorias Predominantes na Investigação em Contabilidade. *XIV Encuentro AECA*.

Farinha, C., Caeiro, S., & Ulisses, A. (2019). Sustainability strategies in Portuguese Higher Education Institutions: Commitments and practices from internal insights. *Sustainability*, *11*(11), 3227. doi:10.3390u11113227

Fleaca, E., Fleaca, B., & Maiduc, S. (2018). Aligning strategy with Sustainable Development Goals (SDGs): Process scoping diagram for entrepreneurial Higher Education Institutions (HEIs). *Sustainability*, *10*(4), 1032. doi:10.3390u10041032

Fonseca, A., Macdonald, A., Dandy, E., & Valenti, P. (2011). The state of sustainability reporting at Canadian universities. *International Journal of Sustainability in Higher Education*, *12*(1), 22–40. doi:10.1108/14676371111098285

Fonseca, L., & Carvalho, F. (2019). T*he Reporting of SDGs by Quality. Environmental, and Occupational Health and Safety-Certified Organizations*, *11*(20), 5797. doi:10.3390u11205797

Gamage, P., & Sciulli, N. (2016). Sustainability Reporting by Australian Universities. *Australian Journal of Public Administration*, *76*(2), 187–203. doi:10.1111/1467-8500.12215

Garde Sánchez, R., Rodriguéz Bolívar, M. P. & López-Hernández, A. M. (2013). Online disclosure of university social responsibility: a comparative study of public and private US universities. *Environmental Education Research, 19*(6), 709–746.

Gunawan, J., Permatasari, P., & Tilt, C. (2019). Sustainable Development Goal disclosures: Do they support responsible consumption and production? *Journal of Cleaner Production*, *246*(10), 118989.

Huber, S., & Bassen, A. (2018). Towards a sustainability reporting guideline in higher education. *International Journal of Sustainability in Higher Education*, *19*(2), 218–232. doi:10.1108/IJSHE-06-2016-0108

Huerta-Riveros, P., & Gaete-Feres, H. (2017). Responsabilidad social universitaria a través de los reportes de sostenibilidad del Global Reporting Initiative: Experiencia de una universidad pública. *Revista Iberoamericana de Educación Superior*, *8*(23), 120–137. doi:10.22201/iisue.20072872e.2017.23.3014

HummelK. (2019). *Reporting on the Sustainable Development Goals – early evidence from Europe*. Available at https://ssrn.com/abstract=3411017, June 2019. doi:10.2139/ssrn.3411017

Kohlbacher, F. (2006). The Use of Qualitative Content Analysis in Case Study Research. *Forum Qualitative Sozialforschung / Forum: Qualitative. Social Research*, *7*(1), 1–30.

KPMG. (2015). *KPMG survey of corporate responsibility reporting 2015*. KPMG.

KPMG (2018). *How to report on the SDGs. What good looks like and why it matters*. kpmg.com/sdgreporting.

Krippendorff, K. (2013). *Content Analysis. An Introduction to Its Methodology* (3rd ed.). Sage Publications.

Larrán Jorge, M., Javier Andrades, F., & Madueño Jesús, H. (2018). An analysis of university sustainability reports from the GRI database: An examination of influential variables. *Journal of Environmental Planning and Management*, *62*(3), 1–26.

Lozano, R. (2011). The state of sustainability reporting in universities. *International Journal of Sustainability in Higher Education*, *12*(1), 67–78. doi:10.1108/14676371111098311

Lozano, R., Llobet, J., & Tideswell, G. (2013a). The process of assessing and reporting sustainability at universities: Preparing the report of the University of Leeds. *Revista Internacional de Tecnología, Sostenibilidad y Humanismo. Desembre*, *8*, 85–112.

Martin, E., & Moneva, J. (2018). Do as I say, not as I do: Drivers of sustainability reporting Spanish universities. *Sociology International Journal*, *2*(5), 411–419. doi:10.15406ij.2018.02.00078

Monteiro, S., Ribeiro, V., & Lemos, K. (2020). Linking corporate social responsibility reporting with the UN Sustainable Development Goals: Evidence from the Portuguese stock market. In Conceptual and Theoretical Approaches to Corporate Social Responsibility, Entrepreneurial Orientation, and Financial Performance (pp. 134–151). IGI Global.

Monteiro, S., Ribeiro, V., & Lemos, K. (forthcoming). Sustainability reporting and the Sustainable Development Goals (SDGs) in Higher Education: a Portuguese university case. In D. Crowther & S. Seifi (Eds.), CSR and Sustainability. Academic Press.

Neary, J., & Osborne, M. (2018). University engagement in achieving Sustainable Development Goals: A synthesis of case studies from the SUEUAA Study. *Australian Journal of Adult Learning*, *58*(3), 336–364.

Ntim, C., Soobaroyen, T., & Broad, M. J. (2017). Governance structures, voluntary disclosures and public accountability. *Accounting, Auditing & Accountability Journal*, *30*(1), 65–118. doi:10.1108/AAAJ-10-2014-1842

ORSIES- Observatório da Responsabilidade Social e Instituições de Ensino Superior. (2018). *Livro Verde sobre Responsabilidade Social e Instituições de Ensino Superior*. Author.

Paletta, A., & Bonoli, A. (2019). Governing the university in the perspective of the United Nations 2030 Agenda: The case of the University of Bologna. *International Journal of Sustainability in Higher Education*, *20*(3), 500–514.

Paletta, A., Fochi, P., Toschi, T., & Ubertini, F. (2020). Adoption of the SDGs as a reporting framework at the Alma Mater Studiorum (University of Bologna) in Italy. In G. Nhamo & V. Mjimba (Eds.), *Sustainable Development Goals and Institutions of Higher Education* (pp. 185–197). Sustainable Development Goals Series.

Romolini, A., Fissi, S., & Gori, E. (2015). Quality disclosure in sustainability reporting: Evidence from universities. *Transylvanian Review of Administrative Sciences*, *11*(44), 196–218.

Rosati, F., & Faria, L. (2019a). Business contribution to the Sustainable Development Agenda: Organizational factors related to early adoption of SDG reporting. *Corporate Social Responsability and Environmental Management*, 1–10. doi:10.1002/csr.1705

Rosati, F., & Faria, L. (2019b). Addressing the SDGs in sustainability reports: The relationship with institutional factors. *Journal of Cleaner Production*, *215*, 1312–1326.

Sassen, R., & Azizi, L. (2017). Voluntary disclosure of sustainability reports by Canadian universities. *Journal of Business Economics*, *88*(1), 97–137.

Sassen, R., & Azizi, L. (2018). Assessing sustainability reports of US universities. *International Journal of Sustainability in Higher Education*, *19*(7), 1158–1184.

Sassen, R., Dienes, D., & Wedemeier, J. (2018). Characteristics of UK higher education institutions that disclose sustainability reports. *International Journal of Sustainability in Higher Education*, *19*(7), 1279–1298.

Schramade, W. (2017). Investing in the UN Sustainable Development Goals: Opportunities for companies and investors. *Journal of Applied Corporate Finance*, *29*, 87–99.

SDSN Australia/Pacific. (2017). *Getting started with the SDGs in universities: A guide for universities, higher education institutions, and the academic sector*. Australia, New Zealand and Pacific Edition. Sustainable Development Solutions Network – Australia/Pacific.

Sepasi, S., Braendle, U., & Rahdari, A. H. (2018b). Comprehensive sustainability reporting in higher education institutions. *Social Responsibility*, *15*(2), 155–170.

Sepasi, S., Rahdari, A., & Rexhepi, G. (2018a). Developing a sustainability reporting assessment tool for higher education institutions: The University of California. *Sustainable Development*, *26*(6), 672–682.

Silva. (2017) *Caracterização do relato de sustentabilidade nas Instituições de Ensino Superior Portuguesas*. Dissertação para obtenção do grau de Mestre em Contabilidade e Controlo de Gestão pela Faculdade de Economia do Porto.

Son-Turan, S., & Lambrechts, W. (2019). Sustainability disclosure in higher education: A comparative analysis of reports and websites of public and private universities in Turkey. *International Journal of Sustainability in Higher Education, 20*(7), 1143–1170.

Tsalis, T. A., Malamateniou, K. E., Koulouriotis, D., & Nikolaou, I. E. (2020). New challenges for corporate sustainability reporting: United Nations' 2030 Agenda for sustainable development and the sustainable development goals. *Corporate Social Responsibility and Environmental Management*, 1–13.

Vale, J., Bertuzi, R., & Monteiro, A. (2020). Social responsibility reporting in Higher Education Institutions: A systematic literature review. In *Conceptual and Theoretical Approaches to Corporate Social Responsibility, Entrepreneurial Orientation, and Financial Performance*. IGI Global.

Vilalta, J. M., Betts, A., & Gomez, V. (2018). Higher Education's role in the 2030 agenda: The why and how of GUNi's commitment to the SDGs. In J. Vilalta, A. Betts & V. Gómez (Eds.), *Sustainable Development Goals: Actors and Implementation. A Report from the International Conference* (pp. 10-13). Barcelona: GUNi.

Wang, X., Yuen, K., Wong, Y., & Li, K. (2020). How can the maritime industry meet Sustainable Development Goals? An analysis of sustainability reports from the social entrepreneurship perspective. *Transportation Research Part D, Transport and Environment*, 78.

Yáñez, S., Uruburu, Á., Moreno, A., & Lumbreras, J. (2018). The Sustainability Report as an essential tool for the holistic and strategic vision of Higher Education Institutions. *Journal of Cleaner Production*. Advance online publication. doi:10.1016/j.jclepro.2018.09.171

Yi, A., Howard, D., & Harun, H. (2017). Sustainability reporting at a New Zealand public university: A longitudinal analysis. *Sustainability, 9*, 1529. doi:10.3390u9091529

Yin, R. K. (2003). *Case study research, design and methods* (3rd ed., Vol. 5). Sage.

Zorio-Grima, A., Sierra-García, L., & Garcia-Benau, M. A. (2018). Sustainability reporting experience by universities: A causal configuration approach. *International Journal of Sustainability in Higher Education, 19*(2), 337–352.

ADDITIONAL READING

Adams, C. (2018). Debate: Integrated reporting and accounting for suitable development across generations by universities. *Public Money & Management, 38*(5), 325–327. doi:10.1080/09540962.2018.1477580

Adams, C., Muir, S., & Hoque, Z. (2014). Measurement of sustainability performance in the public sector. *Sustainability Accounting. Management and Policy Journal, 5*(1), 46–67.

Adhikariparajul, M., Hassan, A., Fletcher, M., & Elamer, A. (2019, January 14th). ((Accepted). Integrated Reporting in UK Higher Education Institutions. *Sustainability Accounting* [Forthcoming]. *Management and Policy Journal.*

Brusca, I., Labrador, M., & Larran, M. (2018). The challenge of sustainability and integrated reporting at universities: A case study. *Journal of Cleaner Production*, *188*, 347–354. doi:10.1016/j.jclepro.2018.03.292

Buchta, K., Jakubiak, M., & Wilczewski, A. (2018). University social responsibility – theory vs. practice. *Research Papers of the Wroclaw University of Economics*, *2018*(520), 22–33.

Chatelain-Ponroy, S., & Morin-Delerm, S. (2016). Adoption of sustainable development reporting by universities. *Accounting, Auditing & Accountability Journal*, *29*(5), 887–918. doi:10.1108/AAAJ-06-2014-1720

Hooey, C., Mason, A., & Triplett, J. (2017). Beyond greening: Challenges to adopting sustainability in Institutions of Higher Education. *The Midwest Quarterly*, (April), 280–291.

Kräusche, K., & Pilz, S. (2018). Integrated sustainability reporting at HNE Eberswalde – a practice report. *International Journal of Sustainability in Higher Education*, *19*(2), 291–312. doi:10.1108/IJSHE-07-2016-0145

Lozano, R. (2006). A tool for a graphical assessment of sustainability in universities (GASU). *Journal of Cleaner Production*, *14*(9-11), 963–972. doi:10.1016/j.jclepro.2005.11.041

Lozano, R., & Huisingh, D. (2011, February). Inter-linking issues and dimensions in sustainability reporting. *Journal of Cleaner Production*, *19*(1-2), 99–107. doi:10.1016/j.jclepro.2010.01.004

Lozano, R., Lukman, R., Lozano, F., Huisingh, D., & Lambrechts, W. (2013b). Declarations for sustainability in higher education: Becoming better leaders, through addressing the university system. *Journal of Cleaner Production*, *48*, 10–19. doi:10.1016/j.jclepro.2011.10.006

Nejati, M., Shafaei, A., Salamzadeh, Y. & Daraei, M. (2011). Corporate social responsibility and universities: A study of top 10 world universities' website. *African Journal of Business Management, 5*(2), 440-447.

Quezada, R. (2011). La responsabilidad social universitaria como desafío para la gestión estratégica de la Educación Superior: El caso de España. *Review of Education*, *355*(Mayo-agosto), 109–133.

Richard, V., Forget, D., & Gonzalez-Bautista, N. (2017). Implementing sustainability in the classroom at Université Laval. In W. Leal Filho, M. Mifsud, C. Shiel, & R. Pretorius (Eds.), *Handbook of Theory and Practice of Sustainable Development in Higher Education* (Vol. 3, pp. 133–146). Springer. doi:10.1007/978-3-319-47895-1_9

Severino-González, P., Villalobos-Antúnez, J., Chamali-Duarte, N., Vergara-Piña, G., & González-Soto, N. (2019). Social responsibility of university student and institutional educational policies. Recognizing the commitments to society. *Opción*, *35*(90), 1171–1197.

Spodarczyk, E., & Szelagowska-Rudza, K. (2019). Socially responsible university from the perspective of a student as a stakeholder – a pilot study report. *Management*, *23*(2), 80–97. doi:10.2478/manment-2019-0020

Vallaeys, F. (2014). University social responsibility: A new university model against commodification. *Ibero-American Journal of Higher Education*, *5*(12), 105–117.

KEY TERMS AND DEFINITIONS

Global Reporting Initiative: Is an international independent standards organization that defines guidelines or standards to support companies and other organizations to disseminate information about their actions and their impact on the environment, human rights, the fight against corruption, etc. It is one of the frameworks most used by organizations when preparing their sustainability reports.

Social Responsibility: Corresponds to business development with a concern for the well-being of society and the environment. Companies are concerned about their impacts on society, in economic, environmental and social terms.

Sustainability Reporting: It consists in disseminating information on the impact of organizations on society, in economic, social and environmental terms. Entities disclose their actions in favour of sustainable development.

Sustainable Development: It corresponds to the development of society and the use of resources without compromising the natural system and the maintenance of the production of new resources.

Sustainable Development Goals: Corresponds to 17 goals defined in the 2030 Agenda of the United Nations, which are interconnected and that seek to contribute to a more sustainable future.

United Nations (UN) 2030 Agenda: At the 2015 United Nations General Assembly, the 17 Sustainable Development Goals which are intended to be achieved by the year 2030 were defined. In this way, the United Nations resolution became known as Agenda 2030.

ENDNOTES

1 Since 2000, GRI has been publishing multiples guidelines for sustainability reporting. In 2013, the GRI G4 was the last version launched before the SDGs. GRI G4 intended to give to the report a higher level of relevant and concentrated information. In 2016 the GRI Standards were published, having a modular, flexible, and inter-related set which starts with the universal standards and can be complete by a variety of other topics in economic, environmental, or social dimensions.

2 For further details about the evolution of UMinho sustainability reports' content, please see Monteiro et al. (forthcoming).

3 As state before, Social Services of Uminho published its first Sustainability reporting in 2018, using GRI guidelines. However, there is no linkage to SDGs issues.

Chapter 14

Organizational Virtuousness and Workplace Well-Being in Organizational Citizenship Behaviors:

A Case Study in Intermunicipal Community of Terras de Trás-os-Montes

Liliana Pimentel
https://orcid.org/0000-0002-9058-9749
Faculty of Economics, University of Coimbra, Portugal

Helena Maria Santos De Oliveira
https://orcid.org/0000-0001-8153-9005
ISCAP, Polytechnic Institute of Porto, Portugal

Sara Vicente Pinheiro
University of Coimbra, Portugal

ABSTRACT

In the current context, strongly marked by globalization, organizations have undergone a period of changes induced by economic, social, and political changes, which have led to changes in the relationships between individuals and society. In this way, organizations, through the implementation of new values and strategies, have established themselves as socially responsible organizations, characterized by their ethical essence. This chapter intends to evaluate the organizational virtuousness, affective well-being at work, and organizational citizenship behaviors, following the studies developed by Cameron et al. and by Rego et al., which relate organizational virtuosity, well-being at work, and behaviors of organizational citizenship. The study was applied in the municipalities of the Intermunicipal Community (CIM) – Trás-os-Montes, and 760 employees were surveyed. Findings indicate that perceptions of organizational virtuousness and well-being at work are related to organizational citizenship behaviors.

DOI: 10.4018/978-1-7998-9410-0.ch014

INTRODUCTION

In the current context, strongly marked by globalization, organizations have undergone a set of economic, social and political transformations that have changed the relationships between individuals and society. The need to respond to these transformations has forced organizations to proceed with the implementation of new values and strategies, in order to build a path that aims to strengthen corporate social responsibility and sustainable development.

The concern with corporate social responsibility has led organizations to proceed with the implementation of an organizational culture that prioritizes the development of high performance, with emphasis on people management. In this way, the last decades have been marked by the growth of investigations directed to understand the different organizational cultures of countries, since organizational values are one of the fundamental elements of these cultures (Tamayo, 2004). Organizational values, present in organizations and perceived through organizational virtuosity and well-being at work, are constantly changing and, therefore, it is important to understand their trend.

The growing trend of virtuous characteristics in individuals and organizations has driven the topic of organizational virtuosity and well-being at work. These themes have assumed a growing importance over the last decades, not only in society, but also, increasingly, in the academic and scientific environment.

In the current context, strongly marked by situations of lack of business ethics and corruption, it is essential to analyze and study how organizations apply policies and practices and what is the impact on society. In this way, the importance of social responsibility has increased, essentially driven by organizational citizenship behaviors. Promoting the well-being of employees and managers is nowadays a priority.

Several studies have focused on the relevance of organizational virtuosity and on organizational environments where virtues such as integrity, forgiveness, trust, optimism and compassion are practiced, disseminated and perpetuated, both individually and collectively (Bright and Exline, 2012; Bright et al., 2006; Cameron, 2003; Cameron et al., 2004; Caza et al., 2004). These studies show that virtuous perceptions and practices are positively related to organizational citizenship behaviors.

In Portugal, this theme has also assumed its importance and several researchers have analyzed the perceptions of organizational virtuosity, relating it with organizational performance, affective well-being and organizational citizenship behaviors (Rego et al., 2010).

Following global and national trends, it is important to understand the true perception that employees develop about organizational values, since these values allow for the creation and permanence of work environments that help the development of people and the organization. This happens because the values present in the organization and perceived by the workers induce their behavior towards the organization.

This study aims to understand whether the organizational citizenship behaviors observed in the public entity Intermunicipal Community (CIM) - Terras de Trás-os-Montes are related to perceptions of organizational virtuosity and work well-being. To this end, we intend to answer the following questions: (1) Are the organizational citizenship behaviors practiced at CIM - Terras de Trás-os-Montes related to perceptions of organizational virtuosity? (2) Are the organizational citizenship behaviors practiced at CIM – Terras de Trás-os-Montes related to well-being at work?; (3) Is the well-being at work felt by workers at CIM - Terras de Trás-os-Montes related to perceptions of organizational virtuosity.

Although studies about integrity and trust within organizations are common in the literature, research combining such aspects as compassion, forgiveness, optimism and corporate social responsibility is yet to be done, namely in public sector entities.

Noticing that the relevance ascribed to character and virtues in the individual and institucional life "has waxed and waned, often depending on broader trends in society and the pervasiveness of corruption and scandals in the private and public sectors,"Wright and Goodstein (2007: 949), asked: "[w]hat is needed to maintain momentum in exploring the role of character and virtude in organizational studies?''. They suggested that more research is necessary to answer the question. Organizational manifestations of virtuousness and its consequences, both for individuals and organizations, remain under-developed theoretically and empirically (Cameron, 2003; Cameron et al,. 2004; Wright and Goodstein, 2007).

This chapter is an answer to the challenges posed by those researchers. We agree with Wright and Goodstein's (2007) suggestion that the exploratory nature of the assessment method employed by Cameron et al. (2004) invites other reserachers to employ such a measure in other contexts, for example, public sector, namely local governments, as is our study, and in relation to other outcomes (e.g., employee loyalty; organizational commitment; and well-being).

Keeping this in mind, the chapter is organized as follows. Section 2 offers an overview of the relevance of the corporate social responsibility to understand the organizational virtuousness concept and its multidimensional nature. Section 3 starts by briefly discussing each construct's definition: organizational virtuousness (OV), affective well-being (AWB) at work and organizational citizenship behaviors (OCB). Making use of a double-source method for collecting data on dependent and independent variables, this chapter hopes to contribute to a better understanding of what is surely a central challenge for many contemporary organizations and their managers: to identify forms of using and developing their human capital in a way that is mutually beneficial for themselves and their employees. Section 4 define the hypotheses of the study. Section 5 explains the methodology and describes the sample. Section 6 discusses the results. Finally, section 7 presents the conclusions and implications of the study for management are presented, as well as the bibliographical references.

CORPORATE SOCIAL RESPONSIBILITY

In the current context, strongly marked by globalization, organizations have undergone a period of change induced by economic, social and political modifications that have originated changes in the relationships between individuals and society. Thus, organizations, through the implementation of new values and strategies, have been building a long way to achieve corporate social responsibility and sustainable development, adapting to existing conditions and resources and available skills.

Given the challenges of sustainability and corporate social responsibility for organizations, whether public or private, accounting plays an important role, as it provides economic, social and environmental information to internal management and external agents, allowing them to make decisions in an effective and conscious manner. The last decades have been largely marked by the theme of corporate social responsibility, which has attracted the attention of researchers and the academic community, so it is important to understand the concept, its evolution and the position of researchers.

The first reference found on corporate social responsibility dates back to 1953 and was led by Howard R. Bowen. In order to define the concept, the author considered two aspects: the first is led by the fact that large organizations are large power centers, and the second is related to how the decisions made by these organizations affect the lives of their employees/leaders in different aspects. Thus, for the pioneer author of the theme, corporate social responsibility is defined as the "obligations of businessmen to adopt policies, make decisions, or follow courses of action that are desirable in relation to the goals and values of our society" (Bowen, 1953: 6). However, and as with most topics, this is a topic that has sparked the interest of other researchers and has therefore changed over time.

Corporate social responsibility, driven by social movements that emerged around the world, gained greater importance in the 1960s. Thus, organizations began to be pressured by society, which defended causes such as consumer rights, the presence of more female employees, the concern with the environment, to adopt socially responsible behaviors that fit with the changes in the social and political environment (Carroll and Shabana, 2010). Thus, one of the most conceptualized definitions of this era was developed by Keith Davis who considered that corporate social responsibility concerns the "decisions and actions that businessmen take for reasons that, at least in part, go beyond the economic or technical interests of the company" (Davis, 1960: 70). For the construction of the definition, the author considered that the obligations borne by businessmen should be equivalent to their power.

Despite the developments around the concept of corporate social responsibility, Friedman criticized the existing definitions up to that time. According to the neoliberalist author, any corporate behavior should be aimed at making profit and safeguarding the interests of the shareholders belonging to the organization (Friedman, 1970).

However, the works carried out on corporate social responsibility continued to emerge, mainly due to the increased ambition of society regarding the moral commitments of organizations, and in this regard, one of the first proposals to analyze the dimensions of corporate social responsibility emerged. In the late 1970s, the researcher Carroll proposed a definition of corporate social responsibility in which he considers the economic, legal, ethical, and discretionary expectations that citizens in general develop toward an organization over a given period of time.

Choosing a different perspective from the previous ones and driven by the importance of transforming the organization's responsible behaviors into business opportunities, Drucker (1984) considers that corporate social responsibility is the ability to transform a social problem into a business opportunity and into benefits such as: productive capacity, increased human capital, higher pay, and more wealth. In this sense, organizations tend to find the balance between commercial success and ethical and environmental values.

The globalization of the economy is a reality that organizations have had to deal with in the 21st century, and because of this, corporate social responsibility has also become a global phenomenon discussed by managers, politicians, trade unionists, and researchers. The European Commission in 2001, in the Green Paper, defines corporate social responsibility as "a concept whereby companies decide, on a voluntary basis, to contribute to a fairer society and a cleaner environment" (EC, 2001, p.4). According to the European Commission, social responsibility can be divided into two categories: the internal, where actions are considered in human capital investment, health and safety measures at work; the external, where the actions and behaviors adopted by the organization should perpetuate to business partners, suppliers, customers, public authorities, etc..

The concept of corporate social responsibility is constantly being updated and therefore consists of an incessant process of seeking higher levels of involvement between organizations and social groups. This process is supported by the World Business Council for Sustainable Development which defines corporate social responsibility as a "continuous process of corporate commitment and ethical behavior that aims to contribute to economic development and, at the same time, drive improvements in the quality of life of its employees and their families, the well-being of the local community, and society in its broadest concept.

The number of companies concerned with corporate social responsibility has increased considerably in recent years, because they consider that the existence of this concern is a competitive advantage (Branco & Rodrigues, 2008). A good relationship with stakeholders is achieved through social commitment, allowing a differentiation in relation to competitors and, simultaneously, improving financial results and performance. These authors also argue that organizations rely on corporate social responsibility with the intention of facing external pressures, because if they do not, they may jeopardize their profitability and survival.

Studies on the topic of corporate social responsibility are numerous and with different approaches. For example, Dobers and Halme (2009) investigated this topic for organizations in developing countries. The authors concluded that given the culture and environment that persists in these countries, corporate social responsibility is approached differently by managers. They go further and, given the weak institutional environment in which these companies operate, present two possible solutions to the problem. The first solution, directed at the government, consists of implementing measures to detect tax fraud, corruption, and unfair competition. The second is directed at the companies themselves, in that they, too, must seek to create a structure that promotes social justice, environmental protection, and poverty reduction.

At the national level, the research on this theme has occurred essentially since 2003. Several authors (Ferreira, 2004; Lourenço et al., 2012; Rego et al., 2007; Roque and Cortez, 2006), who developed studies on corporate social responsibility.

For example, Roque and Cortez (2006) analyzed, for a 5-year period, the reports and accounts of 15 companies listed on Euronext Lisbon. Using the analysis of the relationship between environmental disclosure and financial performance, they concluded that the companies that annually do not disclose environmental information are also those that present the worst financial performance.

Using a different approach, Lourenço et al., (2012), investigated how markets observe Corporate Sustainability Performance, based on the Dow Jones Sustainability Index. The authors developed a set of hypotheses that related market value of net parimony and Corporate Sustainability Performance. The authors concluded that Corporate Sustainability Performance significantly influences stock price over accounting measures such as earnings and net worth book value. They also concluded that investors undervalue large profitable companies when they have low levels of Corporate Sustainability Performance, and therefore this type of analysis should never be conducted on the basis of sustainability alone.

Over the last decades several models of social responsibility have been proposed that show the evolution of the concept and the growing importance of the subject (Carroll, 1991; Wood, 1991). Carroll (1991) considers that the concept covers responsibilities of different types that organizations should have towards society, namely: economic, legal, ethical and discretionary.

Subsequently, Schwartz & Carroll (2003) assume the existence of two limitations: the difficulty in differentiating between ethical and philanthropic actions and the fact that the fulfillment of philanthropic responsibilities is driven by economic ambitions and, therefore, decided to remove this dimension.

After this brief contextual analysis of social responsibility and the business performance model, it is possible to state that a socially responsible organization has ethics at its core, seeks to maximize profits, provide adequate remuneration and integrate employees into the organizational environment. To this end, it adopts virtuous actions and provides well-being at work to its employees that will have consequences in organizational citizenship behaviors.

KEY CONSTRUCTS OF THE MODEL

Organizational Virtuousness

For Aristotle virtues can be described as desires, habits, or actions that produce personal and social good (Aristotle, 1999; Cameron, 2003). Some authors (Peterson and Seligman, 2004) have defined virtues as "core characteristics valued by moral philosophers and religious thinkers" that can be grouped into six main categories, namely, wisdom, courage, humanity, justice, temperance, and transcendence.

Virtuousness refers to the pursuit of the highest aspirations in the human condition (Bright et al., 2006). Recent corporate scandals have prompted a growing interest in the topic as an organizational property/feature, both in the business community and academia. For example, Cameron et al. (2004) developed and validated an instrument for measuring the perceptions of OV. They found a five-factor model comprising organizational optimism, forgiveness, trust, compassion, and integrity. Organizational optimism means that organizational members develop a belief that they will succeed in doing well and doing good, even when faced with major challenges. Organizational forgiveness means that the mistakes are quickly forgiven and used as opportunities for learning in a context characterized by high standards of performance. Organizational trust indicates that courtesy, consideration, and respect govern the organization and that people trust each other and their leaders. Organizational compassion means that people care about each other, and that acts of compassion and concern are common. Organizational integrity indicates that honesty, trustworthiness, and honor pervade the organization.

Statistically significant relationships between perceived OV and performance have been found (Cameron, 2003; Cameron et al., 2004). Virtuousness amplifies positive contexts, attitudes, and behaviors. For example, when observing virtuous behaviors, employees experience positive emotions, which in turn, lead them to replicate virtuous acts, thus subsequently fostering social capital (Baker and Dutton, 2007) and increasing individual and organizational performance (Fredrickson, 2001, 2003; Staw et al., 1993). Virtuousness also buffers the organization and its members from negative contexts, attitudes, feelings, and behaviors, by enhancing resiliency, solidarity, and a sense of efficacy. For example, the exposure to virtuousness may help individuals to absorb misfortune, recover from trauma, and maintain momentum in difficult circumstances (as in the case of downsizing), allowing them and their organizations to overcome problems and difficulties, and to improve performance (Bright et al., 2006; Cameron, 2003; Cameron et al., 2004).

This association between OV and performance may make virtuousness a less foreign construct to the organizational research field (Cameron et al., 2004). As Cameron (2003: 49) pointed out, ''organizational manifestations of OV and its consequences remain underdeveloped theoretically and empirically. This is unfortunate because virtuousness is intimately tied to what is good to and for human

beings, so its omission from scientific investigation leaves a void in understanding the full range of consequential organizational phenomena.''

At the national level, researchers have followed international trends and developed some empirical studies related to organizational virtuosity and its consequences. Rego et al. (2010) studied how organizational virtuosity explains happiness and organizational citizenship behaviors. Based on a survey of fourteen organizations, the researchers suggested that employees who perceive virtuous actions in their organizations are happier at work and therefore engage in more organizational citizenship behaviors.

Rego et al. (2012) developed another empirical study with individuals working in teams in a Portuguese university organization to understand how virtuousness explains affective commitment and the level of power in teams. The authors suggest that in more virtuous organizations, their members developed greater affective commitment that extends to the whole organization and, therefore, both the organization and the teams felt more powerful and more confident in their ability to achieve their goals.

The study shows that organizations whose members have more affective commitment to the team achieve better performance. These relationships occur because in virtuous teams, feelings of gratitude remain, leading employees to feel that they do meaningful work. Thus, it can be seen that they assume positive attitudes and feelings towards the team, showing more interest in contributing to its success. The positive responses from employees lead the organization to feel more empowered, with more power to face challenges and adversities and to achieve goals. As a general rule, the affective commitment of employees within teams produces high levels of energy, initiative, and dedication that are the basis for the potential of teams and the organization.

By relating OV with extra-role performance (i.e., OCB), this chapter hopes to contribute toward making the construct more familiar and relevant to researchers at the interface between business and psychology.

Affective Well-Being

The difficulties in defining the concept of well-being at work are evident in the number of theoretical approaches in the literature. For Warr (2007), well-being at work can be evaluated using three orthogonal axes: anxiety-comfort, depression-pleasure, and tiredness-vitality. In turn, Daniels (2000) uses the same line of thought and adds two new factors: boredom-enthusiasm and anger-serenity. These two authors construct the definition of well-being at work based on the definition of psychological well-being, unlike Van Horn and his team (2004) who used the definition of subjective well-being. They proposed a model that aims at the positive evaluation of the various characteristics present in the work context and that consists of five dimensions: affective, professional, behavioral, cognitive, and psychosomatic.

Despite all the differences, the models mentioned have some similarities, such as: the existence of a concept, well-being at work, which is a multidimensional construct and the existence, always, of reference to positive or negative experiences that occur in the work environment and that, therefore, the affect variable is more important than the others.

After the literature review on the topic, in this work, we focus on AWB, defined as the frequent experience of positive affects and the infrequent experience of negative affects (Daniels, 2000). This study will adopt Warr's definition of well-being at work and follow Daniels' (2000) model to assess it.

The AWB construct is multidimensional and domain specific, and can be measured in relation to the work domain. Daniels (2000) proposed an AWB at work construct comprising five bi-polar dimensions: anxiety–comfort, depression–pleasure, boredom–enthusiasm, tiredness–vigor, and anger–placidity. Daniels (2000) found empirical evidence supporting the fivefactor model. Empirical studies by Rego and his associates (Rego et al., 2010) also support the model, although only after removing several scales. Rego and Cunha (2008a) also found that a second-order factor model, with the five factors loading onto an overall AWB factor, fits the data reasonably well.

Organizational Citizenship Behaviors

The OCB was originally defined as ''individual behavior that is discretionary, not directly or explicitly recognized by the formal reward system, and that in the aggregate promotes the effective functioning of the organization'' (Rego et al., 2010). Observing problems and difficulties in distinguishing (a) between what is and what is not discretionary, and (b) between what is contractual and noncontractual, Organ (1997) suggested that OCB be redefined as contextual performance: ''contributions to the maintenance and enhancement of the social and psychological context that supports task performance'' (Organ, 1997, p. 91). At the heart of the OCB concept resides the idea that organizations depending upon prescribed behavior are fragile social systems (Katz, 1964, cited by Rego et al., 2010). OCB increases social capital and enhances organizational functioning and effectiveness (Rego et al., 2010).

The five dimensions suggested by Organ (1988), mentioned by Rego et al. (2010), and the most frequently examined by researchers (Schnake and Dumler, 2003) are altruism, sportsmanship, courtesy, conscientiousness, and civic virtue. Altruism refers to discretionary behaviors aimed at helping other individuals with an organizationally relevant task or problem. Sportsmanship refers to tolerating the inconveniences and annoyances of organizational life without complaining and filing grievances. Courtesy is about being mindful of how one's action affects other people. Conscientiousness involves employees' behaviors that go beyond minimal requirements in carrying out their tasks. Civic virtue is the responsible participation in the political process of the organization (e.g., contribution to discussions or the involvement in organizational activities to assist and improve the organization). Empirical studies have supported this five-factor structure (e.g., Podsakoff et al., 2000).

A more recent metaanalysis conducted by Hoffman et al. (2007: 562), cited by Rego et al. (2010), also suggested that ''current operationalizations of OCB are best viewed as indicators of a general OCB factor. Thus, there is likely little to be gained through the use of separate dimensional measures as opposed to an overall composite measure''.

In summary, the literature converges that organizational citizenship behaviors influence organizational effectiveness and are likely to "free up resources that can be used for more productive purposes; reduce the need to devote scarce resources to simple maintenance tasks; facilitate the process of coordinating inter- and intra-group activities; enable the organization to attract and retain the best employees, and to adapt more effectively to environmental changes; make organizations more competitive in the knowledge economy in which we live; contribute to improved decision making" (Rego et al., 2010).

HYPOTHESES

This study was directed to the public sector, namely to the local governments that make up the Intermunicipal Community -Terras de Trás-os-Montes.

The choice of this region of the country to conduct the empirical study is related to the lack of studies on this topic for the Trás-os-Montes region. The choice of the public sector is related, on the one hand, to the restructuring process that has been occurring over the past few years and in which the implementation of policies and practices, usually used by private entities, has occurred and, on the other hand, by the lack of studies on the issues addressed for this sector.

The Intermunicipal Community of Terras de Trás-os-Montes is made up of nine municipalities, namely: Alfândega da Fé, Bragança, Macedo de Cavaleiros, Miranda do Douro, Mirandela, Mogadouro, Vila Flor, Vinhais and Vimioso.

Organizational Virtuousness (OV) and Organizational Citizenship Behaviors (OCB)

Perceptions of OV may induce people to adopt more OCB. The existence of positive perceptions of organizational virtuosity by employees promotes greater affective well-being and the adoption of more organizational citizenship behaviors. This phenomenon is explained by the existence of virtuous actions in the organization, which leads employees to perceive meaning in the work they do, to achieve a sense of well-being and, consequently, to engage in organizational citizenship behaviors. In other words, when the employees of an organization experience virtuosity in their work environment, they are induced to feel positive emotions that promote the existence of behaviors that contribute to the benefit of the organization and people (Rego et al., 2010).

The relationship between organizational virtuosity and organizational citizenship behaviors can be based on factors such as organizational identification, psychological contract, positive emotions, perceived organizational support, and reciprocity.

Organizational identification can be described as the degree of identification that the employee feels towards the organization. The way individuals interpret organizational actions induces the way employees see themselves, that is, virtuous behaviors affect individuals' identification with the organization itself. That is, a virtuous individual will identify more with an organization in which virtuous actions occur frequently. The perceptions of organizational virtuosity, through positive spirals, lead to the emergence of more virtuous actions by its employees, resulting, on the one hand, in the increase of the employee's degree of identification with the organization and, on the other hand, in the increase of organizational citizenship behaviors. According to Rego et al. (2010), identification with the organization emerges from the reconciliation of strong personal and organizational values, resulting in higher levels of organizational citizenship behaviors.

The positive emotions experienced by individuals are produced by virtuous actions (Cameron et al., 2004) and are also an important factor in the relationship between organizational virtuosity and organizational citizenship behaviors. People experience positive emotions arising from the relationships that take place within virtuous organizations. These relationships between individuals, when strong and rewarding, create perceptions of meaningful work that further develop positive emotions (Fredrickson, 2003). The existence of organizational virtuosity allows people to feel that their work is rewarded (Ryan and Deci, 2000), they are psychologically and emotionally stronger and more stable.

Perceptions of organizational virtuosity contribute to the construction of psychological contracts between individuals and the organization, based on trust and mutual loyalty and, consequently, to the increase of organizational citizenship behaviors. To quote Rego et al. (2010), the rewards provided to the employee are frankly contingent on their performance, but rather derive from their belonging and participation in the organization. The employee is expected to be loyal to the organization and to have 'love for the shirt'; and the organization is expected to provide job security and individual and family well-being.

The fourth factor that results from perceptions of organizational virtuosity and influences organizational citizenship behaviors is perceived organizational support. There is a relationship between perceived organizational support and organizational citizenship behaviors, mediated by the willingness of employees to assume behaviors/attitudes that will be reciprocated by the organization. In short, in the presence of perceived organizational support, employees act in reciprocity, dedicating themselves more to organizational life and practicing more organizational citizenship behaviors (Rego et al, 2010). The organizational virtuosity observed through perceived organizational support translates the perception of support and value felt by the employee towards the organization, influencing the employee to act accordingly and, therefore, to adopt organizational citizenship behaviors.

Finally, reciprocity emerges as a factor of organizational virtuosity and influences organizational citizenship behaviors. Virtuous organizations create feelings of gratitude in employees towards them, and thus employees behave reciprocally, adopting behaviors that favor both the organization and their colleagues (Rego et al., 2010).

The literature review considers that organizational virtuosity, perceived through factors such as organizational identification, positive emotions, psychological contract, perceived organizational support, and reciprocity, influences organizational citizenship behaviors. Thus, the previous relationship explained can be formalized through the following hypothesis:

Hypothesis 1: Perceptions of OV are related to OCB

Affective Well-Being (AWB) and Organizational Citizenship Behaviors (OCB)

Several authors have suggested that happy employees tend to be more helpful to other people, to be more empathetic and respectful, and to perform OCB (Avey et al., 2008; Miles et al., 2002; Spector and Fox, 2002). Happy individuals, as a general rule, tend not only to adopt altruistic, empathetic, and courteous behaviors, but also to relate more to people who already practice them, in order to foster the happiness already felt (Miles et al., 2002). There is also a greater proximity between individuals, which in turn allows for the strengthening of relationships at all levels, making people more socially integrated (Diener and Seligman, 2002; Fredrickson, 2001). In short, positive emotions allow for the development of sportsmanship (individuals complain less and therefore do not develop resentment), courtesy and altruism, i.e. individuals are more likely to adopt spontaneous behaviors that are extremely important when work problems arise (Avey et al., 2008).

Positive emotions are also particularly important in the process of improving work behavior, as they serve as a driver for this to happen and guide workers that work should not only be based on obtaining a financial reward, but also on gratification and recognition of the work done (Wright and Cropanzano, 2004).

Happiness at work also allows for the development of the worker's physical, intellectual, and social skills through the perception of positive emotions, which reinforce and provide greater possibility for solutions when conflicts arise (Fredrickson, 2001). It is likely that the happiness at work felt by the worker influences how the worker channels the energies arising from happiness into conflict resolution in the work context. Thus, the positive emotions that arise from the perception of happiness at work make the worker more willing to endure inconveniences that may occur in the organizational environment without having to complain about what happened (sportsmanship) and to participate responsibly in the life of the organization (civic virtue).

Through the literature review it is possible to verify the existence of a relationship between well-being at work and organizational citizenship behaviors that can be formalized as follows:

Hypothesis 2: Well-being (AWB) at work is related to organizational citizenship behaviors (OCB).

Organizational Virtuousness (OV) and Affective Well-Being (AWB)

The relationship between organizational virtuosity and well-being at work has been studied by different authors, and for Cameron, one of the main authors on the topic, perceptions of organizational virtuosity trigger some emotions in individuals, namely empathy, enthusiasm, admiration and love (Cameron & Winn, 2012). These positive emotions, due to their characteristics, in addition to being fundamental to achieving success as a manager and organizational excellence, are important in strengthening relationships between individuals (Dutton and Baker, 2007; Fineman, 1999). Strengthened interpersonal relationships make it possible to meet the social needs of individuals, who feel more secure and, consequently, increase well-being at work. In this way, a set of positive spirals is generated that contribute to increased well-being at work (Fredrickson, 2003).

The relationships in an organizational context are characterized by being more rewarding and closer, so that workers feel that their work is meaningful, experiencing more and more positive emotions, causing an increase in workers' well-being (Ryan and Deci, 2000). Virtuous organizations, through the perception of positive emotions, allow their workers to feel psychologically and emotionally stronger and safer, which leads them to face work adversities in a more positive way due to the feeling of control they develop (Spector and Fox, 2002). There is also an increase in respect in the organization, whereby employees respect and are more respected by other employees, contributing to a significant increase in their self-esteem (Ramarajan et al., 2008). Thus, employees in the presence of characteristics such as self-esteem, respect, and more developed emotional and psychological capacity, present and perpetuated by virtuous organizations, increase their well-being. This hypothesis can be formalized as follows:

Hypothesis 3: Perceptions of organizational virtuosity (OV) are related to well-being at work (AWB).

METHOD: PARTICIPANTS AND PROCEDURES

A total of 832 individuals were surveyed, more specifically, 665 employees and 167 team/department leaders. After a more detailed analysis, 72 questionnaires were eliminated, of which 57 belonged to employees and the remaining 15 belonged to team/department leaders, according to the following criteria:

- Individuals who proceeded to fill out the questionnaire inconsistently, such as assigning the maximum or minimum value of the scale for all items;
- Individuals who had been working in the organization for less than a year, since the total number of employees was calculated based on the 2020 forecasted staffing charts, which were assumed to be based on the number of employees working on December 31, 2019.

The final sample is a convenience sample of 760 individuals, two of which 152 are team/department leaders (hierarchical superior) and 608 are employees (subordinate). The questionnaire aimed to collect information regarding how subordinates perceive organizational virtuosity and well-being at work and superiors report organizational citizenship behaviors.

The local governments that participated in the survey are those that make up the Intermunicipal Community - Terras de Trás-os-Montes, namely Alfândega da Fé, Bragança, Macedo de Cavaleiros, Miranda do Douro, Mirandela, Mogadouro, Vila Flor, Vimioso and Vinhais. The size of the municipalities varies between 171 and 366 employees, and the number of employees surveyed varies between

52 and 130 employees. The distribution of the respondents per local authority varies between 18% and 7%, with approximately 59% of the respondents working in the four largest local authorities (Bragança, Mirandela, Macedo de Cavaleiros and Alfândega da Fé) belonging to the Intermunicipal Community - Terras de Trás-os-Montes (Table 1 - Characterization of the total number of employees and respondents per municipality).

Table 1. Characterization of the total number and respondents of employees per municipality

Local Government (LG)	**Total Number of Employees**	**Number of Respondent Employees**			**Distribution of Respondents by LG**
		Employees	*Team Leader / Department*	*Total*	
Alfândega da Fé	208	60	15	75	10%
Bragança	366	116	29	145	19%
Macedo de Cavaleiros	269	92	23	115	15%
Miranda do Douro	181	52	13	65	9%
Mirandela	301	100	25	125	16%
Mogadouro	187	48	12	60	8%
Vila Flor	171	40	10	50	7%
Vimioso	175	44	11	55	7%
Vinhais	203	56	14	70	9%
Total CIM-TTM	**2061**	**608**	**152**	**760**	**100%**

Measurement Instruments

Organizational Virtuousness (OV)

Cameron et al. (2004) proposed the use of 15 six-point Likert scales to measure perceptions of organizational virtuosity consisting of forgiveness, trust, integrity, optimism, and compassion. The scale used by the employees was intended to measure the degree to which each of the items was false (1) or true (6).

The data collected from employees regarding perceptions of organizational virtuosity were subjected to factor analysis. To use this method of data analysis, the assumptions inherent to the applied statistics were validated, i.e. normality (Kolmogorov-Smirnov test and asymmetry and kurtosis analysis) and data validity (Bartlett's Test of Sphericity and KMO test) were assessed.

Factor extraction was performed, using the principal axis factoring method, with Varimax rotation, from which three factors were obtained. The three factors did not allow a simplified and clear factorial structure to be obtained and, as such, based on the diagonal values of the anti-image matrix (they should be greater than 0.5) and on the communalities, some variables were eliminated. The process resulted in the elimination of eleven variables.

The final model was also analyzed using the extraction method used in the first analysis, and resulted in two factors that explain more than 50% of the initial information of the variables and whose degree of reliability (Alpha Cronbach) is higher than 0.7, as advised by Fabrigar et al. (1999). The two factors, called integrity (Factor 1) and optimism (Factor 2), represent two of the five dimensions of organizational virtuosity proposed by Cameron et al., 2004.

Affective Well-Being (AWB)

To measure well-being at work, initially validated and used by Daniels (2000) and in the Portuguese context by Rego et al., (2010), we used a five-dimensional model named: anxiety-comfort, depression-pleasure, tiredness-vitality, boredom-enthusiasm and anger-serenity. The employees (subordinates) of the local authorities of the Intermunicipal Community - Terras de Trás-os-Montes, through a questionnaire, answered 30 items in order to describe the feelings felt during the last three months in the respective local authority, using a seven-point scale in which level (1) corresponds to "never" and level (7) corresponds to always.

The data collected from the employees regarding well-being at work were submitted to an exploratory factor analysis. To proceed with this analysis the assumptions inherent to the applied statistics were validated, i.e., the normality (Kolmogorov-Smirnov test) and validity of the data (Bartlett's Test of Sphericity and KMO test) were evaluated.

In a first stage, factor extraction was performed using the main axis factoring method, with Varimax rotation, from which six factors were obtained (Kaiser Criterion). The six factors did not allow a simplified and clear factorial structure to be obtained and, as such, based on the diagonal values of the anti-image matrix, which should be greater than 0.50, and on the communalities, some variables were eliminated. The process resulted in the elimination of twenty variables.

The final model was also analyzed using the extraction method used in the first analysis, and four factors were obtained which explain more than 50% of the initial information of the variables and whose degree of reliability (Alpha Cronbach) for factor 1 is excellent and for factors 2, 3 and 4 is good, i.e. higher than 0.7, as recommended by Fabrigar et al. (1999). The four factors, named Depression-Pleasure (Factor 1), Boredom-Enthusiasm (Factor 2), Anger-Tranquility (Factor 3), Anxiety-Comfort (Factor 4) represent four of the five dimensions of well-being at work proposed by Daniels (2000).

Organizational Citizenship Behaviors (OCB)

To address organizational citizenship behaviors, a similar methodology was used as for perceptions of organizational virtuosity. A seven-point Likert scale (Konovsky and Organ, 1996) was used, with level (1) corresponding to "This statement does not strictly apply at all to this employee" and level (7) corresponding to "This statement applies completely to this employee," to measure the five dimensions: altruism, sportsmanship, courtesy, conscientiousness, and civic virtue. Superiors rated the 32 items, taken from MacKenzie et al. (1991), Podsakoff et al. (2000) and Smith et al. (1983), reporting the degree to which each of the statements applied to the employee (subordinate) they had previously thought of. For the sportsmanship dimension, items were formulated in the negative (MacKenzie et al., 1991; Podsakoff and MacKenzie, 1994) to control for possible response bias.

The data collected from the hierarchical superiors concerning organizational citizenship behaviors were submitted to an exploratory factor analysis. To perform this analysis, the assumptions inherent to the applied statistics were validated, i.e. the normality (Kolmogorov-Smirnov test) and validity of the data (Bartlett's Test of Sphericity and KMO test) were assessed.

In a first stage, factor extraction was performed using the main axis factoring method, with Varimax rotation, from which six factors were obtained (Kaiser Criterion). A seven-factor model was obtained, all of them with an eigenvalue greater than 1, but with only three consistent factors (Cronbach Alpha). Factor 1 corresponds to a mixture of all dimensions except sportsmanship; factor 2 is composed of a mixture of all dimensions except conscientiousness; and factor 3 is composed of a mixture of all dimensions except, the civic virtue dimension. The model was not satisfactory and, consequently, some items were eliminated, based on the diagonal values of the anti-image matrix, which should be greater than 0.50, and on the communalities.

The final model was also analyzed using the extraction method used in the first analysis, and three factors were obtained that explain more than 50% of the initial information of the variables and whose degree of reliability (Alpha Cronbach) of all factors is good, i.e. higher than 0.7. The three factors, named sportsmanship (Factor 1), conscientiousness (Factor 2), and courtesy (Factor 3), represent three of the five dimensions of well-being at work proposed by the Konovsky and Organ (1996) theoretical model.

RESULTS

In the convenience sample, it is found that 58.95% of the respondents are female and 41.05% of the respondents are male. Regarding age, 73.4% of the individuals are between 36 and 55 years old, 17.8% are over 55 years old, and 8.9% are between 18 and 35 years old.

Descriptive Statistics by Category: Employee (Subordinate) and Team Leader/Department Leader (Higher Hierarchical Level)

With regard to gender, we see that 65.3% of the women hold the position of employee and only 34.7% hold management positions. On the contrary, for the category of team/department leader, 66.4% are male and 33.6% are female, i.e. the higher hierarchical positions are mostly occupied by men.

For the age variable it is possible to verify that the majority (55.9%) of team/department leaders are between 46 and 55 years old, while in the employee category, the majority (44.7%) is between 36 and 45 years old (Table 2).

Table 2. Table showing the percentage of ages by position held in the local government

	Employee		**Team Leader**	
	Percentage	*Cumulative Percentage*	*Percentage*	*Cumulative Percentage*
Between ages 18 and 25	0,8%	0,8%	0%	0%
Between ages 26 and 35	10,2%	11,0%	0%	0%
Between ages 36 and 45	44,7%	55,7%	31,6%	31,6%
Between ages 46 and 55	25,2%	80,9%	55,9%	87,5%
Over 55 years	19,1%	100%	12,5%	100,0%
Total	**100%**		**100,0%**	

With regard to seniority, it can be seen that 40.1% of the team leaders/department leaders have been working in the local authority for more than 20 years and, in the case of the employees, most have been working between 11 and 20 years. We can see a relationship between the seniority of the employee and the position held in the organization, i.e., the older the employee, the larger the position he or she holds and therefore, the team/department leaders are also the individuals who have been in local authority positions the longest.

Finally, it is necessary to analyze the educational qualifications of the two categories. In the category of team/department leader it is possible to verify that 64.5% of the individuals have a college degree and only 28.3% have the minimum compulsory education. In the employee category, 56.4% have a college degree, 34.7% have the minimum compulsory education, and the remaining 8.9% have the 9th grade or less.

Descriptive Analysis of the Primary Data

The following tables present the descriptive analysis of the data for the initial primary variables. The questionnaire items concerning organizational citizenship behaviors (dependent variable), answered by team/department leaders, were typed as CCO_Qi, with i = 1,...,32. The items related to perceptions of organizational virtuosity and well-being at work (independent variables), answered by the employees, were typed as VO_Qj and BET_Qk, with j = 1,...,15 and k = 1,...,30, respectively.

Independent and Primary Variables - Organizational Virtuosity and Well-being at Work

Table 3, presented below, summarizes the data from the analysis to the primary independent variable (organizational virtuosity).

Based on the previous table, it can be seen that most variables have very close mean and median values. Only the variable VO_Q6 is slightly distant from the values. Thus, it is possible to conclude that the distributions are symmetrical and, therefore, it is possible to use the arithmetic mean to describe the center of the distribution.

The normality of the data should be assessed, individually for each variable, using kurtosis and skewness (Field, 2011) in order to avoid distortion of results. With regard to skewness, it can be seen that the distributions of the variables are symmetrical, since the value of the skewness is close to zero for most variables, and there is a standard distribution for all variables under analysis. The kurtosis allows us to understand how the distribution oscillates against a central point that is given by the normal distribution. For the set of variables that measure the perceptions of organizational virtuosity it is verified that the distribution is mesocurtic, since most of the coefficients, in absolute value, are less than 1 and, simultaneously, are also outside the interval]-0.5; 0.5[, except for variable VO_Q8, whose kurtosis value is above 1.

Table 4 summarizes the data from the analysis of the primary independent variable (well-being at work).

Based on the previous table, it can be seen that most of the variables have very close mean and median values. Only the variable BET_Q21 and BET_Q22 are slightly away from the values. Thus, it is possible to conclude that the distributions are symmetrical and, therefore, it is possible to use the arithmetic mean to describe the center of the distribution.

Table 3. Descriptive analysis of the primary variables - Organizational Virtuosity

Primary Variables	Observations	Mean	Standard Deviation	Median	Variance	Skewing	Kurtosis	Tests of Normality					
								Kolmogorov-Smirnov[a]			Shapiro-Wilk		
								Estat.	*df*	*Sig.*	*Estat.*	*df*	*Sig.*
VO_Q1	608	2,99	1,196	3,00	1,43	0,242	-0,761	0,197	608	0,000	0,917	608	0,000
VO_Q2	608	2,87	1,178	3,00	1,39	0,278	-0,746	0,220	608	0,000	0,910	608	0,000
VO_Q3	608	3,14	1,244	3,00	1,55	0,213	-0,739	0,169	608	0,000	0,926	608	0,000
VO_Q4	608	3,68	1,175	4,00	1,38	-0,149	-0,153	0,201	608	0,000	0,929	608	0,000
VO_Q5	608	3,38	1,306	3,00	1,71	0,124	-0,708	0,175	608	0,000	0,935	608	0,000
VO_Q6	608	4,25	1,157	5,00	1,34	-0,441	-0,593	0,246	608	0,000	0,899	608	0,000
VO_Q7	608	3,35	1,372	3,00	1,88	0,070	-0,958	0,163	608	0,000	0,928	608	0,000
VO_Q8	608	3,19	1,352	3,00	1,83	-0,016	-1,060	0,172	608	0,000	0,919	608	0,000
VO_Q9	608	2,90	1,149	3,00	1,32	0,119	-0,746	0,190	608	0,000	0,916	608	0,000
VO_Q10	608	3,68	1,222	4,00	1,49	-0,428	-0,454	0,238	608	0,000	0,912	608	0,000
VO_Q11	608	2,90	1,206	3,00	1,45	0,268	-0,737	0,191	608	0,000	0,918	608	0,000
VO_Q12	608	3,05	1,202	3,00	1,45	0,479	-0,224	0,207	608	0,000	0,917	608	0,000
VO_Q13	608	3,34	1,220	3,00	1,49	0,263	-0,392	0,204	608	0,000	0,930	608	0,000
VO_Q14	608	2,87	1,176	3,00	1,38	0,715	0,275	0,205	608	0,000	0,896	608	0,000
VO_Q15	608	3,08	1,225	3,00	1,50	0,363	-0,412	0,202	608	0,000	0,924	608	0,000

Table 4. Descriptive analysis of the primary variables - Well-being at work

Primary Variables	Observations	Mean	Standard Deviation	Median	Variance	Skewing	Kurtosis	Tests of Normality					
								Kolmogorov-Smirnov[a]			Shapiro-Wilk		
								Estat.	*df*	*Sig.*	*Estat.*	*df*	*Sig.*
BET_Q1	608	4,13	1,526	4,00	2,330	-0,209	-0,513	0,163	608	0,000	0,944	608	0,000
BET_Q2	608	3,94	1,248	4,00	1,558	-0,204	-0,346	0,175	608	0,000	0,938	608	0,000
BET_Q3	608	3,81	1,515	4,00	2,294	-0,054	-0,575	0,192	608	0,000	0,933	608	0,000
BET_Q4	608	3,78	1,364	4,00	1,860	-0,067	-0,886	0,185	608	0,000	0,928	608	0,000

continues on following page

Table 4. Continued

Primary Variables	Observations	Mean	Standard Deviation	Median	Variance	Skewing	Kurtosis	Tests of Normality					
								Kolmogorov-Smirnov[a]			Shapiro-Wilk		
								Estat.	*df*	*Sig.*	*Estat.*	*df*	*Sig.*
BET_Q5	608	3,82	1,416	4,00	2,005	-0,328	-0,722	0,171	608	0,000	0,927	608	0,000
BET_Q6	608	3,29	1,678	4,00	2,814	0,019	-1,291	0,176	608	0,000	0,898	608	0,000
BET_Q7	608	3,68	1,249	4,00	1,560	0,160	-0,581	0,201	608	0,000	0,915	608	0,000
BET_Q8	608	3,36	1,491	3,00	2,222	0,006	-0,893	0,146	608	0,000	0,935	608	0,000
BET_Q9	608	4,14	1,338	4,00	1,790	-0,337	-0,225	0,211	608	0,000	0,930	608	0,000
BET_Q10	608	3,50	1,506	4,00	2,267	-0,099	-0,948	0,147	608	0,000	0,932	608	0,000
BET_Q11	608	3,80	1,561	4,00	2,437	-0,117	-1,048	0,140	608	0,000	0,922	608	0,000
BET_Q12	608	4,02	1,510	4,00	2,281	-0,065	-0,997	0,178	608	0,000	0,934	608	0,000
BET_Q13	608	4,04	1,194	4,00	1,425	-0,385	0,207	0,260	608	0,000	0,902	608	0,000
BET_Q14	608	3,93	1,296	4,00	1,681	0,003	-0,880	0,202	608	0,000	0,922	608	0,000
BET_Q15	608	3,73	1,354	4,00	1,833	0,077	-0,812	0,158	608	0,000	0,937	608	0,000
BET_Q16	608	3,75	1,588	4,00	2,521	-0,098	-1,079	0,146	608	0,000	0,928	608	0,000
BET_Q17	608	3,53	1,546	3,00	2,391	0,252	-0,832	0,173	608	0,000	0,937	608	0,000
BET_Q18	608	3,62	1,499	3,00	2,246	0,165	-0,948	0,165	608	0,000	0,934	608	0,000
BET_Q19	608	3,29	1,579	4,00	2,493	0,203	-0,603	0,186	608	0,000	0,924	608	0,000
BET_Q20	608	3,27	1,532	4,00	2,347	-0,038	-1,037	0,202	608	0,000	0,915	608	0,000
BET_Q21	608	4,19	1,621	5,00	2,628	-0,141	-1,121	0,200	608	0,000	0,921	608	0,000
BET_Q22	608	2,04	1,203	2,00	1,448	1,153	1,123	0,262	608	0,000	0,804	608	0,000
BET_Q23	608	3,85	1,535	4,00	2,356	0,151	-0,955	0,141	608	0,000	0,933	608	0,000
BET_Q24	608	3,97	1,438	4,00	2,068	0,038	-0,948	0,181	608	0,000	0,931	608	0,000
BET_Q25	608	4,11	1,426	4,00	2,035	0,079	-0,959	0,207	608	0,000	0,925	608	0,000
BET_Q26	608	3,70	1,480	4,00	2,190	-0,008	-0,661	0,160	608	0,000	0,947	608	0,000
BET_Q27	608	3,50	1,529	4,00	2,339	0,058	-1,026	0,168	608	0,000	0,930	608	0,000
BET_Q28	608	3,61	1,541	4,00	2,374	0,102	-0,602	0,157	608	0,000	0,946	608	0,000
BET_Q29	608	4,17	1,384	4,00	1,916	-0,015	-0,696	0,174	608	0,000	0,942	608	0,000
BET_Q30	608	3,97	1,332	4,00	1,775	-0,122	-0,470	0,158	608	0,000	0,942	608	0,000

According to some authors (Fabrigar et al., 1999), the normality of the data allows us to analyze which is the most appropriate method for factor extraction, but it is not considered a necessary assumption of factor analysis. Thus, the results of normality tests, Kolmogorov-Smirnov test and Shapiro-Wilk test, allow us to conclude, for significance levels of 0.01 or 0.05, that the distributions of variables are normal.

However, the normality of the data should be assessed, individually for each variable, using kurtosis and skewness (Field, 2011) in order to avoid distortion of results. The study of symmetry and flattening allows us to see if the distribution is symmetric and mesocurtic, respectively, which are necessary but not sufficient conditions for the distribution to be considered normal.

Regarding asymmetry, it can be seen that the distributions of the variables are symmetrical, since the value of the skewness is close to zero for most variables, and there is a standard distribution for all variables under analysis.

Primary Dependent Variable - Organizational Citizenship Behaviors

In Table 5, presented below, the data from the analysis of the primary independent variable (organizational citizenship behaviors) are summarized.

Based on the table above, it can be seen that most variables present very close mean and median values, so it is possible to conclude that the distributions are symmetrical and, therefore, it is possible to use the arithmetic mean to describe the center of the distribution.

Regarding asymmetry, it can be seen that the distributions of the variables are symmetrical, since the value of the skewness is close to zero for most variables, and there is a standard distribution for all variables under analysis, except for the CCO_Q5 variable.

Factor Analysis Results

After the descriptive analysis of the primary data and the verification of the recommendations mentioned in the previous topic, the study began with the application of the exploratory factor analysis. All variables, independent and dependent, follow a non-normal distribution and therefore the principal axis factoring method was used (Fabrigar et al., 1999).

In the initial phase, we proceeded to estimate the correlation matrix, which measures the linear relationship between the variables, through Pearson's correlation coefficients. Based on these coefficients, the existence of correlation between the variables was confirmed, thus concluding that it would be appropriate to use the factorial model, otherwise the existence of common factors would be unlikely.

For each of the variables, two assumptions of the factor analysis application were tested. Through the Bartlett's test of sphericity, the existence of significant correlations between the items was measured, and through the KMO (Kaiser-Meyer-Olkin) test, the adequacy of the sample was evaluated.

Table 6 shows the results of the Bartlett's test of sphericity and the KMO test for the independent variable organizational virtuosity.

The table above shows that the Bartlett's test has a associated significance level of 0.000, a value lower than 0.05, which leads to the rejection of the hypothesis that the matrix of correlations in the sample is the identity, concluding that there is a good correlation between the variables.

The value obtained for the KMO test is 0.916, i.e. greater than 0.90, indicating that the sample size is adequate for the items under analysis, allowing the factor analysis to proceed.

Table 5. Descriptive analysis of the primary variables - Organizational Citizenship Behaviors

Primary Variables	Observations	Mean	Standard Deviation	Median	Variance	Skewing	Kurtosis	Tests of Normality					
								Kolmogorov-Smirnov[a]			Shapiro-Wilk		
								Estat.	*df*	*Sig.*	*Estat.*	*df*	*Sig.*
CCO_Q1	152	5,04	1,667	5,00	2,780	-0,758	-0,048	0,181	152	0,000	0,893	152	0,000
CCO_Q2	152	4,21	1,918	4,00	3,677	-0,133	-0,930	0,183	152	0,000	0,929	152	0,000
CCO_Q3	152	4,95	1,665	5,00	2,773	-0,702	-0,005	0,176	152	0,000	0,893	152	0,000
CCO_Q4	152	3,81	2,080	4,00	4,328	0,164	-1,409	0,222	152	0,000	0,886	152	0,000
CCO_Q5	152	5,59	1,421	6,00	2,018	-1,158	0,768	0,258	152	0,000	0,836	152	0,000
CCO_Q6	152	4,91	1,466	5,00	2,150	-0,388	-0,643	0,199	152	0,000	0,899	152	0,000
CCO_Q7	152	4,93	1,642	5,00	2,698	-0,311	-0,804	0,146	152	0,000	0,914	152	0,000
CCO_Q8	152	4,15	1,741	5,00	3,030	-0,045	-1,043	0,200	152	0,000	0,925	152	0,000
CCO_Q9	152	4,80	1,625	5,00	2,640	-0,603	-0,626	0,188	152	0,000	0,891	152	0,000
CCO_Q10	152	5,18	1,313	5,00	1,723	-0,957	1,063	0,255	152	0,000	0,876	152	0,000
CCO_Q11	152	5,03	1,462	5,00	2,138	-0,702	0,383	0,228	152	0,000	0,899	152	0,000
CCO_Q12	152	4,15	1,639	5,00	2,686	0,257	-1,043	0,219	152	0,000	0,916	152	0,000
CCO_Q13	152	4,83	1,556	3,00	2,421	-0,256	-0,714	0,142	152	0,000	0,930	152	0,000
CCO_Q14	152	3,55	1,918	3,00	3,679	0,286	-1,151	0,192	152	0,000	0,911	152	0,000
CCO_Q15	152	4,18	1,863	4,00	3,472	-0,244	-1,134	0,158	152	0,000	0,919	152	0,000
CCO_Q16	152	4,95	1,641	5,00	2,693	-0,718	-0,060	0,195	152	0,000	0,900	152	0,000
CCO_Q17	152	5,01	1,561	5,00	2,437	-0,837	0,305	0,201	152	0,000	0,895	152	0,000
CCO_Q18	152	4,52	1,798	5,00	3,231	-0,334	-0,999	0,171	152	0,000	0,918	152	0,000
CCO_Q19	152	5,24	1,482	5,00	2,195	-0,800	0,210	0,180	152	0,000	0,897	152	0,000
CCO_Q20	152	4,15	1,844	5,00	3,401	-0,097	-1,075	0,204	152	0,000	0,916	152	0,000
CCO_Q21	152	5,06	1,765	5,00	3,116	-0,706	-0,711	0,197	152	0,000	0,849	152	0,000
CCO_Q22	152	4,88	1,617	5,00	2,613	-0,538	-0,762	0,204	152	0,000	0,902	152	0,000
CCO_Q23	152	3,64	2,037	4,00	4,151	0,131	-1,316	0,218	152	0,000	0,881	152	0,000
CCO_Q24	152	4,24	1,767	4,00	3,122	0,196	-0,965	0,133	152	0,000	0,917	152	0,000
CCO_Q25	152	3,89	2,099	4,00	4,405	0,106	-1,377	0,165	152	0,000	0,905	152	0,000

continues on following page

Table 5. Continued

Primary Variables	Observations	Mean	Standard Deviation	Median	Variance	Skewing	Kurtosis	Tests of Normality					
								Kolmogorov-Smirnov[a]			Shapiro-Wilk		
								Estat.	*df*	*Sig.*	*Estat.*	*df*	*Sig.*
CCO_Q26	152	4,11	1,755	4,50	3,080	0,006	-1,110	0,194	152	0,000	0,923	152	0,000
CCO_Q27	152	4,90	1,841	5,00	3,388	-0,931	-0,158	0,238	152	0,000	0,859	152	0,000
CCO_Q28	152	4,38	1,901	5,00	3,615	-0,269	-1,147	0,233	152	0,000	0,900	152	0,000
CCO_Q29	152	4,88	1,828	5,00	3,342	-0,847	-0,446	0,238	152	0,000	0,852	152	0,000
CCO_Q30	152	4,68	1,815	5,00	3,293	-0,358	-1,135	0,222	152	0,000	0,887	152	0,000
CCO_Q31	152	4,53	1,873	5,00	3,509	-0,095	-1,435	0,182	152	0,000	0,887	152	0,000
CCO_Q32	152	4,789	1,748	5,00	3,055	-0,313	-1,180	0,206	152	0,000	0,886	152	0,000

Table 6. KMO test and Bartlett's Test of Sphericity

Kaiser-Meyer-Olkin Measure of Sampling adequacy		0,916
Test of Sphericity (Bartlett´s Test)	*Approx. Chi-Square*	6883,577
	gl	105
	Sig.	0,000

The data under analysis were subjected to factor analysis, using the principal axis factoring method with Varimax rotation. The principal axis factoring method assumes that each variable has two constituent parts: a part that is common to the factorial structure and another part specific to the variable. The use of rotation allows transforming the retained factors into a more simplified structure, making the identification and interpretation of each factor more understandable. In the Varimax method, the objective is exactly the same as mentioned above; it consists in obtaining a factorial structure in which one and only one of the original variables is strongly associated with a single factor, and little associated with the remaining factors. In order to obtain a valid set of factors, in a first stage only the items with communalities higher than 0.50 were kept, and, subsequently, item by item, the items with lower communalities were eliminated. In the final set of variables, most communalities are higher than 0.60, except one.

Clearly, analyzing the factors after rotation, it is necessary to make improvements, due to the existence of a mixture of dimensions in all factors. Thus, to solve the following problem it is necessary to remove some variables so that the factorial structure becomes more simplified and understandable. The results of the factor analysis, after eliminating some variables, are shown in Table 7.

Table 7. Factor analysis of principal components after varimax rotation

	Factor 1	**Factor 2**
Integrity		
The organization can be described as honest and full of virtues.	0,554	0,529
Honesty and Trust are the "brand image" of this organization.	0,680	0,463
Optimism		
In this organization, we not only care about getting things done right, we care about doing good for each other.	0,474	0,693
In this organization, things are done with a very noble intention.	0,623	0,636
Own Values	1,384	1,378
Variance	34,593%	34,462%
Cronbach's Alpha	0,767	0,849

The previous table also shows the values of Cronbach's alpha coefficients for each factor, allowing us to assess their internal consistency. The analysis of factor reliability is extremely important, because one of the common problems in conducting a study is the existence of multidimensional variables. To analyze the internal reliability of each of the extracted factors and each item, the inter-item and item-total correlations and the alpha coefficient (α) were calculated. The scale used to evaluate the value of the obtained measures was the scale proposed by Hill and Hill (2000).

The results of the analysis of the principal components of the items showed the existence of some strong correlations. Factor 1 and Factor 2 have Cronbach Alpha coefficients higher than 0.7 and 0.8, respectively, and therefore their reliability is reasonable and good. The values corroborate with the minimum accepted value of 0.70 indicated by Hair et al. (2005) .

The two extracted factors were named as explained below:

1. Integrity - The employees of the Intermunicipal Community - Terras de Trás-os-Montes, perceive honest and trustworthy behavior on the part of the organization and colleagues.
2. Optimism - The employees of the Intermunicipal Community - Terras de Trás-os-Montes perceive that the organization and its members adopt behaviors through which they seek, in addition to "doing well", to "do good" driven by the desire to be successful, even in the face of the greatest challenges they may face.

This is followed by a factor analysis of well-being at work. We proceeded to analyze the results of the Bartlett's test of sphericity and the KMO test for the independent variable Table 8.

Table 8. KMO test and Bartlett's Test of Sphericity

Kaiser-Meyer-Olkin Measure of Sampling adequacy		0,938
Test of Sphericity (Bartlett´s Test)	*Approx. Chi-Square*	16662,392
	gl	435
	Sig.	0,000

The table above shows that the Bartlett's test has a associated significance level of 0.000, a value lower than 0.05, which leads to the rejection of the hypothesis that the matrix of correlations in the sample is the identity, concluding that there is a good correlation between the variables.

The value obtained for the KMO test is 0.938, i.e. greater than 0.90, indicating that the sample size is adequate for the items under analysis, allowing the factor analysis to proceed.

The data under analysis were subjected to factor analysis, using the principal axis factoring method with Varimax rotation. The principal axis factoring method assumes that each variable has two constituent parts: a part that is common to the factorial structure and another part specific to the variable. The use of rotation allows transforming the retained factors into a more simplified structure, making the identification and interpretation of each factor more understandable. In the Varimax method, the objective is exactly the same as mentioned above, it consists in obtaining a factorial structure in which one and only one of the original variables is strongly associated with a single factor, and little associated with the remaining factors. In order to obtain a valid set of factors, in a first stage only the items with communalities greater than 0.50 were kept and, subsequently, eliminating variable by variable, the variables with lower communalities were eliminated. In the final set of variables, all have communalities greater than 0.5.

Clearly, analyzing the factor matrix after rotation it is necessary to make improvements, due to the existence of a mixture of dimensions in all factors. Thus, to solve the following problems it is necessary to remove some variables so that the factor structure becomes more simplified and understandable. The results of the factor analysis, after eliminating some variables, are shown in Table 9.

Table 9. Factor analysis of principal components after Varimax Rotation.

	Factor 1	Factor 2	Factor 3	Factor 4
Depression-Pleasure				
Happy	0,762	-0,284	-0,163	-0,323
In a Good Mood	0,601	-0,301	-0,302	-0,419
Content	0,848	-0,246	-0,201	-0,236
Boredom-Enthusiasm				
Boredom	-0,450	0,722	0,162	0,276
Dull	-0,185	0,661	0,288	0,155
Sluggish (without energy)	-0,200	0,746	0,288	0,159
Anger-Tranquility				
Annoyed	-0,233	0,352	0,758	0,155
Annoyed-Incomfortable	-0,195	0,295	0,773	0,159
Anxiety-Comfort				
Anxious / Restless	-0,348	0,149	0,277	0,675
Tense	-0,329	0,305	0,255	0,711
Own Values	2,260	2,072	1,664	1,565
Variance	22,601%	20,719%	16.636%	15.653%
Cronbah´s Apha	0,907	0,858	0,891	0,859

The results of the analysis of the principal components showed the existence of some strong correlations. The value of the Alpha Cronbach of factor 1 is greater than 0.9, indicating that its reliability is excellent and factors 2, 3 and 4 and present Alpha Cronbach coefficients greater than 0.80, and therefore their reliability is good. The values corroborate with the minimum accepted value of 0.70 indicated by (Hair et al., 2005) .

The four factors extracted were named as explained below:

1. Depression-Pleasure - Refers to the positive or negative affections that convey feelings of depression or pleasure in an organizational environment, experienced by the employee of the Intermunicipal Community - Terras de Trás-os-Montes, in the last 3 months.
2. Boredom-Enthusiasm - Encompasses items related to positive or negative emotions that develop boredom or enthusiasm in an organizational environment, felt by the employee of the Intermunicipal Community - Terras de Trás-os-Montes, in the last 3 months.
3. Anger-Tranquility - Comprises a set of positive or negative emotions that arouse anger or tranquility in an organizational environment, felt by the employee of the Intermunicipal Community - Terras de Trás-os-Montes in the last 3 months.
4. Anxiety-Comfort - Comprises positive or negative emotions, experienced in an organizational environment, that develop anxiety or comfort and that have been felt by the employee of the Intermunicipal Community - Terras de Trás-os-Montes in the last 3 months.

Factor Analysis Results – Dependent Variable

In Table 10, below, we analyze the results of the Bartlett's test of sphericity and the KMO test for the dependent variable.

Table 10. KMO test and Bartlett's Test of Sphericity

Kaiser-Meyer-Olkin Measure of Sampling adequacy		0,887
Test of Sphericity (Bartlett´s Test)	*Approx. Chi-Square*	5725,582
	gl	496
	Sig.	0,000

The previous table shows that the Bartlett's test has a associated significance level of 0.000, a value lower than 0.05, which leads to the rejection of the hypothesis that the matrix of correlations in the sample is the identity, concluding that there is a good correlation between the variables.

The value obtained for the KMO test is 0.887, i.e. higher than 0.80, indicating that the sample size is adequate for the items under analysis, allowing for the continuation of the factor analysis.

The data under analysis were subjected to factor analysis, using the principal axis factoring method with Varimax rotation. The principal axis factoring method assumes that each variable has two constituent parts: a part that is common to the factorial structure and another part specific to the variable. The use of rotation allows transforming the retained factors into a more simplified structure, making the identification and interpretation of each factor more understandable In the Varimax method, the objective is exactly the same as mentioned above, it consists in obtaining a factorial structure in which one and only one of the original variables is strongly associated with a single factor, and little associated with the remaining factors. In order to obtain a valid set of factors, in a first stage only the items with communalities greater than 0.50 were kept, and later, eliminating variable by variable, the variables with lower communalities were removed. In the final set of variables, all of them have communalities higher than 0.5. Only one item has a saturation lower than 0.50, since, according to some authors, it is acceptable to select items with saturations equal to or higher than 0.40 if this situation facilitates the interpretation of the new set of variables, factors, and allows achieving a clearer factor structure.

Clearly, analyzing the factor matrix after rotation, improvements are necessary. The first reason for making improvements is the existence of a mixture of dimensions in all factors and the existence of only one variable in factor 4 and 7. The second reason is related to the Cronbach Alpha coefficients, since four of the seven factors have weak or unacceptable values. This way, to solve the following problems it is necessary to remove some variables so that the factorial structure becomes more simplified and understandable. The result of the improvements is shown in Table 11.

The results of the analysis of the principal components showed the existence of some strong correlations. All factors have Cronbach's alpha values higher than 0.80, indicating that their reliability is good. The values corroborate the minimum accepted value of 0.70 indicated by Hair et al. (2005).

The four factors extracted were named as explained below:

1. Sportsmanship - It concerns the discretionary behaviors, adopted by the employees of the Intermunicipal Community - Terras de Trás-os-Montes, which express the willingness to tolerate and avoid complaints by means of mishaps that occur in the workplace, enduring uncomfortable situations without complaining excessively.
2. Conscientiousness - Encompasses the discretionary behaviors adopted by the employees of the Intermunicipal Community - Terras de Trás-os-Montes, which reflect the performance of tasks that go beyond the minimum usually required by the organization.
3. Courtesy - Refers to the discretionary behaviors of employees, adopted by employees of the Intermunicipal Community - Terras de Trás-os-Montes, which are intended to prevent the occurrence of problems that may affect the organization and its members.

Table 11. Factor analysis of principal components, after Varimax Rotation

	Factor 1	Factor 2	Factor 3
Sportsmanship			
It shows ill will towards any change introduced in the management of the organization.	0,753	-0,071	-0,380
Is always finding fault with what the organization does.	0,873	-0,080	-0,065
Only thinks about his own work problems (not those of others).	0,831	-0,062	-0,128
Conscientiousness			
Is always punctual.	0,098	0,820	0,120
Always follows the rules of the organization and the department.	-0,068	0,635	0,268
Keeps the workplace clean.	-0,281	0,734	0,225
Always finishes your work on time.	-0,119	0,563	0,536
Courtesy			
Is concerned about the effects that your actions or decisions may have on others.	-0,266	0,472	0,765
Does not abuse your rights and benefits.	-0,220	0,232	0,849
Own Values	2,245	2,223	1,896
Variance	24,943%	24,704%	21,069%
Cronbah´s Apha	0,875	0,829	0,897

DISCUSSION AND CONCLUSIONS

Organizations, over the last years, have been "forced" to undergo changes influenced by economic, social and political modifications. These changes have affected the relationships between individuals and society, so the theme of corporate social responsibility has become part of the daily life of organizations. The incessant search for higher levels of involvement between organizations and society has resulted in the implementation of new values and strategies that prioritize, in addition to profit maximization, organizational virtuosity, well-being at work, and organizational citizenship behaviors.

The implementation of corporate social responsibility was initially adopted by private organizations and has extended over the past few years to public entities. In this sense, due to the scarcity of studies, for public entities, on the themes mentioned above, we sought to evaluate them in local authorities, specifically in the municipalities belonging to the Intermunicipal Community - Terras de Trás-os-Montes.

From the application of questionnaires to employees and superiors of local authorities, a set of results was obtained about which it is necessary to discuss and draw conclusions. The data analyzed for the convenience sample show two dimensions of organizational virtuosity against a total of five proposed by Cameron et al. (2004), namely integrity and optimism. For well-being at work, four were identified for a total of five dimensions proposed by Daniels (2000), namely depression-pleasure, boredom-enthusiasm, anger-quietness, and anxiety-comfort. Finally, with regard to organizational citizenship behaviors, we verified the existence of three dimensions against a total of five designed by several authors (Konovsky and Organ, 1996; Podsakoff et al., 1994).

The first hypothesis formulated highlights the possibility of the existence of a relationship between perceptions of organizational virtuosity and organizational citizenship behaviors in the Intermunicipal Community - Terras de Trás-os-Montes. The presence of the integrity dimension is indicative that employees perceive honest and trustworthy behaviors, which reinforce mutual trust between individuals. As a result, interpersonal relationships are stronger, teamwork and team spirit and the degree of participation of individuals in the life of these local governments increases. Optimism, in turn, is a trait that influences how employees perceive and act when faced with certain situations (Luthans, 2002), that is, it is considered one of the drivers of good relationships in organizational settings (Gavin and Mason, 2004).

The relationship between the two dimensions of organizational virtuosity present, in the convenience sample, and organizational citizenship behaviors can be explained based on some factors. Two of the factors pointed out are organizational identification and the creation of psychological contracts. The presence of integrity in the Intermunicipal Community - Terras de Trás-os-Montes leads employees to develop a positive image of the organization due to the reduction of the discrepancy between the norms and values idealized and which corroborates the idea of Bakker (2007). In this way, employees feel greater confidence and, simultaneously, increase their degree of identification, decrease the intention to leave the local authority, feel better in their workplace, leading to the adoption of organizational citizenship behaviors. The creation of psychological contracts by these workers can be explained by the sense of trust, honesty, and loyalty that integrity promotes in the organizational environment, driving them to adopt organizational citizenship behaviors, to the extent that employees seek to develop more work than formally stipulated in the contract, corroborating the author Rego et al. (2010).

Fredrickson (2003), based on the amplifying effect of organizational virtuosity, found that, in certain situations, this effect positively enhances the contexts, that is, in the same way that virtuous actions generate more virtuous actions, optimism generates more optimism. In this way, the pessimism associated with a situation fades away and the previously unattainable goals become more easily attainable. This process induces employees to experience positive emotions that ultimately influence more pessimistic employees, strengthen and empower individuals to face unexpected events, leading to the adoption of organizational citizenship behaviors related to the dimension of sportsmanship. Thus, positive emotions are the third factor identified, in the convenience sample, to explain the relationship between organizational virtuosity and organizational citizenship behaviors.

The last factor is reciprocity. Corroborating the idea advocated by Eisenberger et al. (2001), in the presence of organizational virtuosity, employees tend not only to reciprocate the respective local authorities belonging to the Intermunicipal Community, but also the co-workers, through the adoption of organizational citizenship behaviors.

The second hypothesis formulated for this study was the existence of a relationship between well-being at work and organizational citizenship behaviors in the Intermunicipal Community - Terras de Trás-os-Montes.

The existence of the four dimensions in the convenience sample seems to be consistent with the fact that happier employees adopt respectful and empathetic behaviors (courtesy). The presence of well-being at work induces employees to establish closer and stronger relationships, leading to their social integration (Miles et al., 2002). The positive emotions arising from established relationships promote politeness, through concern for the effects their actions may cause on others, conscientiousness, through careful maintenance in the workspace, punctuality, and attendance at work (Avey et al., 2008).

The results obtained for the Intermunicipal Community - Terras de Trás-os-Montes are also consistent with the idea that well-being at work develops the worker's physical, intellectual and social abilities through positive emotions (Fredrickson, 2001). It is expected that well-being at work, associated with positive emotions, will influence workers to direct their energies towards improving their ability to resolve conflicts and tolerate inconveniences that may arise in stressful situations, without the need to complain about what happened (sportsmanship).

The third hypothesis formulated allows for the existence of a relationship between perceptions of organizational virtuosity and well-being at work in the Intermunicipal Community - Terras de Trás-os-Montes. The possibility of the relationship has been addressed by several authors, and for Cameron et al. (2004), organizational virtuosity induces the emergence of emotions such as empathy, enthusiasm, admiration and love. Corroborating with the author, the analysis of the data concerning the Intermunicipal Community seems to indicate that certain positive emotions provided by virtuous actions are fundamental, not only for achieving individual and organizational success, but also for strengthening relationships between individuals (Fredrickson, 2003). Consequently, these relationships contribute to increased well-being at work.

Organizational virtuosity influences well-being at work through the perception of positive emotions, contributing to workers feeling psychologically and emotionally stronger and safer, more optimistic and happier (Spector & Fox, 2002). It is likely that respect gains a new dimension within the organization and that this is reflected in the self-esteem of the workers (Ramarajan et al., 2008). Thus, it is likely that the employees of these local authorities perceive characteristics such as self-esteem, respect, emotional and psychological capacity more developed and, inevitably, increase their well-being at work. This well-being is visible in the increased feeling of comfort, happiness, energy and tranquility, and in the decrease of anxiety, boredom, aggressiveness and tension felt.

This study conducted for the local government of the Intermunicipal Community - Terras de Trás-os-Montes shows how close these local authorities are to the private sector in the implementation of corporate social responsibility, which prioritizes the development of high performance, people management and the enhancement of ethics, corroborating Carroll and Shabana (2010). Using the assessment of perceptions of organizational virtuosity, well-being at work and organizational citizenship behaviors for the different hierarchical levels, it is possible to conclude that the public entities in the convenience sample are giving increasing emphasis to these themes, but clearly this is an unfinished process.

It is also important to mention that the implementation of these practices is relatively recent and, based on some socio-demographic indicators, this is an even more difficult process to implement for the local authorities of the Intermunicipal Community - Terras de Trás-os-Montes. The region is largely marked by a somewhat more conservative culture, which, coupled with the low percentage of younger individuals performing functions in local authorities, and generally more apt to change, are two important aspects to take into consideration when evaluating and discussing these issues.

This study does not analyze the causal relationships between organizational virtuosity and well-being at work and organizational citizenship behaviors, nor the weight of each of the dimensions of the theoretical models, for the Intermunicipal Community - Terras de Trás-os-Montes. Thus, the two topics mentioned above are suggested as future research that should strengthen the literature on this topic for the public sector, which started with this paper.

REFERENCES

Aristotle. (1999). *The Nicomachean ethics (Translated* (W. D. Ross, Ed.). Batoche Books.

Avey, J., Wernsing, T., & Luthans, F. (2008). Can positive employees help positive organizational change?: Impact of psychological capital and emotions on relevant attitudes and behaviors. *The Journal of Applied Behavioral Science, 44*(1), 48–70. doi:10.1177/0021886307311470

Bakker, E. (2007). Integrity and cynicism: Possibilities and constraints of moral communication. *Journal of Agricultural & Environmental Ethics, 20*(1), 119–136. doi:10.100710806-006-9020-y

Bowen, S. A. (1953). *Social Responsabilities of the Businessman.* Harper & Row.

Branco, M. C., & Rodrigues, L. L. (2008). Factors Influencing Social Responsibility Disclosure by Portuguese Companies. *Journal of Business Ethics, 83*(4), 685–701. doi:10.100710551-007-9658-z

Bright, D., Cameron, K., & Caza, A. (2006). The amplifying and buffering effects of virtuousness in downsized organizations. *Journal of Business Ethics, 64*(3), 249–269. doi:10.100710551-005-5904-4

Bright, D., & Exline, J. J. (2012). Forgiveness at four levels: Intrapersonal, relational, organizational and collective.group. In K. Cameron & G. Spreitzer (Eds.), *The Oxford Handbook of Positive Organizational Scholarship* (pp. 244–259). Oxford University Press.

Cameron, K. (2003). Organizational Virtuousness and Performance. In Positive organizational scholarship. Berrett-Koehler.

Cameron, K., Bright, D., & Caza, A. (2004). Exploring the Relationships between Organizational Virtuousness and Performance. *The American Behavioral Scientist, 47*(6), 766–790. doi:10.1177/0002764203260209

Cameron, K., & Winn, B. (2012). Virtuousness in Organizations. In K. S. Cameron & G. Spreitzer (Eds.), *The Oxford Handbook of Positive Organizational Scholarship* (pp. 231–243). Oxford University Press.

Carroll, A., & Shabana, K. (2010). The business case of Corporate Social Responsability: A Review of concepts, Research and Practice. *International Journal of Management Reviews, 12*(1), 85–105. doi:10.1111/j.1468-2370.2009.00275.x

Carroll, A. B. (1991). The pyramid of corporate social responsibility: Toward the moral management of organizational stakeholders. *Business Horizons, 34*(4), 39–48. doi:10.1016/0007-6813(91)90005-G

Caza, A. (2015). Organizational Virtue. Handbook of Virtue Ethics in Business and Management, 1–9. doi:10.1007/978-94-007-6729-4_30-1

Caza, A., Barker, B., & Cameron, K. (2004). Ethics and ethos: The buffering and amplifying effects of ethical behavior and virtuousness. *Journal of Business Ethics*, *52*(2), 169–178. doi:10.1023/B:BUSI.0000035909.05560.0e

CIM-TTM. (2014). *Plano Estratégico de desenvolvimento intermunicipal das Terras de Trás Montes para o período 2014-2020*. Author.

Daniels, K. (2000). Measures of five aspects of affective well-being at work. *Human Relations*, *53*(2), 275–294. doi:10.1177/a010564

Davis, K. (1960). Can business afford to ignore Social Responsabilities? *California Management Review*, *2*(3), 70–76. doi:10.2307/41166246

Diener, E., & Seligman, M. (2002). Very Happy People. *Psychological Science*, *13*(1), 81–84. doi:10.1111/1467-9280.00415 PMID:11894851

Dobers, P., & Halme, M. (2009). Corporate social responsibility and developing countries. *Corporate Social Responsibility and Environmental Management*, *16*(5), 237–249. doi:10.1002/csr.212

Drucker, P. F. (1984). The New Meaning of Corporate Social Responsibility. *California Management Review*, *26*(2), 53–63. doi:10.2307/41165066

Dutton, J. & Baker, W. (2007). Enabling positive social capital in organizations. *Exploring Positive Relationships at Work - Building a Theoretical and Research Foundation*, 325–345.

Europeia, C. (2001). Livro Verde - Promover um quadro europeu para a responsabilidade social das empresas. *Comissão Europeia, COM 366*.

Fabrigar, L., Wegener, D., Maccalum, R., & Strahan, E. (1999). Evaluating the use of exploratory factor analysis in ppsychological research. *Psychological Methods*, *4*(3), 272–299. doi:10.1037/1082-989X.4.3.272

Ferreira, C. (2004). Environmental accounting: The Portuguese case. *Management of Environmental Quality*, *15*(6), 561–573. doi:10.1108/14777830410560665

Field, A. (2011). *Descobrindo a estatística usando o SPSS* (2ª Edição). Artmed Editora.

Fineman, S. (1999). Emotion and organization. In S. Clegg, C. Hardy, & W. Nord (Eds.), *Studying organisation: Theory and method* (pp. 288–310). Sage. doi:10.4135/9781446218556.n11

Fredrickson, B. (2001). The role of positive emotions in positive psychology: The broaden-and-built theory of positive emotions. *The American Psychologist*, *56*(3), 218–226. doi:10.1037/0003-066X.56.3.218 PMID:11315248

Fredrickson, B. (2003). Positive emotions and upward spirals in organizational settings. In K. Cameron, J. Dutton, & R. Quinn (Eds.), *Positive organizational scholarship* (pp. 163–175). Barrett-Koehler.

Friedman, R. E. (1970). The Social Responsibility of Business is to increase its profits. *New York Times Magazine.*

Garriga, E., & Melé, D. (2004). Corporate Social Responsibility Theories: Mapping the Territory. *Journal of Business Ethics, 53*(1), 51–71. doi:10.1023/B:BUSI.0000039399.90587.34

Gavin, J. H., & Mason, R. O. (2004). The virtuous organization: The value of happiness in the workplace. *Organizational Dynamics, 33*(4), 379–392. doi:10.1016/j.orgdyn.2004.09.005

Hair, J. R. J., Anderson, R., Tatham, R., & Black, W. (2005). *Análise Multivariada de Dados*. Bookman.

Hill, M., & Hill, A. (2000). *Investigação por questionários*. Edições Sílabo.

Konovsky, M., & Organ, D. (1996). Dispositional and contextual determinants of organizational citizienship behaviors. *Journal of Organizational Behavior, 17*(3), 253–256. doi:10.1002/(SICI)1099-1379(199605)17:3<253::AID-JOB747>3.0.CO;2-Q

Lourenço, I., Branco, M., Curto, J., & Eugénio, T. (2012). How Does the Market Value Corporate Sustainability Performance? *Journal of Business Ethics, 108*(4), 417–428. doi:10.100710551-011-1102-8

Luthans, F. (2002). Positive organizational behaviour: Developing and managing psychological strengths. *The Academy of Management Executive, 16*(1), 57–72. doi:10.5465/ame.2002.6640181

MacKenzie, S., Podsakoff, P., & Fetter, R. (1991). Organizational citizienship behavior and objective productivity as determinants of managerial on evalutions of salespersons' performance. *Organizational Behavior and Human Decision Processes, 50*(1), 123–150. doi:10.1016/0749-5978(91)90037-T

Miles, D., Borman, W., Spector, P., & Fox, S. (2002). Building an Integrative Model of Extra Role Work Behaviors: A Comparison of Counterproductive Work Behavior with Organizational Citizienship Behavior. *International Journal of Selection and Assessment, 10*(1&2), 51–57. doi:10.1111/1468-2389.00193

Organ, D. W. (1997). Organizational Citiziенship Behavior: It's construct clean-up time. *Human Performance, 10*(2), 85–97. doi:10.120715327043hup1002_2

Park, N., & Peterson, C. (2003). Virtues and organizations. In K. S. Cameron, J. E. Dutton, & R. E. Quinn (Eds.), *Positive organizational scholarship* (pp. 33–47). Barrett-Koehler.

Peterson, C. & Seligman, M. (2004). *Character Strengths and Virtues: A Handbook and Classification.* . doi:10.1176/appi.ajp.162.4.820-a

Podsakoff, P., & MacKenzie, S. (1994). Organizzational citiziенship behaviors and sales unit effectiveness. *JMR, Journal of Marketing Research, 31*(3), 351–363. doi:10.1177/002224379403100303

Podsakoff, P., MacKenzie, S., Paine, J., & Bachrach, D. (2000). Organizational Citizienship Behaviors: A critival review of the theoretical and empirical leterature and suggestions for future research. *Journal of Management, 26*(3), 513–563. doi:10.1177/014920630002600307

Ramarajan, L., Barsade, S., & Burack, O. (2008). The influence of organizational respect on emotional exhaustion in the human services. *The Journal of Positive Psychology, 3*(1), 4–18. doi:10.1080/17439760701750980

Rego, A., & Cunha, M. (2011). *Liderança: A virtude está no meio*. Actual.

Rego, A., Cunha, M., Costa, N., Gonçalves, H. & Cardoso, C. (2007). *Gestão Ética e Socialmente Responsável*. Academic Press.

Rego, A., Ribeiro, N., & Cunha, M. (2010). Perceptions of organizational virtuousness and happiness as predictors of organizational citizenship behaviors. *Journal of Business Ethics, 93*(2), 215–235. doi:10.100710551-009-0197-7

Rego, A., Vitória, A., Magalhães, A., Ribeiro, N., & Cunha, M. (2012). Are authentic leaders associated with more virtuous, committed and potent teams? *The Leadership Quarterly, 24*(1), 61–79. doi:10.1016/j.leaqua.2012.08.002

Roque, V. & Cortez, M. (2006). A divulgação de informação ambiental e a performance financeira das empresas cotadas em Portugal. *Tékhne - Revista de Estudos Politécnicos, 3*(5–6), 119–143.

Ryan, R., & Deci, E. (2000). Self-determination theory and the facilitation of intrinsic motivation, social development, and well-being. *The American Psychologist, 55*(1), 68–78. doi:10.1037/0003-066X.55.1.68 PMID:11392867

Schwartz, M., & Carroll, A. (2014, December). Corporate Social Responsibility: A Three-Domain Approach. *Business Ethics Quarterly*. Advance online publication. doi:10.2307/3857969

Smith, C., Organ, D., & Near, J. (1983). Organizational Citizienship Behavior: Its nature and antecedents. *The Journal of Applied Psychology, 68*(4), 653–663. doi:10.1037/0021-9010.68.4.653

Spector, P., & Fox, S. (2002). An emotion-centered model of voluntary work behavior. *Human Resource Management Review, 12*(2), 269–292. doi:10.1016/S1053-4822(02)00049-9

Staw, B., & Barsade, S. (1993). Affect and managerial performance: A test of the sadder_but_wiser versus happier-and-smarter hypotheses. *Administrative Science Quarterly, 38*(2), 304–331. doi:10.2307/2393415

Tamayo, A. (2004). Introdução. In A. Tamayo (Ed.), *Cultura e saúde nas organizações* (pp. 11–16). Artmed.

Warr, P. (2007). Searching for happiness at work. *The Psychologist, 20*(12), 726–729.

Wood, D. (1991). *Corporate Social Performance Revisited*. Academic Press.

Wright, T. & Cropanzano, R. (2004). The role of psychological well-being in job performance: A fresh look at an age-old quest. *Organizational Dynamics, 33*(4), 338–351. . doi:10.1016/j.orgdyn.2004.09.002

Wright, T., & Goodstein, J. (2007). Character is not "Dead" in Management Research: A review of individual character and organizational-level virtue. *Journal of Management, 33*(6), 928–958. doi:10.1177/0149206307307644

Compilation of References

Abhayawansa, S., Guthrie, J., & Bernardi, C. (2019). Intellectual capital accounting in the age of integrated reporting: A commentary. *Journal of Intellectual Capital*, *20*(1), 2–10. doi:10.1108/JIC-01-2019-223

Accountability. (2008). *AA1000 Assurance Standard*. https://www.accountability.org/standards/aa1000-assurance-standard

Accountancy Europe. (2017). *Enhance the Coordination of Non-Financial Information Initiatives and Frameworks*. Author.

Aceituno, J., Ariza, L., & Sánchez, I. (2013). Is integrated reporting determined by a country's legal system? An exploratory study. *Journal of Cleaner Production*, *44*, 45–55. doi:10.1016/j.jclepro.2012.12.006

Acuti, D., Bellucci, M., & Manetti, G. (2020). Company disclosures concerning the resilience of cities from the Sustainable Development Goals (SDGs) perspective. *Cities (London, England)*, *99*, 102608. doi:10.1016/j.cities.2020.102608

Adams, C. A. & Abhayawansa, S. (2021). Connecting the COVID-19 pandemic, environmental, social and governance (ESG) investing and calls for 'harmonisation' of sustainability reporting. *Critical Perspectives on Accounting*, *102309*, 1-13. doi:10.1016/j.cpa.2021.102309

Adams, C. A. (2015). The International Integrated Reporting Council: A call to action. *Critical Perspectives on Accounting*, *27*, 23–28. doi:10.1016/j.cpa.2014.07.001

Adams, C. A. (2017). Conceptualizing the contemporary corporate value creation process. *Accounting, Auditing & Accountability Journal*, *30*(4), 906–931. doi:10.1108/AAAJ-04-2016-2529

Adhariani, D., & Villiers, C. (2018). *Integrated reporting: perspectives of corporate report preparers and other stakeholders. Sustainability Accounting, Management and Policy Journal*. doi:10.1108/SAMPJ-02-2018-0043

Adhikariparajuli, M., Hassan, A., Fletcher, M., & Elamer, A. A. (2020). Integrated reporting in higher education: Insights from Scotland, Northern Ireland and Wales. *Social Responsibility Journal*. Advance online publication. doi:10.1108/SRJ-01-2019-0031

Adhikariparajul, M., Hassan, A., Fletcher, M., & Elamer, A. A. (2019). Integrated Reporting in UK Higher Education Institutions. *Sustainability Accounting. Management and Policy Journal*, *10*(5), 844–876. doi:10.1108/SAMPJ-03-2018-0093

Adshead, D., Román García, O., Thacker, S., & Hall, J.W. (2021). Infrastructure Strategies for Achieving the Global Development Agendas in Small Islands. *Earth's Future*, *9*(2), e2020EF001699.

Aiyub, K., Arifin, K., Awang, A., & Jahi, J. M. (2009). Environmental Performance in Small and Medium Sized Enterprises (SMEs) Certified to ISO 14001 in the United Kingdom. *International Business Management*, *3*(1), 7–14.

Akcan, S., & Taş, M. A. (2019). Green Supplier Evaluation with SWARA-TOPSIS Integrated Method to Reduce Ecological Risk Factors. *Environmental Monitoring and Assessment*, *191*(12), 736. doi:10.100710661-019-7884-3 PMID:31707512

Aksu, M. H., & Espahbodi, H. (2016). The Impact of IFRS adoption and corporate governance principles on transparency and disclosure: The case of Borsa. Istanbul. *Emerging Markets Finance & Trade*, *52*(4), 1013–1028. doi:10.1080/1540496X.2014.998570

Alazzani, A., & Wan-Hussin, W. N. (2013). Global Reporting Initiative's environmental reporting: A study of oil and gas companies. *Ecological Indicators*, *32*, 19–24. doi:10.1016/j.ecolind.2013.02.019

Albelda, E. (2011). The Role of Management Accounting Practices as Facilitators of the Environmental Management: Evidence from EMAS Organisations. *Sustainability Accounting, Management and Policy Journal*, *2*(1), 76–100. doi:10.1108/20408021111162137

Albu, C. N. (2013). How Relevant is Size for Setting the Scope of The IFRS for SMEs. *Journal of Accounting and Management Information Systems*, *12*(3), 424–439.

Albu, C. N., Albu, N., Pali-Pista, S. F., Gîrbină, M. M., Selimoglu, S. K., Kovács, D. M., & Strouhal, J. (2013). Implementation of IFRS for SMEs in Emerging Economies: Stakeholder Perceptions in the Czech Republic, Hungary, Romania and Turkey. *Journal of International Financial Management & Accounting*, *24*(2), 140–175. doi:10.1111/jifm.12008

Aleixo, A. M., Azeiteiro, U., & Leal, S. (2016). Toward Sustainability Through Higher Education: Sustainable Development Incorporation into Portuguese Higher Education Institutions. *Challenges in Higher Education for Sustainability*, 159–187.

Aleixo, A. M., Azeiteiro, U. M., & Leal, S. (2018a). The implementation of sustainability practices in Portuguese higher education institutions. *International Journal of Sustainability in Higher Education*, *19*(1), 146–178. doi:10.1108/IJSHE-02-2017-0016

Aleixo, A. M., Azeiteiro, U., & Leal, S. (2020). Are the sustainable development goals being implemented in the Portuguese higher education formative offer? *International Journal of Sustainability in Higher Education*, *21*(2), 336–352. doi:10.1108/IJSHE-04-2019-0150

Aleixo, A., Leal, S., & Azeiteiro, U. (2018b). Conceptualization of sustainable higher education institutions, roles, barriers, and challenges for sustainability: An exploratory study in Portugal. *Journal of Cleaner Production*, *172*, 1664–1673. doi:10.1016/j.jclepro.2016.11.010

Al-Htaybat, K., & von Alberti-Alhtaybat, L. (2018). Integrated thinking leading to integrated reporting: Case study insights from a global player. *Accounting, Auditing & Accountability Journal*, *31*(5), 1435–1460. doi:10.1108/AAAJ-08-2016-2680

Alimo-Metcalfe, B. (1995). An investigation of female and male constructs of leadership. *Women in Management Review*, *10*(2), 3–8. doi:10.1108/09649429510146901

Alipour, M., Ghanbari, M., Jamshidinavid, B., & Taherabadi, A. (2019). Does board independence moderate the relationship between environmental disclosure quality and performance? Evidence from static and dynamic panel data. *Corporate Governance: The International Journal of Business in Society*, *19*(3), 580–610. doi:10.1108/CG-06-2018-0196

Almontaser, T., & Gerged, A. (2019). (in press). Exploring corporate Sustainable Development Goals (SDGs) disclosure in a politically unstable environment: Evidence from the Libyan oil industry. *Sustainability, Accounting, Management, and Policy Journal*. Advance online publication. doi:10.1108/SAMPJ-04-2019-0136

Alonso-Almeida, M. del M., Marimon, F., Casani, F., & Rodriguez-Pomeda, J. (2015). Diffusion of sustainability reporting in universities: Current situation and future perspectives. *Journal of Cleaner Production*, *106*, 144–154. doi:10.1016/j.jclepro.2014.02.008

Al-Refaie, A., Al-Tahat, M., & Lepkova, N. (2020). Modelling Relationships Between Agility, Lean, Resilient, Green Practices in Cold Supply Chains Using ISM Approach. *Technological and Economic Development of Economy*, *26*(4), 675–694. doi:10.3846/tede.2020.12866

Altman, E. I., & Sabato, G. (2007). Modelling Credit Risk for SMEs: Evidence from the U.S. Market. *Abacus*, *43*(3), 332–357. doi:10.1111/j.1467-6281.2007.00234.x

Altman, E. I., Sabato, G., & Wilson, N. (2010). The value of non-financial information in SME risk management. *The Journal of Credit Risk*, *6*(2), 95–127. doi:10.21314/JCR.2010.110

Ameli, N., Kothari, S., & Grubb, M. (2021). Misplaced expectations from climate disclosure initiatives. *Nature Climate Change*, *11*(11), 917–924. doi:10.103841558-021-01174-8

Amey, M. & Whooley. (2018, Dec.). Corporate reporting on the SDGs: mapping a sustainable future. *Advisor Perspectives*.

Amosh, H. A. & Khatib, S. F. A. (2021). Ownership structure and environmental, social and governance performance disclosure: the moderating role of the board independence. *Journal of Business and Socio-economic Development*, 1-18. doi:10.1108/JBSED-07-2021-0094

Andreeva, G., Calabrese, R., & Osmetti, S. A. (2016). A comparative analysis of the UK and Italian small businesses using Generalised Extreme Value models. *European Journal of Operational Research*, *249*(2), 506–516. doi:10.1016/j.ejor.2015.07.062

Anh, N. H., Soa, N. L., & Hanh, H. H. (2020). Environmental accounting practices and cost of capital of enterprises in Vietnam. *Cogent Economics & Finance*, *8*(1), 1–17. doi:10.1080/23322039.2020.1790964

Annan-Diab, F., & Molinari, C. (2017). Interdisciplinarity: Practical approach to advancing education for sustainability and for the Sustainable Development Goals. *International Journal of Management Education*, *15*(2), 73–83. doi:10.1016/j.ijme.2017.03.006

Ansari, D., & Holz, F. (2020). Dawud; HOLZ, Franziska. Between stranded assets and green transformation: Fossil-fuel-producing developing countries towards 2055. *World Development*, *130*, 104947. doi:10.1016/j.worlddev.2020.104947

Apud, M. (2020). Ibope: 9 em cada 10 brasileiros mudaram hábitos financeiros na pandemia [Ibope: 9 out of 10 Brazilians have changed financial habits in the pandemic]. *Estadão*. https://einvestidor.estadao.com.br/educacao-financeira/ibope-brasileiros-habitos-financeiros-pandemia

Arimura, T. H., Darnall, N., & Katayama, H. (2011). Is ISO 14001 a Gateway to More Advanced Voluntary Action? The Case of Green Supply Chain Management. *Journal of Environmental Economics and Management*, *61*(2), 170–182. doi:10.1016/j.jeem.2010.11.003

Aristotle. (1999). *The Nicomachean ethics (Translated* (W. D. Ross, Ed.). Batoche Books.

Arowoshegbe, A. O., & Emmanuel, U. (2016). Accounting for social and environmental challenges: A theoretical perspective. *Saudi Journal of Business and Management Studies*, *1*(3), 143–148. doi:10.21276jbms.2016.1.3.10

Asian Development Bank. (1995). *Governance: Sound development management*. Manila, Philippines: Asian Development Bank. https://www.adb.org/sites/default/files/institutional-document/32027/govpolicy.pdf

Atkins, J., & Maroun, W. (2020). The Naturalist's Journals of Gilbert White: Exploring the roots of accounting for biodiversity and extinction accounting. *Accounting, Auditing & Accountability Journal*, *33*(8), 1835–1870. doi:10.1108/AAAJ-03-2016-2450

Avey, J., Wernsing, T., & Luthans, F. (2008). Can positive employees help positive organizational change?: Impact of psychological capital and emotions on relevant attitudes and behaviors. *The Journal of Applied Behavioral Science*, *44*(1), 48–70. doi:10.1177/0021886307311470

Avrampou, A., Skouloudis, A., Iliopoulos, G., & Khan, N. (2019). Advancing the Sustainable Development Goals: Evidence from leading European banks. *Sustainable Development*, *27*(4), 743–757. doi:10.1002d.1938

Azizi, L., Bien, C., & Sassen, R. (2018). Recent trends in sustainability reporting by German universities. *Sustainability Management Forum, 26*, 65–85.

Bajec, P., Tuljak-Suban, D., & Krmac, E. (2015). Do ISO Standards Favour Logistics Provider Efficiency, Competitiveness and Sustainability? A Slovenian Perspective. *International Journal of Logistics Management*, *26*(2), 275–295. doi:10.1108/IJLM-01-2013-0006

Baker, C. R., Cohanier, B., & Pederzoli, D. (2012). Corporate social and environmental reporting in the large retail distribution sector. *Procedia Economics and Finance*, *2*, 209–218. doi:10.1016/S2212-5671(12)00081-0

Baker, H. K., Pandey, N., Kumar, S., & Haldar, A. (2020). A bibliometric analysis of board diversity: Current status, development, and future research directions. *Journal of Business Research*, *108*, 232–246. doi:10.1016/j.jbusres.2019.11.025

Bakker, E. (2007). Integrity and cynicism: Possibilities and constraints of moral communication. *Journal of Agricultural & Environmental Ethics*, *20*(1), 119–136. doi:10.100710806-006-9020-y

Baldarelli, M. G., & Nesheva-Kiosseva, N. (2017). Environmental accounting: Conceptual framework. In S. O. Idowu & R. Schmidpeter (Eds.), *Environmental Accounting and Reporting: Theory and Practice* (pp. 35–67). Springer International Publishing. doi:10.1007/978-3-319-50918-1_2

Baldini, M., Dal Maso, L., Liberatore, G., Mazzi, F., & Terzani, S. (2018). Role of country- and firm-level determinants in environmental, social, and governance disclosure. *Journal of Business Ethics*, *150*(1), 79–98. doi:10.100710551-016-3139-1

Baldwin, R. (2005). Is better regulation smarter regulation? *Public Law*, (Autumn), 485–511.

Bardin, L. (1977). *Análise de Conteúdo*. Lisboa: Edições 70.

Bardin, L. (2016). *Análise de Conteúdo*. Edições 70.

Barker, R., & Eccles, R. G. (2018). Should FASB and IASB Be Responsible for Setting Standards for Nonfinancial Information? SSRN *Electronic Journal*, (October). doi:10.2139/ssrn.3272250

Barrett, M., Watene, K., & McNicholas, P. (2020). Legal personality in Aotearoa New Zealand: An example of integrated thinking on sustainable development. *Accounting, Auditing & Accountability Journal*, *33*(7), 1705–1730. doi:10.1108/AAAJ-01-2019-3819

Barros, A. (2020). PIB tem queda recorde de 9,7% no 2º trimestre, auge do isolamento social [GDP has a record drop of 9.7% in the 2nd quarter, the height of social isolation]. *Agência IBGE Notícias*. https://agenciadenoticias.ibge.gov.br/agencia-noticias/2012-agencia-de-noticias/noticias/28720-pib-tem-queda-recorde-de-9-7-no-2-trimestre-auge-do-isolamento-social

Barros, A. (2021). Inflação acelera em dezembro e chega a 4,52% em 2020, a maior alta desde 2016 [Inflation accelerates in December and reaches 4.52% in 2020, the highest since 2016]. *Agência IBGE Notícias*. https://agenciadenoticias.ibge.gov.br/agencia-noticias/2012-agencia-de-noticias/noticias/29871-inflacao-acelera-em-dezembro-e-chega-a-4-52-em-2020-a-maior-alta-desde-2016

Bartolacci, F., Caputo, A., & Soverchia, M. (2020). Sustainability and financial performance of small and medium sized enterprises: A bibliometric and systematic literature review. *Business Strategy and the Environment*, *29*(3), 1297–1309. doi:10.1002/bse.2434

Baumann-Pauly, D., Wickert, C., Spence, L. J., & Scherer, A. G. (2013). Organizing corporate social responsibility in small and large firms: Size matters. *Journal of Business Ethics*, *115*(4), 693–705. doi:10.100710551-013-1827-7

Baxi, C. V., & Ray, R. S. (2009). Corporate social & environmental disclosures & reporting. *Indian Journal of Industrial Relations*, *44*(3), 355–375.

Bebbington, J., & Unerman, J. (2018). Achieving the United Nations Sustainable Development Goals: An enabling role for accounting research. *Accounting, Auditing & Accountability Journal*, *31*(1), 2–24. doi:10.1108/AAAJ-05-2017-2929

Bebbington, J., & Unerman, J. (2020). Advancing research into accounting and the UN sustainable development goals. *Accounting, Auditing & Accountability Journal*, *33*(7), 1657–1670. doi:10.1108/AAAJ-05-2020-4556

Beck, A. C., Campbell, D., & Shrives, P. J. (2010). Content analysis in environmental reporting research: Enrichment and rehearsal of the method in a British–German context. *The British Accounting Review*, *42*(3), 207–222. doi:10.1016/j.bar.2010.05.002

Begina, Ai. (2017). *Corporate and Social Responsibility Reporting in the Greek banking sector: a comparative approach* (Master diss.). International Hellenic University.

Belt, V. (2002). A female ghetto? Women's careers in call centres. *Human Resource Management Journal*, *12*(4), 51–66. doi:10.1111/j.1748-8583.2002.tb00077.x

Bepari, M. K., & Mollik, A. T. (2016). Stakeholders' interest in sustainability assurance process: An examination of assurance statements reported by Australian companies. *Managerial Auditing Journal*, *31*(6/7), 655–687. doi:10.1108/MAJ-06-2015-1208

Beske, P., Koplin, J., & Seuring, S. (2008). The Use of Environmental and Social Standards by German First-Tier Suppliers of the Volkswagen AG. *Corporate Social Responsibility and Environmental Management*, *15*(2), 63–75. doi:10.1002/csr.136

Betz, M., O'Connell, L., & Shepard, J. M. (1989). Gender Differences in Proclivity for Unethical Behavior. *Journal of Business Ethics*, *8*(5), 321–324. doi:10.1007/BF00381722

Bevilaqua, I. C., Neumann, M., & Faia, V. S. (2020). Fatores que Favorecem e/ou Limitam o Pensamento Integrado na Prática: Uma análise do Conceito, Incorporação e Benefícios apresentados pela Literatura do Relato Integrado. *XX USP International Conference in Accounting*.

Bierema, L. (2005). Women's networks: A career development intervention or impediment? *Human Resource Development International*, *2*(2), 207–224. doi:10.1080/13678860500100517

Bilimoria, D. (2000). Building the business case for women corporate directors. In R. J. Burke & M. C. Mattis (Eds.), *Women on Corporate Boards of Directors: International Challenges and Opportunities* (pp. 25–40). Kluwer Academic. doi:10.1007/978-90-481-3401-4_3

Bilimoria, D., & Huse, M. (1997). A qualitative comparison of the boardroom experiences of US and Norwegian women corporate directors. *International Review of Women and Leadership*, *3*(2), 63–76.

Boiral, O., & Henri, J. F. (2012). Modelling the Impact of ISO 14001 on Environmental Performance: A Comparative Approach. *Journal of Environmental Management*, *99*(May), 84–97. doi:10.1016/j.jenvman.2012.01.007 PMID:22326757

Boiral, O., & Henri, J. F. (2017). Is Sustainability Performance Comparable? A Study of GRI Reports of Mining Organizations. *Business & Society*, *56*(2), 283–317. doi:10.1177/0007650315576134

Borga, F., Citterio, A., Noci, G., & Pizzurno, E. (2009). Sustainability report in small enterprises: Case studies in Italian furniture companies. *Business Strategy and the Environment*, *18*(3), 162–176. doi:10.1002/bse.561

Bos, K., & Gupta, J. (2019). Stranded assets and stranded resources: Implications for climate change mitigation and global sustainable development. *Energy Research & Social Science, 56*, 1-15.

Bouza Garcia, L. (2015). *Participatory Democracy and Civil Society in the EU: Agenda-Setting and Institutionalization.* Palgrave Macmillan. doi:10.1057/9781137436849

Bowen, S. A. (1953). *Social Responsabilities of the Businessman.* Harper & Row.

Boyce, G. (2000). Public discourse and decision making: Exploring possibilities for financial, social and environmental accounting. *Accounting, Auditing & Accountability Journal*, *13*(1), 27–64. doi:10.1108/09513570010316135

Braam, G. J. M., Uit de Weerd, L., Hauck, M., & Huijbregts, M. A. J. (2016). Determinants of corporate environmental reporting: The importance of environmental performance and assurance. *Journal of Cleaner Production*, *129*, 724–734. doi:10.1016/j.jclepro.2016.03.039

Bracci, E., & Maran, L. (2013). Environmental management and regulation: Pitfalls of environmental accounting? *Management of Environmental Quality*, *24*(4), 538–554. doi:10.1108/MEQ-04-2012-0027

Branco, M. C., & Rodrigues, L. L. (2008). Factors Influencing Social Responsibility Disclosure by Portuguese Companies. *Journal of Business Ethics*, *83*(4), 685–701. doi:10.100710551-007-9658-z

Branson, D. M. (2006). *No Seat at the Table: How Corporate Governance and Law Keep Women Out of the Boardroom.* New York University Press.

Bremmers, H., Omta, O., Kemp, R., & Haverkamp, D. J. (2007). Do Stakeholder Groups Influence Environmental Management System Development in the Dutch Agri-Food Sector? *Business Strategy and the Environment*, *16*(3), 214–231. doi:10.1002/bse.480

Briem, C. R., & Wald, A. (2018). Implementing Third-Party Assurance in Integrated Reporting: Motivação das empresas e papel dos auditores. *Accounting, Auditing & Accountability Journal*, *31*(5), 1461–1485. doi:10.1108/AAAJ-03-2016-2447

Bright, D., Cameron, K., & Caza, A. (2006). The amplifying and buffering effects of virtuousness in downsized organizations. *Journal of Business Ethics*, *64*(3), 249–269. doi:10.100710551-005-5904-4

Bright, D., & Exline, J. J. (2012). Forgiveness at four levels: Intrapersonal, relational, organizational and collective.group. In K. Cameron & G. Spreitzer (Eds.), *The Oxford Handbook of Positive Organizational Scholarship* (pp. 244–259). Oxford University Press.

Bringezu, S. (2018). Key strategies to achieve the SDGs and consequences for monitoring resource use. *Managing Water, Soil and Waste Resources to Achieve Sustainable Development Goals: Monitoring and Implementation of Integrated Resources Management*, 11-34.

Broadbridge, A. (2000). Retail managers: Their work stressors and coping strategies. *Journal of Retailing and Consumer Sciences.*

Broadbridge, A. (1999). Retail managers: Stress and the work-family relationship. *International Journal of Retail & Distribution Management*, *27*(9), 374–382. doi:10.1108/09590559910292825

Broadbridge, A. (2008). Senior careers in retailing: An exploration of male and female executives' career facilitators and barriers. *Gender in Management*, *23*(1), 11–35. doi:10.1108/17542410810849105

Brooks, C., & Schopohl, L. (2021). Green accounting and finance: Advancing research on environmental disclosure, value impacts and management control systems [Editorial]. *The British Accounting Review*, *53*(1), 1–5. doi:10.1016/j.bar.2020.100973

Brown, J., & Dillard, J. (2014). Integrated reporting: On the need for broadening out and opening up. *Accounting, Auditing & Accountability Journal*, *27*(7), 1120–1156. doi:10.1108/AAAJ-04-2013-1313

Brown, J., & Fraser, M. (2006). Approaches and perspectives in social and environmental Accounting: An overview of the conceptual landscape. *Business Strategy and the Environment*, *15*(2), 103–117. doi:10.1002/bse.452

Brown, S. (1990). Management of education and social welfare in Scotland: The places of women. *Scottish Journal of Adult Education*, *9*(3), 3–13.

Burke, R. J., Rothstein, M. G., & Bristor, J. M. (1995). Interpersonal networks of managerial and professional women and men: Descriptive characteristics. *Women in Management Review*, *10*(1), 21–27. doi:10.1108/09649429510077458

Burke, S., & Gaughran, W. F. (2007). Developing a framework for sustainability management in engineering SMEs. *Robotics and Computer-integrated Manufacturing*, *23*(6), 696–703. doi:10.1016/j.rcim.2007.02.001

Burrit, R. (1999). Environmental Accounting and Reporting in Victoria, Australia. News Journal of the Asia Pacific Centre for Environmental Accountability, 5(3), 6-11.

Burton, B., Kumar, S., & Pandey, N. (2020). Twenty-five years of The European Journal of Finance (EJF): A retrospective analysis. *European Journal of Finance*, *26*(18), 1–25. doi:10.1080/1351847X.2020.1754873

Busco, C., Malafronte, I., Pereira, J., & Starita, M. G. (2019). The determinants of companies' levels of integration: Does one size fit all? *The British Accounting Review*, *51*(3), 277–298. doi:10.1016/j.bar.2019.01.002

Cai, Y., & Kleiner, H. (1999). Sex discrimination in Hiring: The Glass ceiling. *International Journal: Gender in Management*, *18*(2-4), 51-55.

Calace, D. (2014). Non-Financial Reporting in Italian SMEs: An Exploratory Study on Strategic and Cultural Motivations. *International Journal of Business Administration*, *5*(3), 34–48. doi:10.5430/ijba.v5n3p34

Caldecott. (2016). *Stranded assets: A climate risk challenge*. IDB.

Caldecott, B. (2017). Introduction to special issue: Stranded assets and the environment. *Journal of Sustainable Finance & Investment*, *7*(1), 1–13. doi:10.1080/20430795.2016.1266748

Caldecott, B., Howarth, N., & McSharry, P. (2013). *Stranded assets in Agriculture: Protecting Value from Environment-Related Risks*. Smith School of Enterprise and the Environment, University of Oxford.

Cameron, K. (2003). Organizational Virtuousness and Performance. In Positive organizational scholarship. Berrett-Koehler.

Cameron, K., Bright, D., & Caza, A. (2004). Exploring the Relationships between Organizational Virtuousness and Performance. *The American Behavioral Scientist*, *47*(6), 766–790. doi:10.1177/0002764203260209

Cameron, K., & Winn, B. (2012). Virtuousness in Organizations. In K. S. Cameron & G. Spreitzer (Eds.), *The Oxford Handbook of Positive Organizational Scholarship* (pp. 231–243). Oxford University Press.

Camodeca, R. E., Almici, A., & Sagliaschi, U. (2018). Sustainability Disclosure in Integrated Reporting: Does It Matter to Investors? A Cheap Talk Approach. *Sustainability*, *10*(12), 4393. doi:10.3390u10124393

Campos, A. (2021). Desemprego recua para 13,9% no 4º trimestre, segundo o IBGE [Unemployment drops to 13.9% in the 4th quarter, according to IBGE.] *Agência Brasil*. https://agenciabrasil.ebc.com.br/economia/noticia/2021-02/desemprego-recua-para-139-no-4o-trimestre-segundo-ibge

Campos, R. I. (2021). *Mudanças climáticas e estabilidade financeira: incertezas, riscos e o papel dos bancos centrais* (Master's dissertation). Programa de PósGraduação em Economia da Faculdade de Economia, Administração, Contabilidade e Gestão de Políticas Públicas da Universidade de Brasília.

Caputo, F., Ligorio, L., & Pizzi, S. (2021). The Contribution of Higher Education Institutions to the SDGs—An Evaluation of Sustainability Reporting Practices. *Administrative Sciences*, *11*(3), 97. doi:10.3390/admsci11030097

Carandang, J. C., & Ferrer, R. C. (2020). Effect of environmental accounting on financial performance and firm value of listed mining and oil companies in the Philippines. *Asia-Pacific Social Science Review*, *20*(1), 117–134.

Carbon Tracker Initiative. (2020). *Fault Lines: How diverging oil and gas company strategies link to stranded asset risk*. Author.

Carbon Tracker Initiative. (2021). *The burning need to cut oil dependence in the energy transition*. Author.

Carraro, W., Pereira, M., & Soster, Monise. (2021). *Dicas para manter sua educação financeira em alta* [Tips to keep your financial education high]. Educação Financeira UFRGS [Financial Education UFRGS]. https://lume.ufrgs.br/handle/10183/217579

Carroll, A. B. (1991). The pyramid of corporate social responsibility: Toward the moral management of organizational stakeholders. *Business Horizons*, *34*(4), 39–48. doi:10.1016/0007-6813(91)90005-G

Carroll, A., & Shabana, K. (2010). The business case of Corporate Social Responsability: A Review of concepts, Research and Practice. *International Journal of Management Reviews*, *12*(1), 85–105. doi:10.1111/j.1468-2370.2009.00275.x

Carungu, J., Di Pietra, R., & Molinari, M. (2021). Mandatory vs voluntary exercise on non-financial reporting: Does a normative/coercive isomorphism facilitate an increase in quality? *Meditari Accountancy Research*, *29*(3), 449–476. doi:10.1108/MEDAR-08-2019-0540

Catalyst. (2007a). *2007 Catalyst Census of Women Board Directors*. New York: Catalyst.

Catalyst. (2007b). *2007 Catalyst Census of Women Corporate Officers and Top Earners*. New York: Catalyst.

Caza, A. (2015). Organizational Virtue. Handbook of Virtue Ethics in Business and Management, 1–9. doi:10.1007/978-94-007-6729-4_30-1

Caza, A., Barker, B., & Cameron, K. (2004). Ethics and ethos: The buffering and amplifying effects of ethical behavior and virtuousness. *Journal of Business Ethics*, *52*(2), 169–178. doi:10.1023/B:BUSI.0000035909.05560.0e

CDP. (n.d.). *About Us*. Retrieved from https://www.cdp.net/en/info/about-us

CDP. (n.d.). *Accelerating the Rate of Change: CDP Strategy 2021-2025*. Retrieved from https://cdn.cdp.net/cdp-production/comfy/cms/files/files/000/005/094/original/CDP_STRATEGY_2021-2025.pdf

Cerbasi, G. (2004). *Casais inteligentes enriquecem juntos* [Smart couples get rich together]. Gente.

Cerbasi, G. (2012). *Como organizar sua vida financeira: inteligência financeira pessoal na prática* [How to organize your financial life: personal financial intelligence in practice]. Elsevier.

Ceulemans, K., Lozano, R., & Alonso-Almeida, M. (2015). Sustainability reporting in higher education: Interconnecting the reporting process and organisational change management for sustainability. *Sustainability, 7*(7), 8881–8903. doi:10.3390u7078881

CGMA. (2014). *Integrated thinking: The next step in integrated reporting.* CIMA & American Institute of CPAs. Available in: https://competency.aicpa.org/media_resources/206473-integrated-thinking-the-next-step-in-integrated-re/detail

Chaidali, P., & Jones, M. J. (2017). It's a matter of trust: Exploring the perceptions of Integrated Reporting preparers. *Critical Perspectives on Accounting, 48*(1), 1–20. doi:10.1016/j.cpa.2017.08.001

Chatham, M. D., Larson, R. K., & Vietze, A. (2010). Issues affecting the development of an international accounting standard on financial instruments. *Advances in Accounting, 26*(1), 97–107. doi:10.1016/j.adiac.2010.03.002

Chen, J. C., Cho, C. H., & Patten, D. M. (2014). Initiating disclosure of environmental liability information: An empirical analysis of firm choice. *Journal of Business Ethics, 125*(4), 681-692. doi:10551-013-1939-0 doi:10.1007/s

Cheng, B., Ioannou, I., & Serafeim, G. (2014). Corporate social responsibility and access to finance. *Strategic Management Journal, 35*(1), 22–30. doi:10.1002mj.2131

Cheng, M., Green, W., Conradie, P., Konishi, N., & Romi, A. (2014). The International Integrated Reporting Framework: Key Issues and Future Research Opportunities. *Journal of International Financial Management & Accounting, 25*(1), 90–119. doi:10.1111/jifm.12015

Cherobim, A., & Espejo, M. (Eds.). (2010). *Finanças pessoais: conhecer para enriquecer* [Personal finance: knowing to get rich]. Atlas.

Chiarini, A. (2012). Designing an Environmental Sustainable Supply Chain through ISO 14001 Standard. *Management of Environmental Quality, 24*(1), 16–33. doi:10.1108/14777831311291113

Chiarini, A. (2014). Strategies for Developing an Environmentally Sustainable Supply Chain: Differences Between Manufacturing and Service Sectors: Strategies for Developing an Environmentally Sustainable Supply Chain. *Business Strategy and the Environment, 23*(7), 493–504. doi:10.1002/bse.1799

Chiarini, A. (2017). Environmental Policies for Evaluating Suppliers' Performance Based on GRI Indicators: A Multi-Regression Model for Suppliers' Environmental Performance. *Business Strategy and the Environment, 26*(1), 98–111. doi:10.1002/bse.1907

Cho, C. H., Senn, J., & Sobkowiak, M. (2021). Sustainability at stake during COVID-19: Exploring the role of accounting in addressing environmental crises. *Critical Perspectives on Accounting*, 102327.

Christensen, H. B., Hail, L., & Leuz, C. (2021). Mandatory CSR and sustainability reporting: Economic analysis and literature review. *Review of Accounting Studies, 26*(3), 1176–1248. Advance online publication. doi:10.100711142-021-09609-5

Chung, J., & Tsay, M. Y. (2017). A bibliometric analysis of the literature on open access in Scopus. *Qualitative and Quantitative Methods in Libraries, 4*(4), 821–841.

Churet, C., & Eccles, R. G. (2014). Integrated Reporting, Quality of Management, and Financial Performance. *Journal of Applied Corporate Finance, 26*(1), 56–65. doi:10.1111/jacf.12054

CIM-TTM. (2014). *Plano Estratégico de desenvolvimento intermunicipal das Terras de Trás Montes para o período 2014-2020*. Author.

Climate Disclosure Standards Board. (2019). *CDSB Framework for reporting environmental & climate change information.* Retrieved from https://live-cdsb.pantheonsite.io/sites/default/files/cdsb_framework_2019_v2.2.pdf

Climate Disclosure Standards Board. (n.d.). *About the Climate Disclosure Standards Board.* Retrieved from https://www.cdsb.net/our-story

Committee of Sponsoring Organizations of the Treadway Commission. (2013). *Internal Control – Integrated Framework: Executive Summary.* https://na.theiia.org/standards-guidance/topics/Documents/Executive_Summary.pdf

Confederation National of Goods and Tourism Trade – CNC. (2020) *Pesquisa de endividamento e inadimplência do Consumidor* [Consumer indebtedness and default survey]. http://cnc.org.br/editorias/economia/pesquisas/pesquisa-de-endividamento-e-inadimplencia-do-consumidor-peic-janeiro-2

Cooper, D. R., & Schindler, P. S. (2003). *Métodos de Pesquisa em Administração* (7th ed.). Bookman.

Coraiola, D., Suddaby, R., & Foster, W. M. (2018) Organizational Fields as Mnemonic Communities. Knowledge and Institutions, 13(1), 45-68. doi:10.1007/978-3-319-75328-7_3

Corazza, L. (2017). The standardization of down-Streamed Small Business Social Responsibility (SBSR): SMEs and their sustainability reporting practices. *Information Resources Management Journal*, *30*(4), 39–52. doi:10.4018/IRMJ.2017100103

Corrado, M., Demartini, P., & Dumay, J. (2019). *Assurance on Integrated Reporting: A Critical Perspective*. doi:10.1007/978-3-030-01719-4_11

Coughlan, P., & Coghlan, D. (2002). Action research for operations management. *International Journal of Operations & Production Management*, *22*(2), 220–240. doi:10.1108/01443570210417515

Council of Europe Parliamentary Assembly. (2004). *Environmental accounting as a sustainable development tool.* Retrieved from https://assembly.coe.int/nw/xml/XRef/Xref-XML2HTML-en.asp?fileid=17198&lang=en

Creswell, J. W. (2014). *Investigação Qualitativa e Projeto de Pesquisa: escolhendo entre cinco abordagens.* Penso Editora.

Cross, C., & Linehan, M. (2006). Advancing female careers in the high-tech sector: Empirical evidence from Ireland. *Women in Management Review*, *21*(1), 28–39. doi:10.1108/09649420610643394

CSO. (2006). *Women and Men in Ireland.* Central Statistics Office, Stationery Office.

CSR Europe GRI. (2017). *Member State Implementation of Directive 2014/95/EU: A comprehensive overview of how Member States are implementing the EU Directive on Non-financial and Diversity Information.* https://www.globalreporting.org/resourcelibrary/NFRpublication%20online_version.pdf

Cuesta Lopez, V. (2010). The Lisbon Treaty's provisions on Democratic Principles: A legal framework for Participatory democracy. *European Public Law*, *16*(1), 123–136.

Culpan, O., Marzotto, T., & Demir, N. (2007). Foreign banks: Executive jobs for Turkish Women? *Women in Management Review*, *22*(8), 608–630. doi:10.1108/09649420710836308

Curkovic, S., & Sroufe, R. (2011). Using ISO 14001 to Promote a Sustainable Supply Chain Strategy. *Business Strategy and the Environment*, *20*(2), 71–93. doi:10.1002/bse.671

Daddi, T., Giacomo, M. R., Frey, M., & Iraldo, F. (2018). Analysing the Causes of Environmental Management and Audit Scheme (EMAS) Decrease in Europe. *Journal of Environmental Planning and Management*, *61*(13), 2358–2377. doi:10.1080/09640568.2017.1395316

Daddi, T., Heras-Saizarbitoria, I., Marrucci, L., Rizzi, F., & Testa, F. (2021). The effects of green supply chain management capability on the internalisation of environmental management systems and organisation performance. *Corporate Social Responsibility and Environmental Management*, *28*(4), 1241–125328. doi:10.1002/csr.2144

Dagiliene, L., Frendzel, M., Sutiene, K., & Wnuk-Pel, T. (2020). Wise managers think about circular economy, wiser report and analyze it. Research of environmental reporting practices in EU manufacturing companies. *Journal of Cleaner Production*, *274*, 1–14. doi:10.1016/j.jclepro.2020.121968

Dal Maso, M., Olsen, K. H., Dong, Y., Pedersen, M. B., & Hauschild, M. Z. (2020). Sustainable development impacts of nationally determined contributions: Assessing the case of mini-grids in Kenya. *Climate Policy*, *20*(7), 815–831. doi:10.1080/14693062.2019.1644987

Daniels, K. (2000). Measures of five aspects of affective well-being at work. *Human Relations*, *53*(2), 275–294. doi:10.1177/a010564

Davis, K. (1960). Can business afford to ignore Social Responsabilities? *California Management Review*, *2*(3), 70–76. doi:10.2307/41166246

De Villiers, C., Rinaldi, L., & Unerman, J. (2014). Integrated Reporting: Insights, gaps and an agenda for future research. *Accounting, Auditing & Accountability Journal*, *27*(7), 1–47. doi:10.1108/AAAJ-06-2014-1736

De Villiers, C., & van Staden, C. J. (2006). Can less environmental disclosure have a legitimizing effect? Evidence from Africa. *Accounting, Organizations and Society*, *31*(8), 763–781. doi:10.1016/j.aos.2006.03.001

Del Baldo, M. (2017). The implementation of integrating reporting <IR> in SMEs: Insights from a pioneering experience in Italy. *Meditari Accountancy Research*, *25*(4), 505–532. doi:10.1108/MEDAR-11-2016-0094

DelaPoza, E., Merello, P., Barberá, A., & Celani, A. (2021). Universities' Reporting on SDGs: Using THE Impact Rankings to Model and Measure Their Contribution to Sustainability. *Sustainability*, *13*(4), 2038. doi:10.3390u13042038

Delmas, M. A. (2000). Barriers and Incentives to the Adoption of ISO 14001 by Firms in the United States. *Duke Environmental Law and Policy Forum, 11*(1), 1–38.

Delmas, M. A. (2002). The Diffusion of Environmental Management Standards in Europe and in the United States: An Institutional Perspective. *Policy Sciences*, *35*(1), 91–119. doi:10.1023/A:1016108804453

Delmas, M. A., & Montiel, I. (2009). Greening the Supply Chain: When Is Customer Pressure Effective? *Journal of Economics & Management Strategy*, *18*(1), 171–201. doi:10.1111/j.1530-9134.2009.00211.x

Delmas, M., Montes-Sancho, M. J., & Shimshack, J. P. (2010). Information disclosure policies: Evidence from the electricity industry. *Economic Inquiry*, *48*(2), 483–498. doi:10.1111/j.1465-7295.2009.00227.x

Deloitte. (2015). *Non-financial reporting*. https://www2.deloitte.com/content/dam/Deloitte/lv/Documents/strategy/Non-financial_reporting_2015.pdf

Deloitte. (2016). *IAS 16—Property, Plant and Equipment*. Disponível em: https://www.iasplus.com/en/standards/ias/ias16

Deloitte. (n.d.). *Conceptual Framework for Financial Reporting 2018*. Retrieved from https://www.iasplus.com/en/standards/other/framework

Denedo, M., & Egbon, O. (2021). Africa, from the past to the present: Moving the critical environmental accounting research on Africa forward. *Environmental Accounting*, 265-275.

Dhar, S., & Chowdhury, M. A. F. (2021). Impact of environmental accounting reporting practices on financial performance: Evidence from banking sector of Bangladesh. *International Journal of Asian Business and Information Management*, *12*(1), 24–42. doi:10.4018/IJABIM.20210101.oa2

Di Vaio, A., & Varriale, L. (2020). SDGs and airport sustainable performance: Evidence from Italy on organisational, accounting and reporting practices through financial and non-financial disclosure. *Journal of Cleaner Production*, *249*(10), 119431. doi:10.1016/j.jclepro.2019.119431

Dias, A. (2010). Responsabilidade Social Corporativa – Teorias Predominantes na Investigação em Contabilidade. *XIV Encuentro AECA*.

Diener, E., & Seligman, M. (2002). Very Happy People. *Psychological Science*, *13*(1), 81–84. doi:10.1111/1467-9280.00415 PMID:11894851

Dimaggio, P., & Powell, W. W. (1983). The iron cage revisited: Institutional isomorphism and collective rationality in organizational fields. *American Sociological Review*, *48*(2), 147–160. doi:10.2307/2095101

Dimes, R., & De Villiers, C. (2020). How management control systems enable and constrain integrated thinking. *Meditari Accountancy Research*. Available in: https://papers.ssrn.com/sol3/papers.cfm?abstract_id=3750429

Dincer, B. (2011). do the shareholders really care about corporate social responsibility? *International Journal of Business and Social Science*, *2*(10), 71–76. doi:10.30845/ijbss

Dinh, T., Husmann, A., & Melloni, G. (2021). The role of non-financial performance indicators and integrated reporting in achieving sustainable value creation. Publication for the committee on Economic and Monetary Affairs, Policy Department for Economic, Scientific and Quality of Life Policies, European Parliament.

Directive 2014/95/EU: European Parliament and of the Council. (2014). *Official Journal of the European Union.*

Dobbs, S., & Van Staden, C. (2016). Motivations for corporate social and environmental reporting: New Zealand evidence. *Sustainability Accounting, Management and Policy Journal*, *7*(3), 449–472. doi:10.1108/SAMPJ-08-2015-0070

Dobers, P., & Halme, M. (2009). Corporate social responsibility and developing countries. *Corporate Social Responsibility and Environmental Management*, *16*(5), 237–249. doi:10.1002/csr.212

Driscoll, M. (2004, May 2). The girls ganging up on the old boy network. *The Sunday Times News Review*, p. 9.

Drucker, P. F. (1984). The New Meaning of Corporate Social Responsibility. *California Management Review*, *26*(2), 53–63. doi:10.2307/41165066

Dumay, J., Bernardi, C., Guthrie, J., & Demartini, P. (2016). Integrated reporting: A structured literature review. *Accounting Forum*, *40*(3), 166–185. doi:10.1016/j.accfor.2016.06.001

Dumay, J., Bernardi, C., Guthrie, J., & La Torre, M. (2017). Barriers to implementing the International Integrated Reporting Framework: A contemporary academic perspective. *Meditari Accountancy Research*, *25*(4), 461–480. doi:10.1108/MEDAR-05-2017-0150

Dumay, J., & Dai, T. (2017). Integrated thinking as a cultural control? *Meditari Accountancy Research*, *25*(4), 574–604. doi:10.1108/MEDAR-07-2016-0067

Dutton, J. & Baker, W. (2007). Enabling positive social capital in organizations. *Exploring Positive Relationships at Work - Building a Theoretical and Research Foundation*, 325–345.

Eccles, R. G., Krzus, M. P., & Watson, L. A. (2012). Integrated reporting requires integrated assurance. *Effective auditing for corporates: Key developments in practice and procedures*, 161–178.

Eccles, R. G., & Krzus, M. P. (2014). *The integrated reporting movement: Meaning, momentum, motives, and materiality*. John Wiley & Sons.

Eccles, R., Krzus, M., & Serafeim, G. (2011). Market interest in nonfinancial information. *Journal of Applied Corporate Finance*, *23*(4), 113–127. doi:10.1111/j.1745-6622.2011.00357.x

Ecer, F. (2020, Feb.). Multi-Criteria Decision Making for Green Supplier Selection Using Interval Type-2 Fuzzy AHP: A Case Study of a Home Appliance Manufacturer. *Operational Research.*

Edelman, L. B., Abraham, S. E., & Erlanger, H. S. (1992). Professional construction of law: The inflated threat of wrongful discharge. *Law & Society Review*, *26*(1), 47–83. doi:10.2307/3053836

Ehrich, L. C. (1994). Mentoring and networking for women educators. *Women in Management Review*, *9*(3), 4–10. doi:10.1108/09649429410056281

Ekundayo, G., & Josiah, M. (2020). Environmental accounting disclosure: A critical examination of literature. *British Journal of Economics, Finance and Management Sciences*, *17*(2), 34–45.

El Baz, J., Laguir, I., Marais, M., & Staglianò, R. (2016). Influence of National Institutions on the Corporate Social Responsibility Practices of Small- and Medium-sized Enterprises in the Food-processing Industry: Differences Between France and Morocco. *Journal of Business Ethics*, *134*(1), 117–133. doi:10.100710551-014-2417-z

Elbarkouky, M. M. G., & Abdelazeem, G. (2013). A Green Supply Chain Assessment for Construction Projects in Developing Countries. *WIT Transactions on Ecology and the Environment*, *179*, 1331–1341. doi:10.2495/SC131132

Elmadhoun, I. R. I., & Reddy, G. N. (2019). The role of carbon accounting in managing the climate change risks: Literature review. *International Journal of Scientific Research*, *9*(11), 309–315. doi:10.21275/SR20921154637

Elmuti, D., Minnis, W., & Abebe, M. (2005). Does education have a role in developing leadership skills? *Management Decision*, *43*(7/8), 1018–1031. doi:10.1108/00251740510610017

Equal Opportunity in the Workplace Agency & Catalyst. (2003). *Australian Census of Women Executive Managers and Women Board Directors*. EOWA.

Ernst & Young. (2016). *The road to reliable nonfinancial reporting*. Author.

Ernst & Young. (2017). *EU Directive: Disclosure of non-financial information and diversity information*. https://www.ey.com/Publication/vwLUAssets/Law_NFI_reporting/$FILE/Law%20NFI%20reporting.pdf

Esch, M., Schulze, M., & Wald, A. (2019). The dynamics of financial information and non-financial environmental, social and governance information in the strategic decision-making process. *Journal of Strategy and Management*. doi:10.1108/JSMA05-2018-0043

Espaliú Berdud, C. (2016). A reflection on the participation of individuals in the creation of European Law through the European Citizens´ Initiative and its scope in International Law. *Cuadernos Europeos de Deusto*, *54*, 181–202.

European Commission. (2002). Regulation (EC) No. 1606/2002 of the European Parliament and of the Council of 19 July 2002 on the application of International Accounting Standards. *Official Journal of the European Communities*. L243/1 (11.9.2002). Retrieved from https://eur-lex.europa.eu/legal-content/PT/TXT/?uri=celex%3A32002R1606

European Commission. (2003). Commission Recommendation of 6 May 2003 concerning the definition of micro, small and medium-sized enterprises. *Official Journal of the European Communities*. L124 (20.5.2003). Retrieved from https://eur-lex.europa.eu/legal-content/EN/TXT/?uri=CELEX:32003H0361

European Commission. (2011). *EMAS - Factsheet.* Retrieved October 2020, from https://ec.europa.eu/environment/emas/pdf/factsheets/EMASiso14001_high.pdf

European Commission. (2017). Guidelines on non-financial reporting (methodology for reporting non-financial information). *Official Journal of the European Union. COM 2017/C 215/01.*

European Commission. (2018). *Case Study – Involving Employees in Implementing EMAS.* Retrieved October 2020, from https://ec.europa.eu/environment/emas/pdf/other/Case%20study_Involving%20employees.pdf

European Commission. (2020). *Official Statistics of the European EMAS Helpdesk - Organisations and Sites per Country (September 2020).* Retrieved October 2020, from https://ec.europa.eu/environment/emas/pdf/statistics/EMASStatistics_September%202020_final.pdf

European Commission. (2020a). *Communication from the commission to the European parliament, the council, the European economic and social committee and the committee of the regions: An SME Strategy for a sustainable and digital Europe.* European Commission. Retrieved from https://ec.europa.eu/info/sites/default/files/communication-sme-strategy-march-2020_en.pdf

European Commission. (2020b). *Summary report on the public consultation on the Evaluation and Review of the ePrivacy Directive.* Retrieved from https://ec.europa.eu/info/law/better-regulation/have-your-say/initiatives/12129-Revision-of-Non-Financial-Reporting-Directive/public-consultation_en

European Commission. (n.d.a). *Policy - EMAS Regulation.* Retrieved October 2020, from https://ec.europa.eu/environment/emas/emas_publications/policy_en.htm

European Commission. (n.d.b). *SME Corner.* Retrieved October 2020, from https://ec.europa.eu/environment/emas/join_emas/what_if_i_am_an_sme_en.htm

European Financial Reporting Advisory Group. (2021a). *Current non-financial reporting formats and practices. Efrag, (February), Stream A6 Assessment Report.* Retrieved from https://www.efrag.org/Assets/Download?assetUrl=%2Fsites%2Fwebpublishing%2FSiteAssets%2FEFRAG%2520PTF-NFRS_A6_FINAL.pdf

European Financial Reporting Advisory Group. (2021b). *Proposals for a Relevant and Dynamic Eu Sustainability Reporting Standard-Setting.* Retrieved from https://www.efrag.org/Assets/Download?assetUrl=%2Fsites%2Fwebpublishing%2FSiteAssets%2FEFRAG%2520PTF-NFRS_MAIN_REPORT.pdf

European Investment Bank Group. (2020). *Small and medium enterprises overview.* Retrieved from https://www.eib.org/attachments/thematic/small_and_medium_enterprises_overview_2020_en.pdf

European Parliament. (2013). Directive 2013/34/EU amending Directive 2006/43/EC and repealing Council Directives 78/660/EEC and 83/349/EEC on the annual financial statements, consolidated financial statements and related reports of certain types of undertakings. *Official Journal of the European Union, 2013*(June), L 182/19-75. Retrieved from https://eur-lex.europa.eu/legal-content/EN/TXT/?uri=celex:32013L0034

European Parliament. (2014). Directive 2014/95/EU of the European Parliament and of the Council of 22 October 2014 amending Directive 2013/34/EU as regards disclosure of non-financial and diversity information by certain large undertaking and groups. *Official Journal of the European Union, 2014*(April), 1–9. Retrieved from https://eur-lex.europa.eu/legal-content/EN/TXT/?uri=CELEX:32014L0095

European Parliament. (2021). Proposal for a Directive of the European Parliament and of the Council amending Directive 2013/34/EU, Directive 2004/109/EC, Directive 2006/43/EC and Regulation (EU) No 537/2014, as regards corporate sustainability reporting. *Official Journal of the European Union*, (April), 1–65. Retrieved from https://eur-lex.europa.eu/legal-content/EN/TXT/?uri=CELEX:52016PC0198

Europeia, C. (2001). Livro Verde - Promover um quadro europeu para a responsabilidade social das empresas. *Comissão Europeia, COM 366.*

Eurostat. (2017). *Which Member States have the largest share of EU's GDP?* https://ec.europa.eu/eurostat/web/products-eurostat-news/-/DDN-20180511-1?inheritRedirect=true

Eurostat. (2021). *Environmental protection expenditure accounts (env_ac_epea).* Retrieved from https://ec.europa.eu/eurostat/cache/metadata/en/env_ac_epea_esms.htm

Fabrigar, L., Wegener, D., Maccalum, R., & Strahan, E. (1999). Evaluating the use of exploratory factor analysis in ppsychological research. *Psychological Methods, 4*(3), 272–299. doi:10.1037/1082-989X.4.3.272

Farinha, C., Caeiro, S., & Ulisses, A. (2019). Sustainability strategies in Portuguese Higher Education Institutions: Commitments and practices from internal insights. *Sustainability, 11*(11), 3227. doi:10.3390u11113227

Farrukh, M., Meng, F., Raza, A., & Tahir, M. S. (2020). Twenty-seven years of Sustainable Development Journal: A bibliometric analysis. *Sustainable Development, 28*(6), 1725–1737. doi:10.1002d.2120

Fasan, M. (2013). Annual Reports, Sustainability Reports and Integrated Reports: Trends in Corporate Disclosure. In C. Busco, M. L. Frigo, A. Riccaboni, & P. Quattrone (Eds.), *Integrated Reporting: Concepts and Cases that Redefine Corporate Accountability* (pp. 41–57). Springer International Publishing Switzerland. doi:10.1007/978-3-319-02168-3_3

FASB - Financial Accounting Standards Board. (1980). Qualitative characteristics of accounting information. *Statement of Financial Accounting Concepts, 2,* 1–60.

FEE - Federation of European Accountants. (2016). *EU Directive on disclosure of nonfinancial and diversity information Achieving good quality and consistent reporting.* Author.

Feng, T., Cummings, L., & Tweedie, D. (2017). Exploring integrated thinking in integrated reporting – an exploratory study in Australia. *Journal of Intellectual Capital, 18*(2), 330–353. doi:10.1108/JIC-06-2016-0068

Fernandes, F. (2019). *Informação Financeira e Não Financeira* (Master Thesis). IPC.

Fernandez-Feijoo, B., Romero, S., & Ruiz, S. (2014). Effect of Stakeholders' Pressure on Transparency of Sustainability Reports within the GRI Framework. *Journal of Business Ethics, 122*(1), 53–63. doi:10.100710551-013-1748-5

Fernandez-Feijoo, B., Romero, S., & Ruiz, S. (2015). Multilevel Approach to Sustainability Report Assurance Decisions. *Australian Accounting Review, 25*(75), 346–358. Advance online publication. doi:10.1111/auar.12104

Ferreira, C. (2004). Environmental accounting: The Portuguese case. *Management of Environmental Quality, 15*(6), 561–573. doi:10.1108/14777830410560665

Ferreira, R. Jr, & Rita, L. (2020). Impactos da Covid-19 na Economia: Limites, desafios e políticas [Impacts of Covid-19 on the Economy: limits, challenges and policies]. *Cadernos de Prospecção, 13*(2), 459–476.

FIBRAS. (2020). *Finanças Sustentáveis: Um Panorama.* Projeto Finanças Brasileiras Sustentáveis.

Field, A. (2011). *Descobrindo a estatística usando o SPSS* (2ª Edição). Artmed Editora.

Fijałkowska, J., Zyznarska-Dworczak, B., & Garsztka, P. (2018). *Corporate social-environmental performance versus financial performance of banks in Central and Eastern European countries.* doi:10.3390/su10030772

Fineman, S. (1999). Emotion and organization. In S. Clegg, C. Hardy, & W. Nord (Eds.), *Studying organisation: Theory and method* (pp. 288–310). Sage. doi:10.4135/9781446218556.n11

Flanders, M. L. (1994). *Breakthrough: The Career Woman's Guide to Shattering the Glass Ceiling*. Paul Chapman Publishing.

Fleaca, E., Fleaca, B., & Maiduc, S. (2018). Aligning strategy with Sustainable Development Goals (SDGs): Process scoping diagram for entrepreneurial Higher Education Institutions (HEIs). *Sustainability*, *10*(4), 1032. doi:10.3390u10041032

Fonseca, A., Macdonald, A., Dandy, E., & Valenti, P. (2011). The state of sustainability reporting at Canadian universities. *International Journal of Sustainability in Higher Education*, *12*(1), 22–40. doi:10.1108/14676371111098285

Fonseca, L., & Carvalho, F. (2019). T*he Reporting of SDGs by Quality. Environmental, and Occupational Health and Safety-Certified Organizations*, *11*(20), 5797. doi:10.3390u11205797

Forret, N. L., & Dougherty, T. W. (2001). Correlates of networking behavior for managerial and professional employees. *Group & Organization Management*, *26*(3), 283–311. doi:10.1177/1059601101263004

Forte, C. (2020). O papel da AEF-Brasil na execução da estratégia nacional. In Estratégia Nacional de Educação Financeira (ENEF): em busca de um Brasil melhor [The role of AEF-Brasil in the execution of the national strategy]. Riemma.

Fredrickson, B. (2001). The role of positive emotions in positive psychology: The broaden-and-built theory of positive emotions. *The American Psychologist*, *56*(3), 218–226. doi:10.1037/0003-066X.56.3.218 PMID:11315248

Fredrickson, B. (2003). Positive emotions and upward spirals in organizational settings. In K. Cameron, J. Dutton, & R. Quinn (Eds.), *Positive organizational scholarship* (pp. 163–175). Barrett-Koehler.

Friedman, R. E. (1970). The Social Responsibility of Business is to increase its profits. *New York Times Magazine*.

Fusco, F., & Ricci, P. (2019). Qual é o estoque da situação? Uma análise bibliométrica sobre a pesquisa em contabilidade socioambiental no setor público. *Revista Internacional de Gestão do Setor Público*, 1-38.

Gamage, P., & Sciulli, N. (2016). Sustainability Reporting by Australian Universities. *Australian Journal of Public Administration*, *76*(2), 187–203. doi:10.1111/1467-8500.12215

García-Machado, J. J. (2018). The latest streams in finance research: An updated bibliometric mapping based on co-occurrence data. In *Forum Scientiae Economia* (pp. 7–25). Wydawnictwo Naukowe Akademii WSB.

García-Sánchez, I.-M. (2021). Corporate social reporting and assurance: The state of the art [Información social corporativa y aseguramiento: El estado de la cuestión]. *Revista de Contabilidad-Spanish Accounting Review*, *24*(2), 241–269. doi:10.6018/rcsar.409441

Garde Sánchez, R., Rodriguéz Bolívar, M. P. & López-Hernández, A. M. (2013). Online disclosure of university social responsibility: a comparative study of public and private US universities. *Environmental Education Research, 19*(6), 709–746.

Garriga, E., & Melé, D. (2004). Corporate Social Responsibility Theories: Mapping the Territory. *Journal of Business Ethics*, *53*(1), 51–71. doi:10.1023/B:BUSI.0000039399.90587.34

Gavin, J. H., & Mason, R. O. (2004). The virtuous organization: The value of happiness in the workplace. *Organizational Dynamics, 33*(4), 379–392. doi:10.1016/j.orgdyn.2004.09.005

General Assembly. (1992). Report of the United Nations Conference on Environment and Development. In *Proceedings of the Conference* (Vol. 2). United Nations. Retrieved October 2020, from https://www.un.org/en/development/desa/population/migration/generalassembly/docs/globalcompact/A_CONF.151_26_Vol.I_Declaration.pdf

Generation Foundation. (2013). *Stranded Carbon Assets*. Disponível em: http://genfound.org/media/pdf-generationfoundation-stranded-carbon-assets-v1.pdf

Ghosh, R., & Wolf, S. (2021). Hybrid governance and performances of environmental accounting. *Journal of Environmental Management, 284*(111995), 1–11. doi:10.1016/j.jenvman.2021.111995 PMID:33515840

Gibbs, G. (2009). *Análise de dados qualitativos: coleção pesquisa qualitativa*. Bookman.

Global Reporting Initiative. (n.d.). *About GRI.* Retrieved from https://www.globalreporting.org/about-gri/

Gomes, S. F., Eugénio, T. C. P., & Branco, M. C. (2015). Sustainability reporting and assurance in Portugal. *Corporate Governance: The International Journal of Business in Society, 15*(3), 281–292. doi:10.1108/CG-07-2013-0097

González, P., Sarkis, J., & Adenso-Díaz, B. (2008). Environmental Management System Certification and Its Influence on Corporate Practices: Evidence from the Automotive Industry. *International Journal of Operations & Production Management, 28*(11), 1021–1041. doi:10.1108/01443570810910179

Goodman, K. W. (2015). *Anti-money laundering: an often-overlooked cornerstone of effective compliance*. U.S. S.E.C. https://www.sec.gov/news/speech/anti-money-laundering-an-often-overlooked-cornerstone.html

Goodwin, K., Wiedmann, T., Chen, G., & Teh, S. H. (2021). Benchmarking urban performance against absolute measures of sustainability – A review. *Journal of Cleaner Production, 314*, 128020. doi:10.1016/j.jclepro.2021.128020

GPIF. (2019). *Climate Related Portfolio Risk Assessment. Trucost Analysis Supporting GPIF's Disclosure in line with TCFD Recommendations*. GPIF.

Graafland, J. J. (2018). Ecological Impacts of the ISO14001 Certification of Small and Medium Sized Enterprises in Europe and the Mediating Role of Networks. *Journal of Cleaner Production, 174*(February), 273–282. doi:10.1016/j.jclepro.2017.10.322

Granleese, J. (2004). Occupational pressures in banking: Gender differences. *Women in Management Review, 19*(4), 219–225. doi:10.1108/09649420410541290

Gray, B., Purdy, J. M., & Ansari, S. (2015). From interactions to institutions: Microprocesses of framing and mechanisms for the structuring of institutional fields. *Academy of Management Review, 40*(1), 115–143. doi:10.5465/amr.2013.0299

Gray, R. (1993). Current practice in environmental reporting. *Social and Environmental Accounting, 13*(1), 6–8.

Gray, R., Brennan, A., & Malpas, J. (2014). New accounts: Towards a reframing of social accounting. *Accounting Forum, 38*(4), 258–273.

Gray, R., Owen, D., & Adams, C. (1996). *Accounting & accountability: changes and challenges in corporate social and environmental reporting*. Prentice Hall.

Griffin, M. (2012). *Contabilidade e finanças* [Accounting and Finance. Fundamentals Series]. Série Fundamentos. Saraiva. https://app.minhabiblioteca.com.br/#/books/9788502180161/

Groenfeldt, T. (2017). Customer trust is the key for banks to build long-term relationships. *Forbes*. https://www.forbes.com/sites/tomgroenfeldt/2017/06/16/customer-trust-is-the-key-for-banks-to-build-long-term-relationships/#717e18fb3060

Gros, D., & Schoenmaker, D. (2016). *Too late, too sudden: Transition to a low-carbon economy and systemic risk*. Reports of the Advisory Scientific Committee. European Systemic Risk Board.

Gulluscio, C., Puntillo, P., Luciani, V., & Huisingh, D. (2020). Climate change accounting and reporting: A systematic literature review. *Sustainability, 12*(5455), 1–31. doi:10.3390u12135455

Gunarathne, N., & Senaratne, S. (2017). Diffusion of integrated reporting in an emerging. *Managerial Auditing Journal, 32*(4–5), 524–548. doi:10.1108/MAJ-01-2016-1309

Gunawan, J., Permatasari, P., & Tilt, C. (2019). Sustainable Development Goal disclosures: Do they support responsible consumption and production? *Journal of Cleaner Production*, *246*(10), 118989.

Gunningham, N., & Sinclair, D. (2017). Smart regulation. Regulatory theory: Foundations and applications, 133-148.

Gunter, R., & Stambach, A. (2005). Differences in men and women scientists' perceptions of workplace climate. *Journal of Women and Minorities in Science and Engineering*, *11*(1), 97–115. doi:10.1615/JWomenMinorScienEng.v11.i1.60

Gupta, S., Soni, U., & Kumar, G. (2019). Green Supplier Selection Using Multi-Criterion Decision Making under Fuzzy Environment: A Case Study in Automotive Industry. *Computers & Industrial Engineering*, *136*(October), 663–680. doi:10.1016/j.cie.2019.07.038

Gurarda, S. (2015). Environmental Management Accounting. In U. Akkucuk (Ed.), *Handbook of Research on Developing Sustainable Value in Economics, Finance, and Marketing* (pp. 278–296). IGI Global. doi:10.4018/978-1-4666-6635-1.ch016

Guthrie, J., Manes-Rossi, F., & Orelli, R. L. (2017). Integrated reporting and integrated thinking in Italian public sector organizations. *Meditari Accountancy Research*, *25*(4), 553–573. doi:10.1108/MEDAR-06-2017-0155

Hair, J. R. J., Anderson, R., Tatham, R., & Black, W. (2005). *Análise Multivariada de Dados*. Bookman.

Hakim, C. (2004). *Key Issues in Women's Work: Female Diversity and the Polarisation of Women's Employment*. Glasshouse Press.

Hambrick, D. C. (2007). Upper Echelons Theory: An Update. *Academy of Management Review*, *32*(2), 334–343. doi:10.5465/amr.2007.24345254

Harrigan, K. R. (1981). Numbers and positions of women elected to corporate boards. *Academy of Management Journal*, *24*(3), 619–625.

Harrison, J. L., & Lamberton, G. (2007). Who's doing environmental accounting? A survey of NSW dairy farms. *Rural Society*, *17*(2), 182–195. doi:10.5172/rsj.351.17.2.182

He, R., Luo, L., Shamsuddin, A., & Tang, Q. (2021). Corporate carbon accounting: A literature review of carbon accounting research from the Kyoto Protocol to the Paris Agreement. *Accounting & Finance*, 1-38. doi:10.1111/acfi.12789

Helfaya, A., Whittington, M., & Alawattage, C. (2019). Exploring the quality of corporate environmental reporting: Surveying preparers' and users' perceptions. *Accounting, Auditing & Accountability Journal*, *32*(1), 163–193. doi:10.1108/AAAJ-04-2015-2023

Herath, G. (2005). Sustainable development and environmental accounting: The challenge to the economics and accounting profession. *International Journal of Social Economics*, *32*(12), 1035–1050. doi:10.1108/03068290510630999

Higgins, C., Stubbs, W., Tweedie, D., & McCallum, G. (2019). Journey or toolbox? Integrated reporting and processes of organisational change. *Accounting, Auditing & Accountability Journal*, *32*(6), 1662–1689. doi:10.1108/AAAJ-10-2018-3696

Hill, M., & Hill, A. (2000). *Investigação por questionários*. Edições Sílabo.

Hillman, A. J., Cannella, A. A. Jr, & Harris, I. C. (2002). Women and racial minorities in the boardroom: How do directors differ? *Journal of Management*, *28*(6), 747–763. doi:10.1177/014920630202800603

Hoang, H., & Phang, S. Y. (2021). How does combined assurance affect the reliability of integrated reports and investors' judgments? *European Accounting Review*, *30*(1), 175–195. doi:10.1080/09638180.2020.1745659

Hojnik, J. (2017). Environmental Corporate Reporting under eu Law: Historic Achievement or Just a Moderate Step Forward? *Journal for European Environmental & Planning Law*, *14*(1), 41–75. doi:10.1163/18760104-01401004

Holt, D. (1998). The Perceived Benefits of an Environmental Management Standard. *Business Process Management Journal*, *4*(3), 204–213. doi:10.1108/14637159810224313

Hoque, M. (2017). Why a Company Should Adopt Integrated Reporting? *International Journal of Economics and Financial Issues*, *7*(1), 241–248.

Hossain, M. M. (2010). Environmental accounting challenges of selected manufacturing enterprises in Bangladesh. *Open Journal of Business and Management*, *7*(02), 709–727. doi:10.4236/ojbm.2019.72048

Houdet, J., Ding, H., Quetier, F., Addison, P., & Deshmukh, P. (2020). Adapting double-entry bookkeeping to renewable natural capital: An application to corporate net biodiversity impact accounting and disclosure. *Ecosystem Services*, *45*(101104), 1–14. doi:10.1016/j.ecoser.2020.101104

Hsieh, H., & Shannon, S. (2005). Three Approaches to Qualitative Content Analysis. *Qualitative Health Research*, *15*(9), 1277–1288. doi:10.1177/1049732305276687 PMID:16204405

Huber, S., & Bassen, A. (2018). Towards a sustainability reporting guideline in higher education. *International Journal of Sustainability in Higher Education*, *19*(2), 218–232. doi:10.1108/IJSHE-06-2016-0108

Huerta-Riveros, P., & Gaete-Feres, H. (2017). Responsabilidad social universitaria a través de los reportes de sostenibilidad del Global Reporting Initiative: Experiencia de una universidad pública. *Revista Iberoamericana de Educación Superior*, *8*(23), 120–137. doi:10.22201/iisue.20072872e.2017.23.3014

Hülsmann, S., & Ardakanian, R. (Eds.). (2018). *Managing water, soil, and waste resources to achieve sustainable development goals: monitoring and implementation of integrated resources management*. Springer. doi:10.1007/978-3-319-75163-4

HummelK. (2019). *Reporting on the Sustainable Development Goals – early evidence from Europe*. Available at https://ssrn.com/abstract=3411017, June 2019. doi:10.2139/ssrn.3411017

IASB. (2008). *Exposure draft on an improved conceptual framework for financial reporting: the objective of financial reporting and qualitative characteristics of decision-useful financial reporting information*. IASB.

IEA. (2013). Redrawing The Energy Climate Map. World Energy Outlook Special Report.

IFRS. (2021). *Global sustainability disclosure standards for the financial markets*. Retrieved from https://www.ifrs.org/news-and-events/news/2021/11/global-sustainability-disclosure-standards-for-the-financial-markets/

Instituto Brasileiro de Governança Corporativa. (2009). *Código das melhores práticas de governança corporativa*. https://www.ibgc.org.br/userfiles/files/Codigo_Final_4a_Edicao.pdf

Intergovernmental Panel on Climate Change. (2018). *Global warming of 1,5oC: Summary for policymaker*s. IPCC.

International Auditing and Assurance Standards Board. (2020). *Revised International Standard on Assurance Engagements (ISAE) 3000, Assurance Engagements Other Than Audits or Reviews of Historical Financial Information*. https://www.iaasb.org/publications/international-standard-assurance-engagements-isae-3000-revised-assurance-engagements-other-audits-or-0

International Federation of Accountants (IFAC). (2015). *Creating Value with Integrated Thinking. The Role of Professional Accountants*. Recovered in: https://www.ifac.org/system/files/publications/files/Creating-value-with-iIntegrated-thinking-role-of-accountants.pdf

International Integrated Reporting Council (IIRC). (2011). *Towards Integrated Reporting. Communicating Value in the 21st Century*. Available at https://integratedreporting.org/wp-content/uploads/2011/09/IR-Discussion-Paper-2011_spreads.pdf

International Integrated Reporting Council (IIRC). (2013). *The International <IR> Framework*. Available in: https://integratedreporting.org/wp-content/uploads/2015/03/13-12-08-THE-INTERNATIONAL-IR-FRAMEWORK-2-1.pdf

International Integrated Reporting Council (IIRC). (2013a). *Capitals Background paper for <IR>*. Recuperado em: https://integratedreporting.org/wp-content/uploads/2013/03/IR-Background-Paper-Capitals.pdf

International Integrated Reporting Council (IIRC). (2021). *International <IR> Framework*. IIRC.

International Integrated Reporting Council. (2014a). *A Estrutura Internacional para Relato Integrado*. http://integratedreporting.org/resource/international-ir-framework

International Integrated Reporting Council. (2014b). *Assurance on <IR>: An introduction to the discussion*. http://www.theiirc.org/resources-2/assurance/

International Integrated Reporting Council. (2014c). *Assurance on <IR>: An exploration of issues*. http://www.theiirc.org/resources-2/assurance/

International Integrated Reporting Council. (2014d). *Responses to assurance on <IR>*. http://integratedreporting.org/responses-to-assuranceon-ir/

International Integrated Reporting Council. (n.d.). *The International <IR> Framework*. Retrieved from https://www.integratedreporting.org/wp-content/uploads/2013/12/13-12-08-THE-INTERNATIONAL-IR-FRAMEWORK-2-1.pdf

International Labour Office. (2004). *Breaking Through the Glass-ceiling*. International Labour Office.

International Sustainability Standards Board. (n.d.). *International Sustainability Standards Board*. Retrieved from https://www.ifrs.org/groups/international-sustainability-standards-board/

INTOSAI WGEA. (n.d.). *INTOSAI WGEA research project - Environment and sustainability reporting: Extended outline*. Retrieved from https://www.wgea.org/media/1275/a_2_sc11-sustainability-reporting-eo.pdf

Iraldo, F., Testa, F., & Frey, M. (2009). Is an Environmental Management System Able to Influence Environmental and Competitive Performance? The Case of the Eco-Management and Audit Scheme (EMAS) in the European Union. *Journal of Cleaner Production*, *17*(16), 1444–1452. doi:10.1016/j.jclepro.2009.05.013

Islam, M. S., Tseng, M. L., Karia, N., & Lee, C. H. (2018). Assessing Green Supply Chain Practices in Bangladesh Using Fuzzy Importance and Performance Approach. *Resources, Conservation and Recycling*, *131*(April), 134–145. doi:10.1016/j.resconrec.2017.12.015

ISO. (2015). *Introduction to ISO 14001:2015*. Retrieved October 2020, from https://www.iso.org/files/live/sites/isoorg/files/store/en/PUB100371.pdf

ISO. (2021). *1. ISO Survey 2019 Results - Number of Certificates and Sites per Country and the Number of Sector Overall*. Retrieved October 2020, from https://isotc.iso.org/livelink/livelink?func=ll&objId=21412354&objAction=Open&nexturl=%2Flivelink%2Flivelink%3Ffunc%3Dll%26objId%3D18808772%26objAction%3Dbrowse%26viewType%3D1

ISO. (n.d.). *Certification*. Retrieved October 2020, from https://www.iso.org/certification.html

Jamal, N. M., Wong, W. K., Saat, M. M., & Othman, A. (2020). A Relationship Analysis between Green Supply Chain Practices, Environmental Management Accounting and Performance. *Journal of Sustainability Science and Management*, *15*(3), 113–133.

Jiang, P., Hu, Y. C., Yen, G. F., & Tsao, S. J. (2018). Green Supplier Selection for Sustainable Development of the Automotive Industry Using Grey Decision-Making. *Sustainable Development*, *26*(6), 890–903. doi:10.1002d.1860

Johnson, M. P., & Schaltegger, S. (2016). Two Decades of Sustainability Management Tools for SMEs: How Far Have We Come? *Journal of Small Business Management*, *54*(2), 481–505. doi:10.1111/jsbm.12154

Johnson, S. G., Schnatterly, K., & Hill, A. D. (2013). Board composition beyond independence: Social capital, human capital, and demographics. *Journal of Management*, *39*(1), 232–262. doi:10.1177/0149206312463938

Jones, M. J. (2010). Accounting for the environment: Towards a theoretical perspective for environmental accounting and reporting. *Accounting Forum*, *34*(2), 123–138. doi:10.1016/j.accfor.2010.03.001

Jones, M. J., & Solomon, J. F. (2010). Social and environmental report assurance: Some interview evidence. *Accounting Forum*, *34*(1), 20–31. doi:10.1016/j.accfor.2009.11.002

Jorgensen, S., Mjos, A., & Pedersen, L. J. T. (2021). Sustainability reporting and approaches to materiality: Tensions and potential resolutions. *Sustainability Accounting, Management and Policy Journal*, 1-21. doi:10.1108/SAMPJ-01-2021-0009

Kaminski, I. (2013). *Road from Rio: The Past, the Present and the Future of ISO 14001*. Retrieved October 2020, from https://www.bsigroup.com/LocalFiles/en-GB/iso-14001/whitepapers/BSI-ISO-14001-article-The-ENDS-Report-Road-from-Rio-UK-EN.pdf

Karatzia, A. (2017). *European Citizens´ Initiative. General Court rules of the Commission's obligation to give reason for refusing to register proposals.* http://eulawanalysis.blogspot.com/2017/02/european-citizens-initiative-general.html

Kaya, D., & Koch, M. (2015). Countries adoption of the International Financial Reporting Standard for Small and Medium-sized Entities (IFRS for SMEs) - Early empirical evidence. *Accounting and Business Research*, *45*(1), 93–120. doi:10.1080/00014788.2014.969188

Kellerman, B. (2003). You've come a long way baby – and you've got miles to go. In D. B. Rhode (Ed.), *The Difference 'Difference' Makes*. Stanford University Press.

Kesner, I. F. (1988). Directors' characteristics and committee membership: An investigation of type, occupation, tenure and gender. *Academy of Management Journal*, *31*(1), 66–84.

Khan, S. (2020). Environmental accounting: One dimension of sustainable development. *Jaipuria International Journal of Management Research*, *6*(2), 38–41. doi:10.22552/jijmr/2020/v6/i2/205280

Khosroshahi, H., Azad, N., Jabbarzadeh, A., & Verma, M. (2021). Investigating the level and quality of the information in the environmental disclosure report of a corporation considering government intervention. *International Journal of Production Economics*, *235*, 1–16. doi:10.1016/j.ijpe.2021.108071

Kılıç, M., & Kuzey, C. (2018). Assessing current company reports according to the IIRC integrated reporting framework. *Meditari Accountancy Research*, *26*(2), 305–333. doi:10.1108/MEDAR-04-2017-0138

Kiliç, M., Uyar, A., & Ataman, B. (2014). Preparedness for and perception of IFRS for SMEs: Evidence from Turkey. *Accounting and Management Information Systems*, *13*(3), 492–519.

Kiliç, M., Uyar, A., & Kuzey, C. (2019). The impact of institutional ethics and accountability on voluntary assurance for integrated reporting. *Journal of Applied Accounting Research*, *21*(1), 1–18. doi:10.1108/JAAR-04-2019-0064

Knauer, A., & Serafeim, G. (2014). Attracting Long-Term Investors Through Integrated Thinking and Reporting: A Clinical Study of a Biopharmaceutical Company. *Journal of Applied Corporate Finance*, *26*(2), 57–64. doi:10.1111/jacf.12067

Kohlbacher, F. (2006). The Use of Qualitative Content Analysis in Case Study Research. *Forum Qualitative Sozialforschung / Forum: Qualitative. Social Research*, *7*(1), 1–30.

Kollman, K., & Prakash, A. (2002). EMS-Based Environmental Regimes as Club Goods: Examining Variations in Firm-Level Adoption of ISO 14001 and EMAS in U.K., U.S. and Germany. *Policy Sciences*, *35*(1), 43–67. doi:10.1023/A:1016071810725

Konovsky, M., & Organ, D. (1996). Dispositional and contextual determinants of organizational citizienship behaviors. *Journal of Organizational Behavior*, *17*(3), 253–256. doi:10.1002/(SICI)1099-1379(199605)17:3<253::AID-JOB747>3.0.CO;2-Q

KPMG (2018). *How to report on the SDGs. What good looks like and why it matters.* kpmg.com/sdgreporting.

KPMG. (2015). *KPMG survey of corporate responsibility reporting 2015.* KPMG.

KPMG. (2017). *The road ahead. The KPMG survey of corporate responsibility reporting 2017.* https://home.kpmg/content/dam/kpmg/campaigns/csr/pdf/CSR_Reporting_2017.pdf

KPMG. (2019). *Survey of Integrated Reporting in Japan 2019.* https://www.integratedreporting.org/wp-content/uploads/2020/05/jp-en-integrated-reporting2019.pdf

Krawczyk, P. (2021). Non-Financial Reporting—Standardization Options for SME Sector. *Journal Risk Financial Management*, *14*(9), 417. doi:10.3390/jrfm14090417

Krippendorff, K. (2013). *Content Analysis. An Introduction to Its Methodology* (3rd ed.). Sage Publications.

Kumar, K., Kumari, R., & Kumar, R. (2021). The state of corporate sustainability reporting in India: Evidence from environmentally sensitive industries. *Business and Society Review*, *126*(4), 1–26. doi:10.1111/basr.12247

Lai, A., Melloni, G., & Stacchezzini, R. (2018). Integrated reporting and narrative accountability: The role of preparers. *Accounting, Auditing & Accountability Journal*, *31*(5), 1381–1405. doi:10.1108/AAAJ-08-2016-2674

Laine, M., Jarvinen, J. T., Hyvönen, T., & Kantola, H. (2017). Ambiguity of financial environmental information: A case study of a Finnish energy company. *Accounting, Auditing & Accountability Journal*, *30*(3), 593–619. doi:10.1108/AAAJ-02-2015-1961

Larrán Jorge, M., Javier Andrades, F., & Madueño Jesús, H. (2018). An analysis of university sustainability reports from the GRI database: An examination of influential variables. *Journal of Environmental Planning and Management*, *62*(3), 1–26.

Law Insider Dictionary. (2021). *Environmental Remediation Costs definition.* Retrieved from https://www.lawinsider.com/dictionary/environmental-remediation-costs

Leal, C., & Nascimento, J. (2012). Planejamento financeiro pessoal [Personal financial planning]. *Revista de Ciências Gerenciais, 15*(22), 163-186. https://revista.pgsskroton.com/index.php/rcger/article/view/2101

Leblanc, R., & Gillies, J. (2005). *Inside the Boardroom: How Boards Really Work and the Coming Revolution in Corporate Governance*. John Wiley & Sons.

Lee, K. H., & Kim, J. W. (2009). Current Status of CSR in the Realm of Supply Management: The Case of the Korean Electronics Industry. Supply Chain Management: An International Journal, 14(2), 138–48.

Lee, S. M., Noh, Y., Choi, D., & Rha, J. S. (2017). Environmental Policy Performances for Sustainable Development: From the Perspective of ISO 14001 Certification: The Financial Effects of ISO 14001. *Corporate Social Responsibility and Environmental Management*, *24*(2), 108–120. doi:10.1002/csr.1395

Leibel, E., Hallett, T., & Bechky, B. A. (2018). Meaning at the Source: The Dynamics of Field Formation in Institutional Research. *The Academy of Management Annals*, *12*(1), 154–177. doi:10.5465/annals.2016.0035

Leocádio, M. F. C. (2017). *Relato Integrado em Portugal* (Master Thesis). Economy Faculty of Oporto, Portugal.

Leung, T. C. (2021). Environmental Accounting and Reporting Practices in Asian Countries. In *Routledge Handbook of Environmental Accounting*. Routledge. doi:10.4324/9780367152369-24

Lievrouw, L. A. (1989). The invisible college reconsidered: Bibliometrics and the development of scientific communication theory. *Communication Research*, *16*(5), 615–628. doi:10.1177/009365089016005004

Liff, S., & Ward, K. (2001). Distorted views through the glass ceiling: The construction of women's understandings of promotion and senior management positions. *Gender, Work and Organization*, *8*(1), 19–36. doi:10.1111/1468-0432.00120

Likert, R. (1932). A technique for the measurement of attitudes. *Archives of Psychology*, *22*(140), 1-55. https://legacy.voteview.com/pdf/Likert_1932.pdf

Linehan, M. (2001). Networking for female managers' career development: Empirical evidence. *Journal of Management Development*, *20*(10), 823–829. doi:10.1108/EUM0000000006237

Litjens, R., Bissessur, S., Langendijk, H., & Vergoossen, R. (2012). How Do Preparers Perceive Costs and Benefits of IFRS for SMEs? Empirical Evidence from the Netherlands. *Accounting in Europe*, *9*(2), 227–250. doi:10.1080/17449480.2012.720875

Liu, G., Yin, X., Pengue, W., Benetto, E., Huisingh, D., Schnitzer, H., Wang, Y., & Casazza, M. (2018). Environmental accounting: In between raw data and information use for management practices. *Journal of Cleaner Production*, *197*, 1056–1068. doi:10.1016/j.jclepro.2018.06.194

Liu, Z., & Liu, M. (2021). Quality evaluation of enterprise environmental accounting information disclosure based on projection pursuit model. *Journal of Cleaner Production*, *279*, 1–10. doi:10.1016/j.jclepro.2020.123679

Llanos-Herrera, G., & Merigo, J. M. (2018). Overview of brand personality research with bibliometric indicators. *Kybernetes*, *48*(3), 546–569. doi:10.1108/K-02-2018-0051

Lodhia, S. (2014). Exploring the Transition to Integrated Reporting Through a Practice Lens: An Australian Customer Owned Bank Perspective. *Journal of Business Ethics*, *129*(3), 585–598. doi:10.100710551-014-2194-8

Lopes, S. M. S. (2018). *Relato Integrado no Sector das Águas Minerais Naturais em Portugal: Relação ODS, Métricas e Estrutura* (Master Thesis). Polythetic Institute of Leiria, Portugal.

Lorsch, J. W., & MacIver, E. (1989). *Pawns or Potentates: The Reality of America's Corporate Boards*. Harvard Business School Press.

Lourenço, H. (2017). *A divulgação de informações não financeiras no âmbito da Diretiva 2014/95/UE: O atual grau de cumprimento das empresas cotadas em Portugal.* Tese de Mestrado. Universidade do Porto.

Lourenço, I., Branco, M., Curto, J., & Eugénio, T. (2012). How Does the Market Value Corporate Sustainability Performance? *Journal of Business Ethics*, *108*(4), 417–428. doi:10.100710551-011-1102-8

Lozano, R. (2011). The state of sustainability reporting in universities. *International Journal of Sustainability in Higher Education*, *12*(1), 67–78. doi:10.1108/14676371111098311

Lozano, R., Llobet, J., & Tideswell, G. (2013a). The process of assessing and reporting sustainability at universities: Preparing the report of the University of Leeds. *Revista Internacional de Tecnología, Sostenibilidad y Humanismo. Desembre*, *8*, 85–112.

Luthans, F. (2002). Positive organizational behaviour: Developing and managing psychological strengths. *The Academy of Management Executive*, *16*(1), 57–72. doi:10.5465/ame.2002.6640181

Machado-da-Silva, C. L., Fonseca, V. S., & Crubellate, J. M. (2005). Estrutura, agência e interpretação: Elementos para uma abordagem recursiva do processo de institucionalização. *Revista de Administração Contemporânea*, *9*(1), 9–39. doi:10.1590/S1415-65552005000500002

Machado, M. C., Vivaldini, M., & de Oliveira, O. J. (2020). Production and supply-chain as the basis for SMEs' environmental management development: A systematic literature review. *Journal of Cleaner Production*, *273*, 123141. doi:10.1016/j.jclepro.2020.123141

Macias, H. A., & Farfan-Lievano, A. (2017). Integrated reporting as a strategy for firm growth: Multiple case study in Colombia. *Meditari Accountancy Research*, *25*(4), 605–628. Advance online publication. doi:10.1108/MEDAR-11-2016-0099

MacKenzie, S., Podsakoff, P., & Fetter, R. (1991). Organizational citizienship behavior and objective productivity as determinants of managerial on evalutions of salespersons' performance. *Organizational Behavior and Human Decision Processes*, *50*(1), 123–150. doi:10.1016/0749-5978(91)90037-T

MacLean, R., & Rebernak, R. (2007). Closing the credibility gap: The challenges of corporate responsibility reporting. *Environmental Quality Management*, *16*(4), 1–6. doi:10.1002/tqem.20137

Maniora, J. (2017). Is Integrated Reporting Really the Superior Mechanism for the Integration of Ethics into the Core Business Model? An Empirical Analysis. *Journal of Business Ethics*, *140*(4), 755–786. doi:10.100710551-015-2874-z

Marion, A. (2019). *Coaching financeiro* [Financial Coaching]. Atlas.

Marouen, A. (2015). Environmental Management System: Environmental Impacts and Productivity. *International Journal of Business and Management*, *10*(11), 107. doi:10.5539/ijbm.v10n11p107

Maroun, W. (2017). Assuring the integrated report: Insights and recommendations from auditors and preparers. *The British Accounting Review*, *49*(3), 329–346. doi:10.1016/j.bar.2017.03.003

Maroun, W. (2018). Modifying assurance practices to meet the needs of integrated reporting: The case for 'interpretive assurance'. *Accounting, Auditing & Accountability Journal*, *31*(2), 400–427. doi:10.1108/AAAJ-10-2016-2732

Maroun, W. (2019). Does external assurance contribute to higher quality integrated reports? *Journal of Accounting and Public Policy*, *38*(4), 1–23. doi:10.1016/j.jaccpubpol.2019.06.002

Maroun, W. (2019b). Exploring the rationale for integrated report assurance. *Accounting, Auditing & Accountability Journal*, *32*(6), 1826–1854. doi:10.1108/AAAJ-04-2018-3463

Marques, E., & Correia, J. N. (2016). *Gestão financeira familiar - como as empresas fazem* [Family financial management - how companies do it]. Alta Books.

Martin, E., & Moneva, J. (2018). Do as I say, not as I do: Drivers of sustainability reporting Spanish universities. *Sociology International Journal*, *2*(5), 411–419. doi:10.15406ij.2018.02.00078

Massa, L., Farneti, F., & Scappini, B. (2015). Developing a sustainability report in a small to medium enterprise: Process and consequences. *Meditari Accountancy Research*, *23*(1), 62–91. Advance online publication. doi:10.1108/MEDAR-02-2014-0030

Masson-Delmotte, V., Zhai, P., Pörtner, H. O., Roberts, D., Skea, J., Shukla, P. R., & Pirani, A. (2018). Global Warming of 1.5°C. An IPCC Special Report on the Impacts of Global Warming of 1.5°C above Pre-Industrial Levels and Related Global Greenhouse Gas Emission Pathways, in the Context of Strengthening the Global Response to the Threat of Climate Change. *Sustainable Development, and Efforts to Eradicate Poverty.*

Mata, C., Fialho, A., & Eugenio, T. (2018). A decade of environmental accounting reporting: What we know? *Journal of Cleaner Production, 198*, 1198–1209. doi:10.1016/j.jclepro.2018.07.087

Mathews, M. R. (1995). Social and environmental accounting: A practical demonstration of ethical concern? *Journal of Business Ethics, 14*(8), 663–671. doi:10.1007/BF00871347

Mattis, M. C. (1995). Corporate initiatives for advancing women. *Women in Management Review, 10*(7), 5–14. doi:10.1108/09649429510095872

Maxwell, G., & Ogden, S. (2006). Career development of female managers in retailing: Inhibitors and enablers. *Journal of Retailing and Consumer Services, 13*(2), 111–120. doi:10.1016/j.jretconser.2005.08.010

Medel, F., García, L., Enriquez, S., & Anido, M. (2011). Reporting models for corporate sustainability in SMEs. *Environmental Science and Engineering*, 407–418. doi:10.1007/978-3-642-19536-5_32

Medina-Serrano, R., González-Ramírez, M., Gascó-Gascó, J. L., & Llopis-Taverner, J. (2019, December). Prácticas de Evaluación de Proveedores Sostenibles a Través de La Cadena de Suministro. *Dirección y Organización*, (69), 13–26. doi:10.37610/dyo.v0i69.558

Meintjes, C., & Grobler, A. F. (2014). Do public relations professionals understand corporate governance issues well enough to advise companies on stakeholder relationship management? *Public Relations Review, 40*(1), 161–170. doi:10.1016/j.pubrev.2013.10.003

Melloni, G., Galio, A., & Perego, P. (2017). Saying more with less? Disclosure conciseness, completeness, and balance in Integrated Reports. *Journal of Accounting and Public Policy, 36*(3), 220–238. doi:10.1016/j.jaccpubpol.2017.03.001

Melnyk, S. A., Sroufe, R. P., & Calantone, R. (2003). Assessing the Impact of Environmental Management Systems on Corporate and Environmental Performance. *Journal of Operations Management, 21*(3), 329–351. doi:10.1016/S0272-6963(02)00109-2

Member State Implementation of Directive 2014/95/EU: A Comprehensive Overview of How Member States are Implementing the EU Directive on Non-financial and Diversity Information. (2017).

Mendes, J. (2011). Participation and the role of law after Lisbon Treaty: A legal vie on article 11 TUE. *Common Market Law Review, 48*, 1849–1878.

Meseguer-Sánchez, V., Abad-Segura, E., Belmonte-Ureña, L.J., & Molina-Moreno, V. (2020). Examining the research evolution on the socio-economic and environmental dimensions of university social responsibility. *International Journal of Environmental Research and Public Health, 17*(13), 1-30.

Meseguer-Sánchez, V., Gálvez-Sánchez, F. J., López-Martínez, G., & Molina-Moreno, V. (2021). Corporate Social Responsibility and Sustainability. A Bibliometric Analysis of Their Interrelations. *Sustainability, 13*(4), 1636. doi:10.3390u13041636

Meyer, J. W., & Rowan, B. (1977). Institutionalized organizations: Formal structure as myth and ceremony. *American Journal of Sociology, 83*(1), 340–363. doi:10.1086/226550

Miklosik, A., & Evans, N. (2021). Environmental sustainability disclosures in annual reports of mining companies listed on the Australian Stock Exchange (ASX). *Heliyon, 7*(7), 1–11. doi:10.1016/j.heliyon.2021.e07505 PMID:34296012

Miles, D., Borman, W., Spector, P., & Fox, S. (2002). Building an Integrative Model of Extra Role Work Behaviors: A Comparison of Counterproductive Work Behavior with Organizational Citizienship Behavior. *International Journal of Selection and Assessment*, *10*(1&2), 51–57. doi:10.1111/1468-2389.00193

Mills, C. W. (1956). *The Power Elite*. Oxford University Press.

Mio, C., Marco, F., & Pauluzzo, R. (2016). Internal application of IR principles: Generali's Internal Integrated Reporting. *Journal of Cleaner Production*, *139*, 204–218. doi:10.1016/j.jclepro.2016.07.149

Mishra, A. R., Rani, P., Pardasani, K. R., & Mardani, A. (2019). A Novel Hesitant Fuzzy WASPAS Method for Assessment of Green Supplier Problem Based on Exponential Information Measures. *Journal of Cleaner Production*, *238*(November), 117901. doi:10.1016/j.jclepro.2019.117901

Moneva, J. M., & Cuellar, B. (2009). The value relevance of financial and non-financial environmental reporting. *The Official Journal of the European Association of Environmental and Resource Economists*, *44*(3), 441–456. doi:10.100710640-009-9294-4

Monteiro, S., Ribeiro, V., & Lemos, K. (2020). Linking corporate social responsibility reporting with the UN Sustainable Development Goals: Evidence from the Portuguese stock market. In Conceptual and Theoretical Approaches to Corporate Social Responsibility, Entrepreneurial Orientation, and Financial Performance (pp. 134–151). IGI Global.

Monteiro, S., Ribeiro, V., & Lemos, K. (forthcoming). Sustainability reporting and the Sustainable Development Goals (SDGs) in Higher Education: a Portuguese university case. In D. Crowther & S. Seifi (Eds.), CSR and Sustainability. Academic Press.

Morrow, D., & Rondinelli, D. (2002). Adopting Corporate Environmental Management Systems. *European Management Journal*, *20*(2), 159–171. doi:10.1016/S0263-2373(02)00026-9

Nawrocka, D. (2008). Environmental Supply Chain Management, ISO 14001 and RoHS. How Are Small Companies in the Electronics Sector Managing? *Corporate Social Responsibility and Environmental Management*, *15*(6), 349–360. doi:10.1002/csr.176

Nawrocka, D., Brorson, T., & Lindhqvist, T. (2009). ISO 14001 in Environmental Supply Chain Practices. *Journal of Cleaner Production*, *17*(16), 1435–1443. doi:10.1016/j.jclepro.2009.05.004

Neary, J., & Osborne, M. (2018). University engagement in achieving Sustainable Development Goals: A synthesis of case studies from the SUEUAA Study. *Australian Journal of Adult Learning*, *58*(3), 336–364.

Needles, B. E., Jr., Frigo, M. L., Powers, M., & Shigaev, A. (2016). Integrated reporting and sustainability reporting: An exploratory study of high performance companies. In M. J. Epstein, F. Verbeeten, & S. K. Widener (Eds.), Performance Measurement and Management Control: Contemporary Issues (pp. 41-81). Emerald Group Publishing Limited. doi:10.1108/S1479-351220160000031019

Nishitani, K. (2010). Demand for ISO 14001 Adoption in the Global Supply Chain: An Empirical Analysis Focusing on Environmentally Conscious Markets. *Resource and Energy Economics*, *32*(3), 395–407. doi:10.1016/j.reseneeco.2009.11.002

Nishitani, K., Haider, M. B., & Kokubu, K. (2020). Are third-party assurances preferable to third-party comments for promoting financial accountability in environmental reporting? *Journal of Cleaner Production*, *248*, 1–13. doi:10.1016/j.jclepro.2019.119199

Ntim, C., Soobaroyen, T., & Broad, M. J. (2017). Governance structures, voluntary disclosures and public accountability. *Accounting, Auditing & Accountability Journal*, *30*(1), 65–118. doi:10.1108/AAAJ-10-2014-1842

O'Dwyer, B., & Unerman, J. (2016). Fostering rigor in accounting for social sustainability. *Accounting, Organizations and Society*, *49*, 32–40. doi:10.1016/j.aos.2015.11.003

Oberthür, S., & Kelly, C. R. (2008). EU Leadership in International Climate Policy: Achievements and Challenges. *The International Spectator*, *43*(3), 35–50. doi:10.1080/03932720802280594

Official Journal of the European Union. (2014). *Directive 2014/95/EU of the European Parliament and of the Council of 22 October 2014 amending Directive 2013/34/EU as regards disclosure of non-financial and diversity information by certain large undertakings and groups*. Retrieved from https://eur-lex.europa.eu/legal-content/EN/TXT/PDF/?uri=CELEX:32014L0095&from=FI

Ogden, S. M., McTavish, D., & McKean, L. (2006). Clearing the way for gender balance in the management of the UK financial services industry: Enablers and barriers. *Women in Management Review*, *21*(1), 40–54. doi:10.1108/09649420610643402

Oliveira. (2019). Bibliometric method for mapping the state-of-the-art and identifying research gaps and trends in literature: An essential instrument to support the development of scientific projects. In Scientometrics Recent Advances. IntechOpen.

Oliver, J., Vesty, G., & Brooks, A. (2016). Conceptualising integrated thinking in practice. *Managerial Auditing Journal*, *31*(2), 228–248. doi:10.1108/MAJ-10-2015-1253

Omazić, M.A., Vukić, N.M., & Ivanović, I.R., Marszalek, V., & Rösler, B. (2020). The analysis of the state of non-financial reporting in Croatia in 2017 and 2018 [Analyse des stands der nichtfinanziellen berichterstattung in Kroatien in den jahren 2017 und 2018]. *Socijalna Ekologija*, *29*(2), 207.

Onat, N. C., Kucukvar, M., Halog, A., & Cloutier, S. (2017). Systems thinking for life cycle sustainability assessment: A review of recent developments, applications, and future perspectives. *Sustainability*, *9*(5), 706. doi:10.3390u9050706

Opferkuch, K., Caeiro, S., Salomone, R., & Ramos, T. B. (2021). Circular economy in corporate sustainability reporting: A review of organisational approaches. *Business Strategy and the Environment*, *30*(8), 4015–4036. doi:10.1002/bse.2854

Oprisor, T. (2015). Auditing integrated reports: Are there solutions to this puzzle? *Procedia Economics and Finance*, *25*, 87–95. doi:10.1016/S2212-5671(15)00716-9

Orcos, R., & Palomas, S. (2019). The Impact of National Culture on the Adoption of Environmental Management Standards: The Worldwide Diffusion of ISO 14001. *Cross Cultural & Strategic Management*, *26*(4), 546–566. doi:10.1108/CCSM-10-2018-0168

Oreskes, N. (2004). The Scientific Consensus on Climate Change. *Science*, *306*(5702), 1686–1686. doi:10.1126cience.1103618 PMID:15576594

Organ, D. W. (1997). Organizational Citizienship Behavior: It's construct clean-up time. *Human Performance*, *10*(2), 85–97. doi:10.120715327043hup1002_2

ORSIES- Observatório da Responsabilidade Social e Instituições de Ensino Superior. (2018). *Livro Verde sobre Responsabilidade Social e Instituições de Ensino Superior*. Author.

Ortiz-Martínez, E., & Marín-Hernández, S. (2021). European SMEs and non-financial information on sustainability. *International Journal of Sustainable Development and World Ecology*, *00*(00), 1–13. doi:10.1080/13504509.2021.1929548

Owen, G. (2013). Integrated Reporting: A Review of Developments and their Implications for the Accounting Curriculum. *Accounting Education*, *22*(4), 340–356. doi:10.1080/09639284.2013.817798

Paletta, A., & Bonoli, A. (2019). Governing the university in the perspective of the United Nations 2030 Agenda: The case of the University of Bologna. *International Journal of Sustainability in Higher Education*, *20*(3), 500–514.

Paletta, A., Fochi, P., Toschi, T., & Ubertini, F. (2020). Adoption of the SDGs as a reporting framework at the Alma Mater Studiorum (University of Bologna) in Italy. In G. Nhamo & V. Mjimba (Eds.), *Sustainable Development Goals and Institutions of Higher Education* (pp. 185–197). Sustainable Development Goals Series.

Palludeto, A. W. A., & Felipini, A. R. (2019). Panorama da literatura sobre a financeirização (1992-2017): Uma abordagem bibliométrica. *Economia e Sociedade*, *28*(2), 313–337. doi:10.1590/1982-3533.2019v28n2art02

Papaterra. (2019). *Stranded assets e o petróleo no Brasil. Trabalho de Conclusão de Curso - Monografia da Escola Superior de Guerra. Curso de Altos Estudos de Política e Estratégia (CAEPE).* Escola Superior de Guerra.

Parker, L., Guthrie, J., & Gray, R. (1998). Accounting and management research: Passwords from the gatekeepers. *Accounting, Auditing & Accountability Journal*, *11*(4), 371–406.

Parker, L., Guthrie, J., & Gray, R. (1998). Accounting and management research: Passwords from the gatekeepers. *Accounting, Auditing & Accountability Journal*, *11*(4), 371–406. doi:10.1108/09513579810231420

Park, N., & Peterson, C. (2003). Virtues and organizations. In K. S. Cameron, J. E. Dutton, & R. E. Quinn (Eds.), *Positive organizational scholarship* (pp. 33–47). Barrett-Koehler.

Paternostro, S. (2020). Integrated Reporting and Social Disclosure: True Love or Forced Marriage? A Multidimensional Analysis of a Contested Concept. In L. Songini, A. Pistoni, P. Baret, & M. H. Kunc (Eds.), Non-Financial Disclosure and Integrated Reporting: Practices and Critical Issues (pp. 107-146). Emerald Publishing Limited. doi:10.1108/S1479-351220200000034008

Perego, P., Kennedy, S., & Whiteman, G. (2016). A lot of icing but little cake? Taking integrated reporting forward. *Journal of Cleaner Production, 136*(July), 53–64. doi:10.1016/j.jclepro.2016.01.106

Peterson, C. & Seligman, M. (2004). *Character Strengths and Virtues: A Handbook and Classification.* . doi:10.1176/appi.ajp.162.4.820-a

Petrushenko, Y., Vadym, A., Vorontsova, A., & Ponomarenko, O. (2020). Sustainable development goals as a tool for strategic planning in communities: A bibliometric analysis of research. In *E3S Web of Conferences* (Vol. 202, p. 03005). EDP Sciences. doi:10.1051/e3sconf/202020203005

Pfeiffer, A., Hepburn, C., Vogt-Schilb, A., & Caldecott, B. (2018). Committed emissions from existing and planned power plants and asset stranding required to meet the Paris Agreement. Alexander Pfeiffer, Cameron Hepburn, Adrien Vogt-Schilb and Ben Caldecott. *Environmental Research Letters*, *13*(5), 054019. doi:10.1088/1748-9326/aabc5f

Pieter de Groen, W., Alcidi, C., Simonelli, F., Campmas, A., Salvo, M., Musmeci, R., Oliinyk, I., & Tadi, S. (2020). *Study on the non-financial reporting directive.* https://www.ceps.eu/download/publication/?id=32991&pdf=EV0220277ENN.en_.pdf

Pizzi, S., Caputo, A., Corvino, A., & Venturelli, A. (2020). Management research and the UN sustainable development goals (SDGs): A bibliometric investigation and systematic review. *Journal of Cleaner Production*, *276*, 124033. doi:10.1016/j.jclepro.2020.124033

Podsakoff, P., & MacKenzie, S. (1994). Organizzational citizienship behaviors and sales unit effectiveness. *JMR, Journal of Marketing Research*, *31*(3), 351–363. doi:10.1177/002224379403100303

Podsakoff, P., MacKenzie, S., Paine, J., & Bachrach, D. (2000). Organizational Citizienship Behaviors: A critival review of the theoretical and empirical leterature and suggestions for future research. *Journal of Management*, *26*(3), 513–563. doi:10.1177/014920630002600307

Powell, G. N., Butterfield, D. A., & Parent, J. D. (2002). Gender and managerial stereotypes: Have the times changed? *Journal of Management, 28*(2), 177–193. doi:10.1177/014920630202800203

Prajogo, D., Chowdhury, M., Yeung, A. C. L., & Cheng, T. C. E. (2012). The Relationship between Supplier Management and Firm's Operational Performance: A Multi-Dimensional Perspective. *International Journal of Production Economics, 136*(1), 123–130. doi:10.1016/j.ijpe.2011.09.022

Prajogo, D., Tang, A. K., & Lai, K. H. (2014). The Diffusion of Environmental Management System and Its Effect on Environmental Management Practices. *International Journal of Operations & Production Management, 34*(5), 565–585. doi:10.1108/IJOPM-10-2012-0448

Prasad, D. S., Pradhan, R. P., Gaurav, K., Chatterjee, P. P., Kaur, I., Dash, S., & Nayak, S. (2018). Analysing the Critical Success Factors for Implementation of Sustainable Supply Chain Management: An Indian Case Study. *Decision (Washington, D.C.), 45*(1), 3–25.

Prieditis, M. (2002). Participation in the legislative process and new information technology. *Proceedings of the Fifth Congress of the European Association of Legislation in Athens, Nomos*, 99-118.

Prinsloo, A., & Maroun, W. (2020). An exploratory study on the components and quality of combined assurance in an integrated or a sustainability reporting setting. *Sustainability Accounting. Management and Policy Journal, 12*(1), 1–29.

Purcell, J. (1996). Human Resource Bundles of Best Practise: A Utopian Cul-de-sac? Paper presented in the ESRC/BUIRA seminar series on the Contribution of HR Strategy to Business Performance.

PwC. (2017). *The non-financial reporting regulations What do they mean in practice?* PwC.

Qian, W., Tilt, C., & Belal, A. (2021). Social and environmental accounting in developing countries: Contextual challenges and insights. *Accounting, Auditing & Accountability Journal, 34*(5), 1021–1050. doi:10.1108/AAAJ-03-2021-5172

Quarchioni, S., Ruggiero, P., & Damiano, R. (2020). *Flows of information and meaning: a vocabulary approach to integrated thinking and reporting*. Meditari Accountancy Research. doi:10.1108/MEDAR-01-2020-0677

Radaelli, C. M. (2005). Diffusion without convergence: How political context shapes the adoption of regulatory impact assessment. *Journal of European Public Policy, 12*(5), 924–943. doi:10.1080/13501760500161621

Radhouane, I., Nekhili, M., Nagati, H., & Pache, G. (2018a). Customer-related performance and the relevance of environmental reporting. *Journal of Cleaner Production, 190*, 315–329. doi:10.1016/j.jclepro.2018.04.149

Radhouane, I., Nekhili, M., Nagati, H., & Pache, G. (2018b). The impact of corporate environmental reporting on customer-related performance and market value. *Management Decision, 56*(7), 1630–1659. doi:10.1108/MD-03-2017-0272

Rajan, R., Dhir, S., & Sushil. (2020). Alliance termination research: A bibliometric review and research agenda. *Journal of Strategy and Management, 13*(3), 351–375. doi:10.1108/JSMA-10-2019-0184

Raman, R., Vinuesa, R., & Nedungadi, P. (2021). Bibliometric Analysis of SARS, MERS, and COVID-19 Studies from India and Connection to Sustainable Development Goals. *Sustainability, 13*(14), 7555. doi:10.3390u13147555

Ramarajan, L., Barsade, S., & Burack, O. (2008). The influence of organizational respect on emotional exhaustion in the human services. *The Journal of Positive Psychology, 3*(1), 4–18. doi:10.1080/17439760701750980

Rayn, M. K., Haslam, S. A., & Postmes, T. (2007). Reaction to Glass cliff: Gender differences in the explanations for the precariousness of women's leadership positions. *Journal of Organizational Change Management, 20*(2), 182–197. doi:10.1108/09534810710724748

Razak, A. A., Rowling, M., White, G., & Mason-Jones, R. (2016). Public Sector Supply Chain Management: A Triple Helix Approach to Aligning Innovative Environmental Initiatives. *Foresight and STI Governance*, *10*(1), 43–52. doi:10.17323/1995-459x.2016.1.43.52

Rees, T. (1992). *Women and the Labour Market*. Routledge.

Rego, A., Cunha, M., Costa, N., Gonçalves, H. & Cardoso, C. (2007). *Gestão Ética e Socialmente Responsável*. Academic Press.

Rego, A., & Cunha, M. (2011). *Liderança: A virtude está no meio*. Actual.

Rego, A., Ribeiro, N., & Cunha, M. (2010). Perceptions of organizational virtuousness and happiness as predictors of organizational citizenship behaviors. *Journal of Business Ethics*, *93*(2), 215–235. doi:10.100710551-009-0197-7

Rego, A., Vitória, A., Magalhães, A., Ribeiro, N., & Cunha, M. (2012). Are authentic leaders associated with more virtuous, committed and potent teams? *The Leadership Quarterly*, *24*(1), 61–79. doi:10.1016/j.leaqua.2012.08.002

Reimsbach, D., Hahn, R., & Gürtürk, A. (2017). Integrated Reporting and Assurance of Sustainability Information: An Experimental Study on Professional Investors' Information Processing. *European Accounting Review*, 1–23.

Renda, A. (2006). *Impact Assessment in the EU: The State of the Art and the Art of the State*. CEPS.

Reports of the Advisory Scientific Committee No 6. (2016). *Too late, too sudden: Transition to a low-carbon economy and systemic risk*. ESRB Advisory Scientific Committee.

Rezaee, Z., & Tuo, L. (2017). Voluntary disclosure of non-financial information and its association with sustainability performance. *Advances in Accounting*, *39*, 47–59. doi:10.1016/j.adiac.2017.08.001

Ribeiro, V. P., Aibar-Guzman, C., Aibar-Guzman, B., & Monteiro, S. (2016). Determinants of environmental accounting and reporting practices in Portuguese local entities. *Corporate Communications*, *21*(3), 352–370. doi:10.1108/CCIJ-11-2015-0071

Richard, G., & Odendaal, E. (2021). Credibility-enhancing mechanisms, other than external assurance, in integrated reporting. *The Journal of Management and Governance*, *25*(1), 61–93. doi:10.100710997-020-09509-x

Rivera-Arrubla, Y. A., Zorio-Grima, A., & García-Benau, M. A. (2017). Integrated Reports: Disclosure level and explanatory factors. *Social Responsibility Journal*, *13*(1), 155–176. doi:10.1108/SRJ-02-2016-0033

Roberts, L., Hassan, A., Elamer, A., & Nandy, M. (2021). Biodiversity and extinction accounting for sustainable development: A systematic literature review and future research directions. *Business Strategy and the Environment*, *30*(1), 705–720. doi:10.1002/bse.2649

Rodríguez-Gutiérrez, P., Guerrero-Baena, M. D., Luque-Vílchez, M., & Castilla-Polo, F. (2021). An approach to using the best-worst method for supporting sustainability reporting decision-making in SMEs. *Journal of Environmental Planning and Management*, *0*(0), 1–24. doi:10.1080/09640568.2021.1876003

Romolini, A., Fissi, S., & Gori, E. (2015). Quality disclosure in sustainability reporting: Evidence from universities. *Transylvanian Review of Administrative Sciences*, *11*(44), 196–218.

Roque, V. & Cortez, M. (2006). A divulgação de informação ambiental e a performance financeira das empresas cotadas em Portugal. *Tékhne - Revista de Estudos Politécnicos, 3*(5–6), 119–143.

Rosati, F., & Faria, L. (2019a). Business contribution to the Sustainable Development Agenda: Organizational factors related to early adoption of SDG reporting. *Corporate Social Responsability and Environmental Management*, 1–10. doi:10.1002/csr.1705

Rosati, F., & Faria, L. (2019b). Addressing the SDGs in sustainability reports: The relationship with institutional factors. *Journal of Cleaner Production*, *215*, 1312–1326.

Rosener, J. B. (1995). *America's Competitive Secret: Utilizing Women as a Management Strategy*. Oxford University Press.

Rosener, J. B. (1995). *America's Competitive Secret: Women Managers*. Oxford University Press.

Rounaghi, M. M. (2019). Economic analysis of using green accounting and environmental accounting to identify environmental costs and sustainability indicators. *International Journal of Ethics and Systems*, *35*(4), 504–512. doi:10.1108/IJOES-03-2019-0056

Rousseau, J.-J. (1772). *Considerations sur le gouvernement de Pologne*. Academic Press.

Rozenberg, J., Vogt-Schilb, A., & Hallegatte, S. (2020). Instrument choice and stranded assets in the transition to clean capital. *Journal of Environmental Economics and Management*, *100*, 102183. doi:10.1016/j.jeem.2018.10.005

Rudebusch. (2021). Climate Change Is a Source of Financial Risk. FRBSF Economic Letter, 3, 1-6.

Rupley, K. H., Brown, D., & Marshall, S. (2017). Evolution of corporate reporting: From stand-alone corporate social responsibility reporting to integrated reporting. *Research in Accounting Regulation*, *29*(2), 172–176. doi:10.1016/j.racreg.2017.09.010

Ryan, R., & Deci, E. (2000). Self-determination theory and the facilitation of intrinsic motivation, social development, and well-being. *The American Psychologist*, *55*(1), 68–78. doi:10.1037/0003-066X.55.1.68 PMID:11392867

Sadiq, M., & Tran, Q. H. (2020). Environmental Accounting Perception: Evidence from Nigerian Pharmaceutical Firms. *Systematic Reviews in Pharmacy*, *11*(1), 303–309. doi:10.5530rp.2020.1.39

Sáez-Martín, A., López-Hernandez, A. M., & Caba-Perez, C. (2017). Access to public information: A scientometric study of legal versus voluntary transparency in the public sector. *Scientometrics*, *113*(3), 1697–1720. doi:10.100711192-017-2541-5

SAICA. (2015). *Integrated thinking: An exploratory survey*. Disponível em: https://www.saica.co.za/Portals/0/Technical/Sustainability/SAICAIntegratedThinkingLandscape.pdf

Salim, H. K., & Padfield, R. (2017). Environmental Management System in the Food and Beverage Sector: A Case Study from Malaysia. *Chemical Engineering Transactions*, *56*(April), 253–258.

Salim, H. K., Padfield, R., Lee, C. T., Syayuti, K., Papargyropoulou, E., & Tham, M. H. (2018). An Investigation of the Drivers, Barriers, and Incentives for Environmental Management Systems in the Malaysian Food and Beverage Industry. *Clean Technologies and Environmental Policy*, *20*(3), 529–538. doi:10.100710098-017-1436-8

Santos, B. M., Godoy, L. P., & Campos, L. M. S. (2019). Performance Evaluation of Green Suppliers Using Entropy-TOPSIS-F. *Journal of Cleaner Production*, *207*(January), 498–509. doi:10.1016/j.jclepro.2018.09.235

Sarkar, S. H., & Ahmed, R. (2020). Corporate environmental accounting disclosure in Bangladesh. *International Journal of Applied Business and Economic Research*, *18*(1), 39–53.

Sassen, R., & Azizi, L. (2017). Voluntary disclosure of sustainability reports by Canadian universities. *Journal of Business Economics*, *88*(1), 97–137.

Sassen, R., & Azizi, L. (2018). Assessing sustainability reports of US universities. *International Journal of Sustainability in Higher Education*, *19*(7), 1158–1184.

Sassen, R., Dienes, D., & Wedemeier, J. (2018). Characteristics of UK higher education institutions that disclose sustainability reports. *International Journal of Sustainability in Higher Education*, *19*(7), 1279–1298.

Saygin. (2019). Power sector asset stranding effects of climate policies. Energy Sources, Part B: Economics, Planning, and Policy, 14(4), 99-124.

Schaltegger, S. (2018). Linking Environmental Management Accounting: A Reflection on (Missing) Links to Sustainability and Planetary Boundaries. *Social and Environmental Accountability Journal*, *38*(1), 19–29. doi:10.1080/0969160X.2017.1395351

Schaltegger, S., Etxeberria, I. A., & Ortas, E. (2017). Innovating corporate accounting and reporting for sustainability – Attributes and challenges. *Sustainable Development*, *25*(2), 113–122. doi:10.1002d.1666

Schoon, I., Martin, P., & Ross, A. (2006). Career transitions in times of social change. His and her story. *Journal of Vocational Behavior*, *70*(1), 78–96. doi:10.1016/j.jvb.2006.04.009

Schramade, W. (2017). Investing in the UN Sustainable Development Goals: Opportunities for companies and investors. *Journal of Applied Corporate Finance*, *29*, 87–99.

Schwartz, M., & Carroll, A. (2014, December). Corporate Social Responsibility: A Three-Domain Approach. *Business Ethics Quarterly*. Advance online publication. doi:10.2307/3857969

Scur, G., & Barbosa, M. E. (2017). Green Supply Chain Management Practices: Multiple Case Studies in the Brazilian Home Appliance Industry. *Journal of Cleaner Production*, *141*(January), 1293–1302. doi:10.1016/j.jclepro.2016.09.158

SDSN Australia/Pacific. (2017). *Getting started with the SDGs in universities: A guide for universities, higher education institutions, and the academic sector*. Australia, New Zealand and Pacific Edition. Sustainable Development Solutions Network – Australia/Pacific.

SEI, IISD, ODI, E3G, & UNEP. (2020). *The Production Gap Report: 2020 Special Report*. Authors.

Sellami, Y. M., & Gafsi, Y. (2018). What drives developing and transitional countries to adopt the IFRS for SMEs? An institutional perspective. *Journal of Corporate Accounting & Finance*, *29*(2), 34–56. doi:10.1002/jcaf.22331

Senn, J., & Giordano-Spring, S. (2020). The limits of environmental accounting disclosure: Enforcement of regulations, standards and interpretative strategies. *Accounting, Auditing & Accountability Journal*, *33*(6), 1367–1393. doi:10.1108/AAAJ-04-2018-3461

Sen, S., & von Schickfus, M.-T. (2020). Climate policy, *stranded assets*, and investors' expectations. *Journal of Environmental Economics and Management*, *100*, 102277. doi:10.1016/j.jeem.2019.102277

Sepasi, S., Braendle, U., & Rahdari, A. H. (2018b). Comprehensive sustainability reporting in higher education institutions. *Social Responsibility*, *15*(2), 155–170.

Sepasi, S., Rahdari, A., & Rexhepi, G. (2018a). Developing a sustainability reporting assessment tool for higher education institutions: The University of California. *Sustainable Development*, *26*(6), 672–682.

Serafeim, G. (2014). Turning a profit while doing good: aligning sustainability with corporate performance. *Governance Studies*, 3-9.

Shehata, M. (2020). Do non-financial information reduce the fraud risk management? *Journal of Behavioural Economics Finance Entrepreneurship Accounting and Transport*, *8*(1), 14–25. doi:10.12691/jbe-8-1-3

Shields, J. F., & Shelleman, J. M. (2020). SME sustainability dashboards: An aid to manage and report performance. *Journal of Small Business Strategy*, *30*(2), 106–114.

Shimbar, A. (2021). Environment-related stranded assets: An agenda for research into value destruction within carbon-intensive sectors in response to environmental concerns. *Renewable & Sustainable Energy Reviews*, *144*, 111010. doi:10.1016/j.rser.2021.111010

Shojaei, P., & Bolvardizadeh, A. (2020). Rough MCDM Model for Green Supplier Selection in Iran: A Case of University Construction Project. *Built Environment Project and Asset Management*, *10*(3), 437–452. doi:10.1108/BEPAM-11-2019-0117

Sial, M., Zheng, C., Khuong, N., Khan, T., & Usman, M. (2018). Does firm performance influence corporate social responsibility reporting of Chinese listed companies? *Sustainability*, *10*(7), 112. doi:10.3390u10072217

Silva. (2017) *Caracterização do relato de sustentabilidade nas Instituições de Ensino Superior Portuguesas*. Dissertação para obtenção do grau de Mestre em Contabilidade e Controlo de Gestão pela Faculdade de Economia do Porto.

Simnett, R., & Huggins, A. L. (2015). Integrated reporting and assurance: Where can research add value? *Sustainability Accounting. Management and Policy Journal*, *6*(1), 29–53.

Simnett, R., Vanstraelen, A., & Chua, W. F. (2009). Assurance on sustainability reports: An international comparison. *The Accounting Review*, *84*(3), 937–967. doi:10.2308/accr.2009.84.3.937

Simnett, R., Zhou, S., & Hoang, H. (2016). Assurance and other credibility enhancing mechanisms for integrated reporting. In *Integrated Reporting: a new accounting disclosure* (pp. 269–286). Palgrave Macmillan UK. doi:10.1057/978-1-137-55149-8_14

Simões. (2020, September/December). Panorama da Literatura sobre a Governança Corporativa: Uma Análise Bibliométrica das Bases Teóricas e Abordagens mais Utilizadas em Artigos Revista de Gestão. *Finanças e Contabilidade*, *10*(3), 62–82.

Singh, A., Panackal, N., & Venkataramani, B. (2016). Environmental Accounting for Industrial Growth and Sustainable Development in India: An Overview and Theoretical Framework. In R. Dubey & A. Gunasekaran (Eds.), *Strategic Management of Sustainable Manufacturing Operations* (pp. 74–94). IGI Global. doi:10.4018/978-1-5225-0350-7.ch004

Singh, P. (2008). Empirical assessment of ISO 9000 related management practices and performance relationships. *International Journal of Production Economics*, *113*(1), 40–59. doi:10.1016/j.ijpe.2007.02.047

Singh, V., Terjesen, S., & Vinnicombe, S. (2008). Newly appointed directors in the boardroom: How do women and men differ? *European Management Journal*, *26*(1), 48–58. doi:10.1016/j.emj.2007.10.002

Singh, V., & Vinnicombe, S. (2004). Why so few women directors on UK corporate boards: Evidence and theoretical explanations. *Corporate Governance*, *12*(4), 479–488. doi:10.1111/j.1467-8683.2004.00388.x

Skouloudis, A., Malesios, C., & Dimitrakopoulos, P. G. (2019). Corporate biodiversity accounting and reporting in mega-diverse countries: An examination of indicators disclosed in sustainability reports. *Ecological Indicators*, *98*, 888–901. doi:10.1016/j.ecolind.2018.11.060

Smith, C. (1993). BS 7750 and Environmental Management. *Journal of the Society of Dyers and Colourists*, *109*(9), 278–279. doi:10.1111/j.1478-4408.1993.tb01574.x

Smith, C., Organ, D., & Near, J. (1983). Organizational Citizienship Behavior: Its nature and antecedents. *The Journal of Applied Psychology*, *68*(4), 653–663. doi:10.1037/0021-9010.68.4.653

Smith, S. (2014). Integrated reporting, corporate governance, and the future of the accounting function. *International Journal of Business and Social Science*, *5*(10), 58–63.

Solovida, G. T., & Latan, H. (2017). Linking environmental strategy to environmental performance: Mediation role of environmental management accounting. *Sustainability Accounting, Management and Policy Journal, 8*(5), 595–619. doi:10.1108/SAMPJ-08-2016-0046

Solum, S., Jacob, V. F., Levitt, M., & the Harvard Law School Forum on Corporate Governance. (2021). *The SEC's Upcoming Climate Disclosure Rules.* Retrieved from https://corpgov.law.harvard.edu/2021/09/01/the-secs-upcoming-climate-disclosure-rules/

Sommer, B., & Hain, A. (2017). Europe as a Green Leader? A Brief Evaluation of Both the European Union's Climate and Energy Policy and Common Agricultural Policy. *Culture, Practice & Europeanization, 2*(2), 33–45. doi:10.5771/2566-7742-2017-2-33

Son-Turan, S., & Lambrechts, W. (2019). Sustainability disclosure in higher education: A comparative analysis of reports and websites of public and private universities in Turkey. *International Journal of Sustainability in Higher Education, 20*(7), 1143–1170.

Soriya, S. & Rastogi, P. (2021). A systematic literature review on integrated reporting from 2011 to 2020. *Journal of Financial Reporting and Accounting*, 1-22. doi:10.1108/JFRA-09-2020-0266

Spector, P., & Fox, S. (2002). An emotion-centered model of voluntary work behavior. *Human Resource Management Review, 12*(2), 269–292. doi:10.1016/S1053-4822(02)00049-9

Speziale, M. (2019). Theoretical Perspectives on Purposes and Users of Integrated Reporting: A Literature Review. In S. O. Idowu & M. D. Baldo (Eds.), *Integrated Reporting: Antecedents and Perspectives for Organizations and Stakeholders* (pp. 13–60). Springer International Publishing Switzerland. doi:10.1007/978-3-030-01719-4_2

Stacchezzini, R., Florio, C., Sproviero, A. F., & Corbella, S. (2019). An intellectual capital ontology in an integrated reporting context. *Journal of Intellectual Capital, 20*(1), 83–99. doi:10.1108/JIC-05-2018-0090

Staw, B., & Barsade, S. (1993). Affect and managerial performance: A test of the sadder_but_wiser versus happier-and-smarter hypotheses. *Administrative Science Quarterly, 38*(2), 304–331. doi:10.2307/2393415

Stechemesser, K., & Guenther, E. (2012). Carbon accounting: A systematic literature review. *Journal of Cleaner Production, 36*, 17–38. doi:10.1016/j.jclepro.2012.02.021

Steinhofel, E., Galeitzke, M., Kohl, H., & Orth, R. (2019). Sustainability reporting in German manufacturing SMEs. *Procedia Manufacturing, 33*, 610–617. doi:10.1016/j.promfg.2019.04.076

Stekelorum, R., Laguir, I., & Elbaz, J. (2018). CSR disclosure and sustainable supplier management: A small to medium-sized enterprises perspective. *Applied Economics, 50*(46), 5017–5030. doi:10.1080/00036846.2018.1466990

Stolowy, H., & Paugam, L. (2018). The expansion of non-financial reporting : an exploratory study The expansion of non- fi nancial reporting : an exploratory study. *Accounting and Business Research ISSN, 48*(5), 525–548. doi:10.1080/00014788.2018.1470141

Stubbs, W., & Higgins, C. (2014). Integrated Reporting and internal mechanisms of change. *Accounting, Auditing & Accountability Journal, 27*(7), 1068–1089. doi:10.1108/AAAJ-03-2013-1279

Sturm, A., & Upasena, S. (1998). *ISO 14001 – Implementing an Environmental Management System.* Ellipson Management Consultants.

Sustainability Accounting Standards Board. (n.d.). *Standards Overview.* Retrieved from https://www.sasb.org/standards/

Sweileh, W. M. (2020). Bibliometric analysis of scientific publications on "sustainable development goals" with emphasis on "good health and well-being" goal (2015–2019). *Globalization and Health*, *16*(1), 1–13. doi:10.118612992-020-00602-2 PMID:32723366

Szymansk, H. (2011). *A entrevista na educação: a prática reflexive*. B. L. Livro, Ed.

Tamanini, M., Zorzeto, G., & Neves, A. (2020). A pandemia do coronavírus e os grupos vulneráveis: uma análise quanto à eficácia dos atos do governo brasileiro [The coronavirus pandemic and vulnerable groups: an analysis of the effectiveness of the Brazilian government's actions]. *Revista da Seção Judiciária do Rio de Janeiro, 24*(49), 112-135. http://177.223.208.8/index.php/revistasjrj/article/view/463

Tamayo, A. (2004). Introdução. In A. Tamayo (Ed.), *Cultura e saúde nas organizações* (pp. 11–16). Artmed.

Tarquinio, L., & Posadas, S. (2020). Exploring the term "non-financial information: An academics' view. *Meditari Accountancy*, *28*(5), 727–749. doi:10.1108/MEDAR-11-2019-0602

Tavares. (2019). *Política energética em um contexto de transição: a construção de um regime de baixo carbono* (Doctoral thesis). Universidade Federal do Rio de Janeiro, Instituto de Economia, Programa de Pós-Graduação em Economia da Indústria e Tecnologia.

Taygashinova, K., & Akhmetova, A. (2019). Accounting for environmental costs as an instrument of environmental controlling in the company. *Management of Environmental Quality*, *30*(1), 87–97. doi:10.1108/MEQ-08-2017-0088

Teixeira, L. M. S., & Ribeiro, M. S. (2014). Estudo bibliométrico sobre as características da contabilidade ambiental em periódicos nacionais e internacionais [Bibliometric study about environmental accounting caracteristics in nacional and international journals]. *Revista de Gestão Social e Ambiental*, *8*(1), 20. doi:10.24857/rgsa.v8i1.810

Tharenou, P. (2005). Women's advancement in management. In R. J. Burke & M. C. Mattis (Eds.), *Supporting Women's Career Advancement: Challenges and Opportunities*. Edward Elgar Publishing. doi:10.4337/9781845423452.00009

The Institute of Internal Auditors. (2009). *Código de ética*. http://www.iiabrasil.org.br/new/images/down/02_IPPF_Intro_Cod_de_etica.pdf

Thomson, I. (2015). "But does sustainability need capitalism or an integrated report" a commentary on "The International Integrated Reporting Council: A story of failure" by Flower, J. *Critical Perspectives on Accounting*, *27*, 18–22. doi:10.1016/j.cpa.2014.07.003

Tirado-Valencia, P., Cordobés-Madueño, M., Ruiz-Lozano, M., & De Vicente-Lama, M. (2019). Integrated thinking in the reporting of public sector enterprises: A proposal of contents. *Meditari Accountancy Research*, *28*(3), 435–453. doi:10.1108/MEDAR-03-2019-0458

Toledo, A. (2020). Educação financeira: por que precisamos dela? [Financial education: why do we need it?]. In Estratégia Nacional de Educação Financeira (ENEF): Em busca de um Brasil melhor. Riemma.

Torugsa, N. A., O'Donohue, W., & Hecker, R. (2012). Capabilities, Proactive CSR and Financial Performance in SMEs: Empirical Evidence from an Australian Manufacturing Industry Sector. *Journal of Business Ethics*, *109*(4), 483–500. doi:10.100710551-011-1141-1

To, W. M., & Lee, P. K. C. (2014). Diffusion of ISO 14001 Environmental Management System: Global, Regional and Country-Level Analyses. *Journal of Cleaner Production*, *66*(March), 489–498. doi:10.1016/j.jclepro.2013.11.076

Traves, J., Brockbank, A., & Tomlinson, F. (1997). Careers of women managers in the retail industry. *Service Industries Journal*, *17*(1), 133–154. doi:10.1080/02642069700000007

Tregidga, H. & Laine, M. (2021). On crisis and emergency: Is it time to rethink long-term environmental accounting? *Critical Perspectives on Accounting*, *102311*, 1-11. doi:10.1016/j.cpa.2021.102311

Tsalis, T. A., Malamateniou, K. E., Koulouriotis, D., & Nikolaou, I. E. (2020). New challenges for corporate sustainability reporting: United Nations' 2030 Agenda for sustainable development and the sustainable development goals. *Corporate Social Responsibility and Environmental Management*, 1–13.

U.S. Environmental Protection Agency. (1995). *An Introduction to Environmental Accounting As A Business Management Tool: Key Concepts and Terms*. Retrieved from https://www.epa.gov/sites/production/files/2014-01/documents/busmgt.pdf

United Nations Climate Change. (n.d.). *The Paris Agreement*. Retrieved from https://unfccc.int/process-and-meetings/the-paris-agreement/the-paris-agreement

United Nations Environment Programme Finance Initiative. (n.d.). *About Us*. Retrieved from https://www.unepfi.org/about/

University of Cambridge Institute for Sustainability Leadership. (2020). *Bank 2030: Accelerating the transition to a low carbon economy*. CISL.

Urban, B., & Govender, D. P. (2012). Empirical Evidence on Environmental Management Practices. *The Engineering Economist*, *23*(2), 209–215.

Useem, M. (1984). *The Inner Circle: Large Corporations and the Rise of Business Political Activity in the US and UK*. Oxford University Press.

Uyar, A., & Güngörmüş, A. H. (2013). Perceptions and knowledge of accounting professionals on IFRS for SMEs: Evidence from Turkey. *Research in Accounting Regulation*, *25*(1), 77–87. doi:10.1016/j.racreg.2012.11.001

Vale, J., Bertuzi, R., & Monteiro, A. (2020). Social responsibility reporting in Higher Education Institutions: A systematic literature review. In *Conceptual and Theoretical Approaches to Corporate Social Responsibility, Entrepreneurial Orientation, and Financial Performance*. IGI Global.

Value Reporting Foundation. (2021). *<IR> Reporters*. http://examples.integratedreporting.org/reporters?start=A&page=1

Value Reporting Foundation. (n.d.). *Understanding the Value Reporting Foundation*. Retrieved from https://www.valuereportingfoundation.org/

Van Der Ploeg, . (2020). Race to burn the last ton of carbon and the risk of stranded assets. *European Journal of Political Economy*, *64*, 101915. doi:10.1016/j.ejpoleco.2020.101915

Van der Veldt, D. (1997). Case Studies of ISO 14001: A New Business Guide for Global Environmental Protection. *Environmental Quality Management*, *7*(1), 1–19. doi:10.1002/tqem.3310070102

Van Eck, N., & Waltman, L. (2010). Software survey: VOSviewer, a computer program for bibliometric mapping. *Scientometrics*, *84*(2), 523–538. doi:10.100711192-009-0146-3 PMID:20585380

Van Gossum, P., Arts, B., & Verheyen, K. (2010). From "smart regulation" to "regulatory arrangements". *Policy Sciences*, *43*(3), 245–261. doi:10.100711077-010-9108-0

Van Staden, C., Kern, T., McGuigan, N., & Wild, S. (2011). Social and Environmental Accounting and Accountability [Guest Editorial]. *Accounting Forum*, *35*(3), 127–129. doi:10.1016/j.accfor.2011.07.003

Vardon, M., Burnett, P., & Dovers, S. (2016). The accounting push and the policy pull Balancing environment and economic decisions. *Ecological Economics*, *124*, 145–152. doi:10.1016/j.ecolecon.2016.01.021

Venter, E. R., Stiglingh, M., & Smit, A. R. (2017). Integrated Thinking and the Transparency of Tax Disclosures in the Corporate Reports of Firms. *Journal of International Financial Management & Accounting*, *28*(3), 394–427. doi:10.1111/jifm.12064

Venturelli, A., Caputo, F., Cosma, S., Leopizzi, R., & Pizzi, S. (2017). Directive 2014/95/EU: Are Italian Companies Already Compliant? *Sustainability*, *9*(81-19).

Vereckey, B. (2021). *5 forces remaking accounting*. Retrieved from https://mitsloan.mit.edu/ideas-made-to-matter/5-forces-remaking-accounting

Vilalta, J. M., Betts, A., & Gomez, V. (2018). Higher Education's role in the 2030 agenda: The why and how of GUNi's commitment to the SDGs. In J. Vilalta, A. Betts & V. Gómez (Eds.), *Sustainable Development Goals: Actors and Implementation. A Report from the International Conference* (pp. 10-13). Barcelona: GUNi.

Villiers, C., & Sharma, U. (2018). A critical reflection on the future of financial, intellectual capital, sustainability and integrated reporting. *Critical Perspectives on Accounting*, *70*, 101999. Advance online publication. doi:10.1016/j.cpa.2017.05.003

Volkova, A., Kulakova, T. A., & Aliaskarova, Z. A. (2017). "Smart Regulation" and Models of Civil Cooperation in the Context of Global Challenges. In Globalization and its Socio-Economic Consequences (pp. 2903-2910). Academic Press.

Voss, R., Quaas, M. F., Stiasny, M. H., Hänsel, M., Stecher Justiniano Pinto, G. A., Lehmann, A., Reusch, T. B. H., & Schmidt, J. O. (2019). Ecological-economic sustainability of the Baltic cod fisheries under ocean warming and acidification. *Journal of Environmental Management*, *238*, 110–118. doi:10.1016/j.jenvman.2019.02.105 PMID:30849595

Wang, X., Yuen, K., Wong, Y., & Li, K. (2020). How can the maritime industry meet Sustainable Development Goals? An analysis of sustainability reports from the social entrepreneurship perspective. *Transportation Research Part D, Transport and Environment*, 78.

Warr, P. (2007). Searching for happiness at work. *The Psychologist*, *20*(12), 726–729.

Weber, O. (2014). The financial sector's impact on sustainable development. *Journal of Sustainable Finance & Investment*, *4*(1), 1–8. doi:10.1080/20430795.2014.887345

WEF. (n.d.). *World Economic Forum Global Risks. Perception Survey 2019-2020*. Author.

Weidman, S. M., Welsh, C. N., & Bonino, L. N. (1994). Accounting for environmental remediation costs. *Business & Professional Ethics Journal*, *13*(1/2), 147–163. doi:10.5840/bpej1994131/23

Wellalage, N. H. & Kumar, V. (2021). Environmental performance and bank lending: Evidence from unlisted firms. *Business Strategy & the Environment*, 1-21. doi:10.1002/bse.2804

Wensen, K., Van Broer, W., Klein, J., & Knopf, J. (2011). The state of play in sustainability reporting in the EU. *European Union*.

White, G. B. (2016). *Sustainability Reporting: Getting Started* (K. A. Merchant, Ed.; 2nd ed.). Business Expert Press.

Wiengarten, F., Pagell, M., & Fynes, B. (2013). ISO 14000 Certification and Investments in Environmental Supply Chain Management Practices: Identifying Differences in Motivation and Adoption Levels between Western European and North American Companies. *Journal of Cleaner Production*, *56*(October), 18–28. doi:10.1016/j.jclepro.2012.01.021

Wilson, C., & Caldecott, B. (2021). *Breaking the Bond: Primary Markets and Carbon-Intensive Financing*. University of Oxford Smith School of Enterprise and the Environment Working Paper 21-05.

Wirth, L. (2004). Breaking through the Glass Ceiling: Women in Management. International Labour Office.

Withers, M. C., Hillman, A. J., & Cannella, A. A. Jr. (2012). A multidisciplinary review of the director selection literature. *Journal of Management, 38*(1), 243–277. doi:10.1177/0149206311428671

Wood, D. (1991). *Corporate Social Performance Revisited.* Academic Press.

World Economic Forum. (2022). *The Global Risks Report 2022.* Retrieved from https://www3.weforum.org/docs/WEF_The_Global_Risks_Report_2022.pdf

World Intellectual Capital Initiative (WICI). (2013). *Connectivity: Background Paper for <IR>*. Disponível em: https://integratedreporting.org/wp-content/uploads/2013/07/IR-Background-Paper-Connectivity.pdf

Wright, T. & Cropanzano, R. (2004). The role of psychological well-being in job performance: A fresh look at an age-old quest. *Organizational Dynamics, 33*(4), 338–351. . doi:10.1016/j.orgdyn.2004.09.002

Wright, T., & Goodstein, J. (2007). Character is not "Dead" in Management Research: A review of individual character and organizational-level virtue. *Journal of Management, 33*(6), 928–958. doi:10.1177/0149206307307644

Wu, J. Z., Santoso, C. H., & Roan, J. (2017). Key Factors for Truly Sustainable Supply Chain Management: An Investigation of the Coal Industry in Indonesia. *International Journal of Logistics Management, 28*(4), 1196–1217. doi:10.1108/IJLM-07-2014-0103

Xanthaki, H. (2013). European Union legislative quality after the Lisbon Treaty: The Challenges of smart regulation. *Statute Law Review, 35*(1), 66–80. doi:10.1093lr/hmt005

Yáñez, S., Uruburu, Á., Moreno, A., & Lumbreras, J. (2018). The Sustainability Report as an essential tool for the holistic and strategic vision of Higher Education Institutions. *Journal of Cleaner Production.* Advance online publication. doi:10.1016/j.jclepro.2018.09.171

Yen, A., Hirst, E., & Hopkins, P. (2007). A Content analysis of the comprehensive income exposure draft comment letters. *Research in Accounting Regulation, 19*(1), 53–79. doi:10.1016/S1052-0457(06)19003-7

Yi, A., Howard, D., & Harun, H. (2017). Sustainability reporting at a New Zealand public university: A longitudinal analysis. *Sustainability, 9*, 1529. doi:10.3390u9091529

Yin, R. K. (2003). *Case study research, design and methods* (3rd ed., Vol. 5). Sage.

Yiridoe, E. K., & Marett, G. E. (2004). Mitigating the High Cost of ISO 14001 EMS Standard Certification: Lessons from Agribusiness Case Research. *International Food and Agribusiness Management Review, 7*, 37-62.

Younis, H., & Sundarakani, B. (2019). The Impact of Firm Size, Firm Age and Environmental Management Certification on the Relationship between Green Supply Chain Practices and Corporate Performance. *Benchmarking, 27*(1), 319–346. doi:10.1108/BIJ-11-2018-0363

Zahid, R. M. A., & Simga-Mugan, C. (2018). An Analysis of IFRS and SME-IFRS Adoption Determinants: A Worldwide Study. *Emerging Markets Finance & Trade, 55*(2), 391–408. doi:10.1080/1540496X.2018.1500890

Zaklan, A., Wachsmuth, J., & Duscha, V. (2021, July 3). The EU ETS to 2030 and beyond: Adjusting the cap in light of the 1.5°C target and current energy policies. *Climate Policy, 21*(6), 778–791. Advance online publication. doi:10.1080/14693062.2021.1878999

Zhang, J., Li, Y., Liu, C., Qu, Z., Li, F., Yang, Z., Jiang, L., & Fu, J. (2019). Application of set pair analysis in a comprehensive evaluation of water resource assets: A case study of Wuhan City, China. *Water (Basel), 11*(8), 1718. doi:10.3390/w11081718

Zhou, S., Simnett, R., & Green, W. (2017). Does integrated reporting matter to the capital market? *Abacus*, *53*(1), 94–132. doi:10.1111/abac.12104

Zhu, Q., Cordeiro, J., & Sarkis, J. (2012). International and Domestic Pressures and Responses of Chinese Firms to Greening. *Ecological Economics*, *83*(November), 144–153. doi:10.1016/j.ecolecon.2012.04.007

Zhu, Q., Tian, Y., & Sarkis, J. (2012). Diffusion of Selected Green Supply Chain Management Practices: An Assessment of Chinese Enterprises. *Production Planning and Control*, *23*(10–11), 837–850. doi:10.1080/09537287.2011.642188

Zilahy, G. (2017). Environmental Management Systems—History and New Tendencies. In *Encyclopedia of Sustainable Technologies* (pp. 23–31). Elsevier. doi:10.1016/B978-0-12-409548-9.10529-9

Zilber, T. B. (2002). Institutionalization as an interplay between actions, meanings, and actors: The case of a rape crisis center in Israel. *Academy of Management Journal*, *45*(1), 234–254.

Zilber, T. B. (2006). The work of the symbolic in institutional processes: Translations of rational myths in Israeli high tech. *Academy of Management Journal*, *49*(2), 281–303. doi:10.5465/amj.2006.20786073

Zimon, D. (2020). Iso 14001 and the Creation of SSCM in the Textile Industry. *International Journal of Qualitative Research*, *14*(3), 739–748. doi:10.24874/IJQR14.03-06

Zimon, D., & Madzik, P. (2019). Impact of Implementing ISO 14001 Standard Requirements for Sustainable Supply Chain Management in the Textile Industry. *Fibres & Textiles in Eastern Europe*, *27*(6), 8–14. doi:10.5604/01.3001.0013.4462

Zorio-Grima, A., Sierra-García, L., & Garcia-Benau, M. A. (2018). Sustainability reporting experience by universities: A causal configuration approach. *International Journal of Sustainability in Higher Education*, *19*(2), 337–352.

Zrnic, A., Starcevic, D. P., & Mijoc, I. (2020). Evaluating environmental accounting and reporting: The case of Croatian listed manufacturing companies. *Pravni Vjesnik (Journal of Law, Social Sciences and Humanities)*, *36*(1), 47-63. doi:10.25234/pv/8089

Zscheischler, J., & Seneviratne, S. I. (2017). Dependence of Drivers Affects Risks Associated with Compound Events. *Science Advances*, *3*(6), e1700263. doi:10.1126ciadv.1700263 PMID:28782010

Zucker, L. G. (1987). Institutional theories of organization. *Annual Review of Sociology*, *13*(1), 443–464. doi:10.1146/annurev.so.13.080187.002303

About the Contributors

Teresa Eugénio received her post-doctoral degree in Management (2019) at Lisbon School of Economics and Management (ISEG), University of Lisbon. She is PhD in Management, specialization in Accounting at University Institute of Lisbon (ISCTE - Portugal). She is Associate Professor of Accounting, Auditing and Social Responsibility at Department of Management and Economics, School of Technology and Management of Leiria at Polytechnic of Leiria (ESTG/ IPL, Portugal). She is an integrated member and vice-coordenator of CARME (Centre of Applied Research in Management and Economics). She published different books and papers in national and International journals as well as presentations in several conferences, and has participated in several research projects. She is international associate member of CSEAR, St Andrews University, UK, and member of the Coordination of GRUDIS (Portuguese Network of Accounting Research). Her research interests include Financial Accounting, Social and Environmental Accounting; Sustainability Education; Sustainable Development Goals; Non-Financial Reporting Information and Auditing.

Graça Azevedo has a PhD in Management (scientific field: Accounting) at the Institute of Higher Labor and Enterprise – ISCTE. She has published and presented several articles in national and international journals and conferences as well. She is a member of the Editorial Board of international journal "Accounting and Finance Research". She currently teaches courses in scientific field of Financial Accounting at the Institute of Higher Learning in Accounting and Administration of the University of Aveiro. Her research interests are related to financial reporting, impression management strategies and international accounting standards.

Ana Fialho is a Professor at the University of Évora, Management Department, specialized in financial accounting, financial and non-financial reporting, social and environmental accounting and reporting and responsible for several Accounting courses. Her main areas of research have included the study of reporting in private and public firms, impression management strategies, sustainability reporting regulation and practices. She is integrated member of Center for Advanced Studies in Management and Economics of the University of Évora (CEFAGE) and over the last 20 years she has been involved in research projects, creation of Management and Accounting study plans and management tasks as subdirector of Social Sciences School of the University of Évora. She is co-authors of many articles, conference presentations and other research works (http:orcid.org/0000-0003-0612-1535).

* * *

Fábio Albuquerque is Coordinator Professor in Accounting at Instituto Politécnico de Lisboa. PhD in Accounting and Financial Economics.

Marta Almeida received her PhD in Accounting and Finance from the University of Manchester, MSc. in Finance and Bachelor in Management from Nova School of Business and Economics. Is currently Assistant Professor at Nova School of Business and Economics. Previously, she has worked as a controller in project finance and as financial analyst in the tourism, manufacturing, and banking industries.

Isabelle Bevilaqua is a doctoral student in accounting at the University of Brasília, Brazil.

Wendy Carraro is a Post-Doctor in the area of Entrepreneurship, Strategy, Planning and Innovation in Business at the University of Porto. PhD in Development Economics from the PPGE of the Federal University of Rio Grande do Sul. Adjunct Professor of the Postgraduate Program in Controllership and Accounting at the Federal University of Rio Grande do Sul.

Valter da Silva Faia received his Ph.D. in Management from State University of Maringá (Brazil). Currently, he is professor of management accounting at State University of Maringá and research in the areas of sales force management, management controls, and individual and organizational performance.

Juliano Faria is Professor of the Accounting Course at the Federal University of Sergipe, Campus Itabaiana. Doctor in Industrial Engineering. Master in Accounting. Specialist in management accounting with an emphasis on controllership. Bachelor of Accounting Sciences. He works as an ad hoc evaluator of national and international journals and scientific events, in addition to developing research in the area of accounting auditing, environmental disclosure, Stranded Assets, management accounting, controllership and research in accounting education. Member of GIEC (UFS_DCCI), Brasfi and GpS.

Leonor Fernandes Ferreira is professor at Nova School of Business and Economics and Universidade do Algarve. She has taught undergraduate, postgraduate and MBA courses, as well as professional education in the field of financial reporting and analysis, financial management, corporate taxation, and accounting history. Currently, she coordinates the curricular internship program at Nova SBE. She authored the books Financial Reporting in Portugal (Routledge, 1994) and Modelos de Avaliação de Empresas e Utilidade da Informação Contabilística (Lusíada, 2008) and contributed with chapters to The Millers' European Accounting Guide (Aspen, 2003) and Transnational Accounting (Palgrave, 2001), among others. Her research has been published in several journals, such as Accounting History, Certified Accountant, Ciência e Técnica Fiscal, Entreprises et Histoire, Tourism and Management Studies, Revista Brasileira de Gestão de Negócios, Revista Iberoamericana de Contabilidad de Gestión, Revisores & Empresas, and Revista de Contabilidade e Comércio. She is a member of the Portuguese Accounting Standardization Committee and a member of the jury for admission to the Portuguese Order of Auditors.

Sónia Gomes is a PhD in Production Engineering from the Federal University of Santa Catarina and Master in Controllership and Accounting from the University of São Paulo. She is currently Technical Vice-President of the Regional Accounting Council and editor of CRCBA's Revisa Diálogo Contábil. President of the Academic Teaching Council of the Federal University of Bahia (2018 and 2019), being a full professor at the Faculty of Accounting Sciences/UFBA; professor and researcher of the Postgraduate

Programs Stricto Sensu in Administration (NPGA) and Accounting at UFBA. Research in the areas: climate change disclosure and accounting education for sustainability. Conducts consultancy and courses and lectures in the area of corporate sustainability and topics related to the Brazilian Accounting Standards.

Kennedy D. Gunawardena is Professor of Accounting Information Systems at the department of Accounting, University of Sri Jayewardenepura, Sri Lanka. In addition to being the professor of Accounting in his department, He is the chairman of the Board of studies and Coordinator of the PhD in management program of Management, faculty of Management studies and commerce in the same university. He teaches Cost and Management Accounting, Accounting Information Systems, Artificial Neural Networks for Accounting and Research Methodology in Accounting and finance for the undergraduates and Postgraduates students.

Kátia Lemos is a PhD in Accounting by University of Santiago de Compostela Professor at the Department of Accounting and Taxation School of Management Institute Polytechnic of Cávado and Ave.

Ana Morais is a full professor in accounting. She is the Associate Dean of ISEG - Lisbon School of Economics & Management. She has a PhD in Accounting from ISCTE. Her research interests include IFRS, non-financial information and audit. Her research has been published in academic journals.

Marguit Neumann graduated in Accounting (2003), Specialist in Management Accounting (2006), Master's in Development (2007). Doctorate at Development Economics from the Federal University of Rio Grande do Sul (UFRGS) and Sciences Economics from the University of Grenoble / France (2011). She is currently an effective professor at the State University of Maringá (UEM) in the Department of Accounting Sciences.

Cátia Nunes is an Investigator Specialist for UBS Switzerland AG handling whistleblowing and non-whistleblowing investigations for the firm. Cátia has also compliance and operational risk control experience in the Investment Bank sector and worked as an auditor at KPMG in Financial Services. She received her Master's degree in Management from Nova School of Business and Economics in 2019 and her Bachelor's degree in Economics from ISEG, Lisbon School of Economics & Management in 2016.

Liliana Pimentel is a PhD in Management, specialization in Accounting, in 2012/03/23 by University of Coimbra, Faculty of Economics, Master in Economy in 2006/07/26 by University of Coimbra, Faculty of Economics and Degree in Economy in 2002/02/27 by the same University. Assistant Professor in University of Coimbra, Faculty of Economics. Trainer in Foundation for Studies and Training in Local Government (FEFAL), Trainer in Portuguese Professional Association of Certified Accountants (OCC) and Auxiliary Researcher in University of Coimbra Centre for Business and Economics Research. In her curriculum Ciência Vitae the most frequent terms in the context of scientific, technological and artistic-cultural output are: capital structure; family firms; Earnings Quality; Environmental Accounting; Corporate Social Responsability; Public Sector and Local Government Accounting; leverage; non-family firms; Risk; Firm Valuation; Market Value Added; Residual Income Model; Valuation; Linear Information Model; Persistence; Predictability; Relevance; Accruals Quality; Informativeness; Smoothness; Timeliness.

Maria Albertina Barreiro Rodrigues is Chairman of the Audit Committees of entities from the public sector. Professor at Universidade Europeia and at Lisbon Accounting and Business School, ISCAL – IPL. PhD in Management. Specialist in Financial Audit. Holds a post-graduation in Strategic Management. The research areas of interest are corporate reporting, sustainability, integrated reporting, audit, ethics, and higher education. Financial and Statutory Auditor and Certified Accountant. Participated in International Conferences and published academic articles in journals and in the proceedings of said conferences. Has a professional career of over 25 years, 10 of which were spent as a Financial Manager in a multinational organisation. Was a member of committees from the Portuguese Institute of Statutory Auditors: Audit - Family and Profession Committee and Code of Ethics Monitoring Committee.

Paula Santos is Accounting Professor at Instituto Politécnico de Lisboa. Member of the Portuguese Accounting Standards Committee. Author of technical and scientific books and articles.

Monise Soster has a Bachelor's Degree in Administration and Accounting Sciences from the Federal University of Rio Grande do Sul.

André Luis Souza is a PhD in Industrial Engineering from the Graduate Program in Industrial Engineering - PEI/UFBA (2012-2016). Master in Business Administration (2010-2012) and Specialist in Business Finance (2009-2010), both from the Graduate Center in Business Administration - NPGA/EAUFBA. Specialist in Accounting, Expertise and Auditing at Universidade Norte do Paraná, UNOPAR (2017-2018); MBA in Public Accounting and Fiscal Responsibility from Centro Universitário Internacional UNINTER (PR) (2019-2020). Bachelor of Accounting Sciences from Visconde de Cairu Foundation (2004-2007). Graduated in Law at Faculdade 2 de Julho (2020-Current). Director of Administration and Planning at IFBA/Campus de Salvador (2020-Current). Permanent Professor of the Professional Masters in Intellectual Property and Technology Transfer for Innovation PROFNIT/IFBA. Professor and Researcher at the Federal Institute of Bahia - IFBA and Former Advisor to the Adjunct Board of Higher Education - DAES/IFBA. IFBA Productivity Researcher (2019-2020). He was a member of the Research Ethics Committee at the IFBA. He is the Lead Researcher of the Research Laboratory in Finance, Asset Valuation and Sustainability-LABFINVAS at the IFBA / Salvador Campus; member of the research group Governance for Sustainability and Low Carbon Management (GPS/EAUFBA) and Management Accounting Research Laboratory at FCC/UFBA. Area of interest and research: Intellectual Property Management and Technology Transfer for Innovation; Patent Valuation, Technologies (Asset Valuation) and Economic-Financial Feasibility Study; Technology Transfer Management Models. Corporate Finance; Corporate Initiatives in Climate and Sustainability. Carbon Market. Costs management. Accounting.

Viviane Torinelli is Guest Professor at FDC- Fundação Dom Cabral. Member of the Former Research and Innovation Committee of the Green Finance Program Brazil-UK; Member of the Innovation Chamber for Sustainability of the Salvador Climate Change Panel - R100 and C40; Co-founder of BRASFI, the alliance of Brazilian researchers around the theme of finance and sustainable investments. Co-founder of WeESG. Doctoral student with a focus on sustainable finance and ESG investment, at the School of Business Administration of the Federal University of Bahia (NPGA-UFBA), with master's progression. MBA in Project Management completed at Universidade Salvador (Unifacs; 2009), in a closed class organized for Ford Motor Company. Graduated in Business Administration from the Federal University of Paraná (UFPR; 2003), having attended the last 6 months at Universitat Passau, Germany. Professional with 20 years of experience in risk management, audit, control and compliance of conduct, in addition to M&A (company valuation and advice on corporate transactions). Practice in Brazil and abroad in the three lines of risk management defense: controller´s office, risk management and internal audit. In addition, experience as an external auditor in big-four, consultant in mergers and acquisitions and due diligence processes. Currently an analyst at the Central Bank of Brazil (BCB), the regulator of the National Financial System. Approved PMP, 2013; CISA, 2007; Six Sigma Green Belt, 2007; CIA, 2006.

Derya Ucoglu received her MSc in Financial Economics (Accounting and Finance Path) from İstanbul Bilgi University and her Ph.D. in Accounting and Finance from Marmara University in 2012. After working in different financial institutions, she has been an Assistant Professor of Accounting at Istanbul Bilgi University since 2014. She has published articles mainly on financial reporting. She currently focuses on the changes in the recognition and presentation of financial statement items due to the pandemic and disclosures regarding COVID-19 and the impact of technological innovations on finance, accounting, and auditing.

Index

K

L

M

N

O

P

S

T

U

W

Z

Ensure Quality Research is Introduced to the Academic Community

Become an Evaluator for IGI Global Authored Book Projects

The overall success of an authored book project is dependent on quality and timely manuscript evaluations.

Applications and Inquiries may be sent to:
development@igi-global.com

Applicants must have a doctorate (or equivalent degree) as well as publishing, research, and reviewing experience. Authored Book Evaluators are appointed for one-year terms and are expected to complete at least three evaluations per term. Upon successful completion of this term, evaluators can be considered for an additional term.

If you have a colleague that may be interested in this opportunity, we encourage you to share this information with them.

www.ingramcontent.com/pod-product-compliance
Lightning Source LLC
LaVergne TN
LVHW060414300926
849162LV00006B/1425